Ergebnisse der Mathematik und ihrer Grenzgebiete

3. Folge · Band 1

A Series of Modern Surveys in Mathematics

Albrecht Fröhlich

Galois Module Structure of Algebraic Integers

Springer-Verlag
Berlin Heidelberg New York Tokyo 1983

Albrecht Fröhlich
Imperial College London and
Robinson College Cambridge

AMS-MOS (1980) Classification numbers: 12 A 55, 12 A 57, 12 A 70

ISBN 3-540-11920-5 Springer-Verlag Berlin Heidelberg New York Tokyo
ISBN 0-387-11920-5 Springer-Verlag New York Heidelberg Berlin Tokyo

Library of Congress Cataloging in Publication Data
Fröhlich, A. (Albrecht), 1916 –
Galois module structure of algebraic integers.
(Ergebnisse der Mathematik und ihrer Grenzgebiete; 3. Folge, v. 1)
Bibliography: p. Includes index.
1. Algebraic number theory. 2. Galois theory. 3. Integral
representations. I. Title. II. Series.
QA247.F76. 1983. 512′74. 82-19588
ISBN 0-387-11920-5 (U.S.)

Printed in Germany
Typesetting: Dipl.-Ing. Schwarz' Erben KG, A-3910 Zwettl
Printing and binding: Konrad Triltsch, D-8700 Würzburg
2141/3140-543210

To my wife Ruth

Acknowledgement

My thanks are due to the Springer-Verlag for their patience and cooperation, to Mrs. Bunn for typing the whole book with her customary perfection, to Steve Ullom and Jan Brinkhuis and to the Bordeaux collective for proof-reading the typescript, and to Martin Taylor who read the manuscript draft and who made many suggestions leading to improvements of the presentation.

Table of Contents

Introduction

In this volume we present a survey of the theory of Galois module structure for rings of algebraic integers. This theory has experienced a rapid growth in the last ten to twelve years, acquiring mathematical depth and significance and leading to new insights also in other branches of algebraic number theory. The decisive take-off point was the discovery of its connection with Artin L-functions. We shall concentrate on the topic which has been at the centre of this development, namely the global module structure for tame Galois extensions of numberfields – in other words of extensions with trivial local module structure. The basic problem can be stated in down to earth terms: the nature of the obstruction to the existence of a free basis over the integral group ring ("normal integral basis"). Here a definitive pattern of a theory has emerged, central problems have been solved, and a stage has clearly been reached when a systematic account has become both possible and desirable. Of course, the solution of one set of problems has led to new questions and it will be our aim also to discuss some of these.

We hope to help the reader early on to an understanding of the basic structure of our theory and of its central theme, and to motivate at each successive stage the introduction of new concepts and new tools. For this reason we shall often state theorems as soon as their statement has become meaningful to the reader, even though the techniques for the proof have as yet not become available. In particular the main theorems (such as Theorem 5 and Theorem 9), which provide the basic orientation, require for their proof a lot of preparation in the way of a whole range of rather technical new methods, sophisticated new ideas and intermediary results. This really takes up, directly or indirectly, a considerable part of this whole volume. There is thus naturally a large gap between the original formulation and the completion of the proof. On the other hand we shall not delay proofs of subsidiary results or the derivation of consequences of the main theorems, if these are quick and straightforward.

We shall endeavour to include proofs, or at least outlines of proofs, in particular where these contain important ideas and new tools. An exception are those results which lie on the borderline of the subject and for which suitable references can be provided.

In Chapter I we trace the background and history of the subject and then develop the basic notions, giving the appropriate definitions and stating many of the results. It is hoped that this chapter can usefully be read by itself, as an introduction to the whole theory. The subsequent chapters are arranged by separate topics, going deeper and more systematically into these. Chapter II contains the required theory of the class group, the generalized group determinants and the group logarithm. In Chapter III the notions of resolvents and of Galois Gauss sums are developed and the proof of the first main theorem, Theorem 5, is completed. Next Chapter IV deals mainly with congruences for Galois Gauss sums and with logarithms and includes the complete proof of the second main theorem, Theorem 9. These four chapters form the core of the book. Chapter V deals with the problem of root number values, which arises out of the basic theory in the earlier chapters. Chapter VI on the other hand contains an account of interesting new developments, as yet incomplete, on what is best called "relative Galois module structure" for tame extensions. Finally in an appendix we shall give a brief outline of some recent work, which falls under the general heading of Galois module structure of rings of integers, but which is not properly included in this volume.

There are separate notes attached to most chapters containing subsidiary material which may be of interest.

Notation and Conventions

1. General. All rings have identities, acting as identity maps on modules, and preserved by all ring homomorphisms. The multiplicative group of invertible elements of a ring S is denoted by S^*, the group ring over S of a group Γ by $S\Gamma$, the ring of n by n matrices by $\mathrm{M}_n(S)$. Thus $\mathrm{M}_n(S)^* = \mathrm{GL}_n(S)$. The centre of S is $\mathrm{cent}(S)$.

We shall use standard symbols as follows

$$\mathbb{N} = \text{set of natural numbers,}$$

$$\mathbb{Z} = \text{ring of rational integers,}$$

$$\mathbb{Q} = \text{field of rational numbers,}$$

$$\mathbb{Q}_p = \text{field of } p\text{-adic rationals,}$$

$$\mathbb{R} = \text{field of real numbers,}$$

$$\mathbb{C} = \text{field of complex numbers,}$$

$$\mathbb{H} = \text{quaternion division algebra over } \mathbb{R},$$

$$\mathbb{F}_q = \text{finite field of } q \text{ elements.}$$

The Galois group of a Galois extension E/F of fields will be denoted by $\mathrm{Gal}(E/F)$, or by special symbols to be introduced; the action of $\mathrm{Gal}(E/F)$ on E is usually written exponentially: $x \mapsto x^\gamma$, for $x \in E$, $\gamma \in \mathrm{Gal}(E/F)$. The separable closure of a field F will be denoted by F^c, and we shall always write $\mathrm{Gal}(F^c/F) = \Omega_F$; this is the *absolute Galois group* of F. It is important to interpret $\mathbb{Q}^c$ as the algebraic closure of $\mathbb{Q}$ in $\mathbb{C}$. For, we shall repeatedly have to consider elements in $\mathbb{Q}^c$, such as Gauss sums, which are in the first place given as complex numbers (see in particular the remark early in Chap. I, §5).

If E/F is a field extension of finite degree, $N_{E/F}$ stands for the norm, $t_{E/F}$ for the trace.

The field of primitive m-th roots of unity over a field F of characteristic zero is denoted by $F(m)$.

2. Number Fields and Local Fields. A *number field* (or global field) K is here always a subfield of $\mathbb{Q}^c$ of finite degree over $\mathbb{Q}$. A *local field* F is always a subfield of $\mathbb{Q}_p^c$, an extension of finite degree of $\mathbb{Q}_p$ – where we also allow p to be "infinite", i.e. $\mathbb{Q}_p^c = \mathbb{C}$ and $\mathbb{Q}_p = \mathbb{R}$. The ring of integers of a number field or non-Archimedean local field F is denoted by $\mathfrak{o}_F$. Thus in the "global case", $\mathfrak{o}_F$ is the ring of algebraic integers in F, in the local case it is the valuation ring. Moreover for a non-Archimedean local field F, the symbol $\mathfrak{p}_F$ stands for the valuation ideal, i.e., the maximal ideal of $\mathfrak{o}_F$; we shall often refer to it as the "prime ideal". $\mathfrak{D}_F$ is the absolute different of F (i.e. over $\mathbb{Q}$, or $\mathbb{Q}_p$, respectively).

A *prime divisor* $\mathfrak{p}$ of a number field K is an equivalence class of nontrivial valuations of K. Thus $\mathfrak{p}$ is either *finite*, i.e., comes from a (non zero) prime ideal of $\mathfrak{o}_K$, and corresponds to an embedding of K into $\mathbb{Q}_p^c$, for some finite p, or $\mathfrak{p}$ is *infinite*, corresponding to an embedding of K into $\mathbb{C}$. If the latter factorizes via $\mathbb{R}$, $\mathfrak{p}$ is *real*, otherwise it is *complex*. The term *localization* is always used in the sense of local completion. The completion of K at $\mathfrak{p}$ (or of $\mathfrak{o}_K$ at $\mathfrak{p}$, when $\mathfrak{p}$ is finite) is denoted by $K_\mathfrak{p}$, and by $\mathfrak{o}_{K,\mathfrak{p}}$ respectively. More generally, if V is a finite dimensional K-space, or – in the case of finite $\mathfrak{p}$ – a finitely generated $\mathfrak{o}$-module, then $V_\mathfrak{p}$ stands for the completion of V at $\mathfrak{p}$. For infinite $\mathfrak{p}$ we make the following convention: If V is a finitely generated $\mathfrak{o}$-module, then $V_\mathfrak{p} = (V \otimes_\mathfrak{o} K)_\mathfrak{p}$, i.e., is the completion of $V \otimes_\mathfrak{o} K$. *Caution* If L is a number field, containing K, and $\mathfrak{p}$ a prime divisor of K, then $L_\mathfrak{p}$ is the completion at $\mathfrak{p}$. This is not in general a field, but a product of fields. We must distinguish it from the completion of L at a prime divisor $\mathfrak{P}$ of L; in fact $L_\mathfrak{p} = L \otimes_K K_\mathfrak{p} \cong \prod_\mathfrak{P} L_\mathfrak{P}$ (product over all prime divisors of L above $\mathfrak{p}$). Accordingly we refer to $L_\mathfrak{p}$ as a *semilocal* completion. Similarly we use the symbol $\mathfrak{o}_{L,\mathfrak{p}} = \mathfrak{o}_L \otimes_{\mathfrak{o}_K} \mathfrak{o}_{K,\mathfrak{p}} \cong \prod_\mathfrak{P} \mathfrak{o}_{L,\mathfrak{P}}$. For $\mathfrak{p}$ finite, this is the integral closure of $\mathfrak{o}_{K,\mathfrak{p}}$ in $L_\mathfrak{p}$. We shall also use the symbol $V_\mathfrak{p}$ for K-spaces of arbitrary dimension, as defined by $V \otimes_K K_\mathfrak{p}$. If $\dim_K(V)$ is finite this is consistent with our original definition, but in general $V_\mathfrak{p}$ will not be complete. Thus $(\mathbb{Q}^c)_\mathfrak{p} = \mathbb{Q}^c \otimes_K K_\mathfrak{p}$. This is of course also the direct limit of the algebras $L_\mathfrak{p}$, as L runs over the number fields containing K.

There are many different, but related aspects, associated with the notions of a prime divisor, a completion, a prime ideal, and we shall move freely from one to the other, without the impediment of pedantic notations. Thus, unless there is danger of confusion, we shall not distinguish symbolically between a finite prime divisor $\mathfrak{p}$ of K, the associated prime ideal of $\mathfrak{o}_K$, or the associated prime ideal of $\mathfrak{o}_{K,\mathfrak{p}}$, or the corresponding embedding of K in $K_\mathfrak{p}$.

The idele group of a number field K will be denoted by $\mathfrak{J}(K)$, the group of unit ideles, i.e., of ideles u whose components $u_\mathfrak{p}$ are units (i.e., lie in $\mathfrak{o}_{K,\mathfrak{p}}^*$), for all finite prime divisors $\mathfrak{p}$, will be denoted by $\mathfrak{U}(K)$, and we shall also use the symbol $\mathfrak{U}_\mathfrak{p}(K)$ for $\mathfrak{o}_{K,\mathfrak{p}}^*$, when considered as a subgroup of $\mathfrak{J}(K)$. The multiplicative group K^* is viewed as embedded diagonally in $\mathfrak{J}(K)$, and $\mathfrak{J}(K)/K^*$ is the idele classgroup. We shall actually mainly work in the direct limit $\mathfrak{J}(\mathbb{Q}^c)$ of the idele groups $\mathfrak{J}(E)$, and we use the symbol $\mathfrak{U}(\mathbb{Q}^c)$ for the limit of the $\mathfrak{U}(E)$. Here E runs through the algebraic number fields and the limit is taken with respect to the inclusion maps $E \subset E'$. We shall often have to consider ideles, given in an indirect way, whose field of definition is neither needed, nor easy to find – from this point of view the group $\mathfrak{J}(\mathbb{Q}^c)$ is then the appropriate object – and a similar remark applies to groups of fractional ideals (see below). We shall again have to deal with semilocal components. Thus if $\mathfrak{p}$ is a finite prime divisor of K, then for any number field $E \supset K$, $\mathfrak{U}_\mathfrak{p}(E) = \prod_\mathfrak{P} \mathfrak{U}_\mathfrak{P}(E)$, $\mathfrak{P}$ running through the prime divisors of E, above $\mathfrak{p}$. In the limit we then write $\mathfrak{U}_\mathfrak{p}(\mathbb{Q}^c)$. Locally we write $\mathbb{Z}_p^c$ for the integral closure of $\mathbb{Z}_p$ in $\mathbb{Q}_p^c$, and put $\mathfrak{U}(\mathbb{Q}_p^c) = (\mathbb{Z}_p^c)^*$.

If F is a number field, or a non-Archimedean local field, a *fractional ideal in* F is always a non-zero fractional ideal of $\mathfrak{o}_F$; the group of these will be denoted by $\mathfrak{I}(F)$;

if K is a number field we may sometimes identify $\mathfrak{J}(K) = \coprod \mathfrak{J}(K_\mathfrak{p})$ (restricted product over the finite prime divisors). We shall view a fractional ideal essentially as determined by the prescribed values, – at all finite prime divisors, i.e., non-Archimedean valuations, in the global case, and at the given valuation, in the non-Archimedean local case. The fractional ideals which will occur will usually be the contents of an idele in $\mathfrak{J}(\mathbb{Q}^c)$, globally, or be given by an element of $\mathbb{Q}_p^c$ locally. Again the actual field of definition is rarely obvious, and often not relevant. We shall accordingly view $\mathfrak{J}(F)$ as a subgroup of $\mathfrak{J}(\mathbb{Q}^c)$ (global case) or of $\mathfrak{J}(\mathbb{Q}_p^c)$ (local case) – the latter groups being defined as the direct limit or union of the $\mathfrak{J}(F)$ (with F running through the number fields, or the local fields over $\mathbb{Q}_p$, respectively). There are instances when a fractional ideal $\mathfrak{a}$ in F should be viewed as a subset of F, e.g. in the use of the symbol $\mathfrak{o}_F/\mathfrak{a}$ for the residue class ring of $\mathfrak{o}_F$ modulo an ideal $\mathfrak{a}$. This is however always obvious from the context. The only place where care is needed under our convention is the definition of the *absolute norm*. We shall always denote the absolute norm (with respect to F) of $\mathfrak{a} \in \mathfrak{J}(F)$ by $\mathbf{N}_F\mathfrak{a}$. If $\mathfrak{a}$ is an ideal of $\mathfrak{o}_F$ then $\mathbf{N}_F\mathfrak{a}$ is the cardinality of $\mathfrak{o}_F/\mathfrak{a}$; for fractional ideals this extends by multiplicativity.

3. Group Representations. We consider representations $T\colon \Gamma \to \mathrm{GL}_n(A)$ of a finite group Γ over a field A of characteristic zero, and in this context we use the classical language of character theory. Thus the *character* χ associated with T is the function $\Gamma \to A$ given by $\chi(\gamma) = \operatorname{trace} T(\gamma)$ and, by definition, $n = \deg(\chi)$, the degree of χ or of T. At the same time we use the fact that the character uniquely describes the equivalence class of the representation T. (Only once will we have to consider representations in characteristic p, the appropriate definitions to be given then).

The characters of degree one will be called *Abelian* characters. These are homomorphism $\Gamma \to \mathrm{GL}_1(A) = A^*$, i.e., at this level the character "coincides" with the representation. Writing Γ^{ab} for the maximal Abelian quotient of Γ, the Abelian characters of course factorize through Γ^{ab}. They form an Abelian group under (pointwise) multiplication: $\chi\phi(\gamma) = \chi(\gamma)\phi(\gamma)$ for all $\gamma \in \Gamma$. If χ has finite order n in this group, we shall simply say that χ is of finite order, and write $n = \text{order}\ (\chi)$.

We add characters by the rule $(\chi + \phi)(\gamma) = \chi(\gamma) + \phi(\gamma)$ for all $\gamma \in \Gamma$. The additive group of functions $\Gamma \to A$ generated by the characters in this way will be denoted by $R_\Gamma(A)$. Its elements are the *virtual characters*. $R_\Gamma(A)$ is a commutative ring, with multiplication $\chi\phi(\gamma) = \chi(\gamma)\phi(\gamma)$ for all $\gamma \in \Gamma$. From a K-theoretic point of view, $R_\Gamma(A) = \mathfrak{K}_0(A\Gamma)$, the Grothendieck group of finitely generated $A\Gamma$-modules. The term "irreducible character" will be used in the absolute sense, i.e., only if A is algebraically closed. $R_\Gamma(A)$ is then additively the free Abelian group on the irreducible characters. Usually we shall be concerned with representations over such an algebraically closed field, and if in this case the choice of A is immaterial, we shall simply write $R_\Gamma(A) = R_\Gamma$. The various $R_\Gamma(A)$ are of course isomorphic, but sometimes we shall actually need the fact that the values $\chi(\gamma)$ are particular elements of A. If in doubt always interpret R_Γ as $R_\Gamma(\mathbb{Q}^c)$. In certain situations we shall have to consider both $R_\Gamma(\mathbb{Q}^c) = R_\Gamma$ and $R_\Gamma(\mathbb{Q}_p^c)$ and the various isomorphisms between them. We shall then also write $R_\Gamma(\mathbb{Q}_p^c) = R_{\Gamma,p}$.

In our use of brackets we have sacrificed consistency to intelligibility and easy reading. A profusion of symbols will demand brackets, to keep their individual meanings separate. On the other hand elsewhere in the same logical, but simpler notational context, these brackets may become superfluous and even confusing, and so are better omitted. Nowhere should this pragmatic use of notation lead to misunderstanding.

Further permanent symbols will be introduced in the text. A list of some of these, together with the places where they are defined, will be given at the end of the book. The meaning of other letters will be more or less fixed within each section, but not necessarily between sections.

In any given chapter, the first ordinal of an equation refers to the section, and the second one gives the consecutive numbering within this section. Back references within each chapter use the same numbering system; back references to other chapters are preceded by the chapter number. Thus within Chap. II, Eq. (2.5) is the 5th in §2, and will be referred to as such within this chapter. If a back reference to it is required in Chap. IV, this will be given as (II.2.5). The same system is used for propositions and lemmas. Theorems however are numbered consecutively throughout the book; a list of their place will be given at the end.

I. Survey of Results

§1. The Background

The questions covered by the heading "Galois module structure of algebraic integers" are quite classical in their origin. The power, depth and interest of the theory however stem from its recently discovered connection with arithmetic invariants, associated with representations of Galois groups, in particular those coming from the functional equation of the Artin L-function. Since this discovery, the subject which previously had lain rather barren, has developed rapidly. It now has a rich and powerful body of theorems, which can effectively be used to provide quite explicit, concrete information, and which at the same time also lead to new insight into other branches of algebraic number theory, such as the Stickelberger relations, the Galois module structure of ideal classgroups, and the embedding problem. In particular one must mention here the study of the arithmetic nature of local and global root numbers and Galois Gauss sums, and indeed this can be taken as a second main topic of the present volume. The aim of the present section is to outline in approximate – but not strict – chronological order the several strands of research and their interaction, which led to the development of the theory in the first place.

To formulate the basic problem, recall first the *normal basis theorem* for a Galois extension E/F (always of finite degree) of fields. This asserts that E as an F-vector space has a basis of form $\{a^\gamma\}$, called a *normal basis*, where a is fixed in E and γ runs over the Galois group $\mathrm{Gal}(E/F) = \Gamma$, i.e. that E is a free $F\Gamma$-module, necessarily of rank one. Now let $N/\mathbb{Q}$ be a Galois extension, with $\mathfrak{o}_N$ the ring of algebraic integers in N. The analogous question is then: Does $\mathfrak{o}_N$ have a $\mathbb{Z}$-basis of the form $\{a^\gamma\}$, γ running over $\Gamma = \mathrm{Gal}(N/\mathbb{Q})$, and a fixed in $\mathfrak{o}_N$ – a *normal integral basis* (abbreviated to NIB), i.e. is $\mathfrak{o}_N$ free over $\mathbb{Z}\Gamma$. The *existence problem for a* NIB (in a slightly more general context indicated below) and the *nature of the obstructions* which intervene, form the central theme of this report.

The "canonical" first attack on such a problem is via localization – and by this we shall mean transition to local completions. So for the moment we consider a Galois extension E/F of non-Archimedean local fields with Galois group Δ. A theorem first stated by E. Noether (cf. [No]) part of which goes back to Speiser (cf. [Sp]) asserts that the ring $\mathfrak{o}_E$ of integers in E (the valuation ring of E) is free over $\mathfrak{o}_F\Delta$, i.e. has an $\mathfrak{o}_F$-basis $\{a^\delta\}$ (δ running through Δ) if, and only if, E/F is *tame*, i.e. at most tamely ramified. Returning now to a Galois extension $N/\mathbb{Q}$ as above, and completing at a rational prime p, it follows that $\mathfrak{o}_{N,p}$ is free over $\mathbb{Z}_p\Gamma$ if, and only if,

$N/\mathbb{Q}$ is tame at p, i.e., p is at most tamely ramified in N. Now consider all primes p simultaneously. We shall call a finitely generated $\mathbb{Z}\Gamma$-module X *locally free* if, for all p, X_p is free over $\mathbb{Z}_p\Gamma$. Noether's theorem then implies that $\mathfrak{o}_N$ is locally free as $\mathbb{Z}\Gamma$-module if, and only if, $N/\mathbb{Q}$ is tame (i.e. is tame at all p). Tame ramification is thus clearly necessary for the existence of a NIB, and the global question we have to answer is: *Is it sufficient*, and if not, *what are the global obstructions*. We shall note below that both the local and the global theory extend naturally to relative Galois extensions N/K, with $\mathrm{Gal}(N/K) = \Gamma$, replacing the base field $\mathbb{Q}$ by some other number field K, but retaining $\mathbb{Z}\Gamma$ as the operator domain.

The only classical result on global structure is due to Hilbert ([Hi] Satz 132). Extending Hilbert's result slightly, we have the theorem that if $N/\mathbb{Q}$ has Abelian Galois group Γ, and is tame, then there exists a NIB. The crucial point is here, that according to the Kronecker-Weber theorem, N will be a subfield of a field $\mathbb{Q}(e^{2\pi i/m})$ with m square free. Once this has been observed the proof is trivial: The trace of $e^{2\pi i/m}$ in N will generate a NIB. Still working over $\mathbb{Q}$, Martinet ([Ma1]) more recently established in the tame case the existence of a NIB, when $\mathrm{Gal}(N/\mathbb{Q}) = D_{2p}$, the dihedral group of order $2p$, p an odd prime. But he then went on to show that when $\mathrm{Gal}(N/\mathbb{Q}) = H_8$, the quaternion group of order 8, each of the two possible structures of locally free $\mathbb{Z}H_8$-modules (of rank one of course) occurs in the form $\mathfrak{o}_N$. In particular he found examples for $N/\mathbb{Q}$ tame, which do not have a NIB (cf. [Ma2]).

For the moment we shall leave the global problem in this unsatisfactory state, it had reached about 12 years ago, and return to Noether's ramification criterion for the module structure of rings of integers. We shall for the time being state everything in local terms, considering a Galois extension E/F of non-Archimedean local fields with Galois group Δ. The equivalence of the two properties, namely (i) $\mathfrak{o}_E$ being projective over $\mathfrak{o}_F\Delta$, and (ii) E/F being tame, is only one aspect of a general pattern. Vaguely one can say that the "wilder" the ramification of E/F, the further removed is $\mathfrak{o}_E$ from being $\mathfrak{o}_F\Delta$-projective. To provide a more quantitative measure, it is here that our first arithmetic representation invariants appear. These are the module conductors and module resolvents whose theory was developed in [F3], [F4], [F14] and extended in [Ne1], [Ne2]. They are "ideal like" objects, associated with integral representations of the Galois group (in the present context representations over local rings of integers). They reflect simultaneously the ramification of E/F and the module properties of $\mathfrak{o}_E$ over $\mathfrak{o}_F\Delta$, and can be used to establish stronger versions of Noether's theorem. In the case when E/F is tame, they only depend on the character χ, associated with the given integral representation of Δ. We thus obtain in particular ideals $\mathfrak{c}(E/F, \chi)$, the module conductors. These objects then lead to the proof of a conjecture of E. Noether, which arose out of her theorem, and which we shall now explain.

Some years before Noether's paper, and in connection with his own work on L-functions, E. Artin had generalized the concept of a conductor from Abelian to general characters χ of (say here local) Galois groups. Denote it by $\mathfrak{f}(E/F, \chi)$, cf. [Art1] (see (5.26) for the definition in the relevant case). Artin obtained a product formula

(1.1) $$\mathfrak{d}(E/F) = \prod_{\chi} \mathfrak{f}(E/F, \chi)^{\deg(\chi)}$$

(product over the irreducible characters χ) for the relative discriminant $\mathfrak{d}(E/F)$. On the other hand if E/F is tame, one sees easily that $\mathfrak{d}(E/F)$ is the square of the classical "group determinant" for a local NIB, and thus has a module theoretic decomposition – and this can be shown in fact to be given in terms of the module conductors by

$$\mathfrak{d}(E/F) = \prod_{\chi} \mathfrak{c}(E/F, \chi)^{\deg(\chi)}. \tag{1.2}$$

Noether had conjectured that the two decompositions – the arithmetic and the module theoretic one – coincide, and indeed this is so, by a fundamental result on module conductors which asserts that in the tame case

$$\mathfrak{c}(E/F, \chi) = \mathfrak{f}(E/F, \chi). \tag{1.3}$$

In a slightly different formulation this will be proved here (see Theorem 24). We mention in passing that in the general case the ratio between module conductors and Artin conductors has some very interesting applications (cf. [Ne2]).

Now we return to a Galois extension N/K of number fields, with $\mathrm{Gal}(N/K) = \Gamma$. The basefield K may now be different from $\mathbb{Q}$. Both Artin conductors $\mathfrak{f}(N/K, \chi)$, and module conductors $\mathfrak{c}(N/K, \chi)$, are defined in the global situation, and in particular $\mathfrak{f}(N/K, \chi)$ is a fractional ideal of $\mathfrak{o}_K$. Denote its class in the ideal classgroup $\mathrm{Cl}(\mathfrak{o}_K)$ of $\mathfrak{o}_K$ (one often says "of K") by $(\mathfrak{f}(N/K, \chi))_{\mathfrak{o}_K}$. If χ is real valued and realizable over K, then one can easily deduce that the class $(\mathfrak{c}(N/K, \chi))_{\mathfrak{o}_K}$ is a square. In view of (1.3) this led the author to a conjecture that for all Galois extensions N/K – not only tame ones –

$$(\mathfrak{f}(N/K, \chi))_{\mathfrak{o}_K} \quad \textit{is a square for real valued } \chi \tag{1.4}$$

(cf. [F3]), and this conjecture turned out to have some rather startling consequences.

We can restate (1.4) in terms of quadratic characters ψ of the ideal classgroup, i.e., in terms of homomorphisms $\psi: \mathrm{Cl}(\mathfrak{o}_K) \to \pm 1$. Viewing such a ψ as a function on ideals, we see that (1.4) is equivalent with

$$\begin{cases} \psi(\mathfrak{f}(N/K, \chi)) = 1, & \textit{for every real valued character} \\ \chi, \textit{ and for every quadratic character } \psi \textit{ of } \mathrm{Cl}(\mathfrak{o}_K). \end{cases} \tag{1.5}$$

In this form the conjecture appears as what Serre has called a "*parity question*": to determine values of numerical arithmetic invariants with possible domain ± 1.

The character ρ of the regular representation of $\mathrm{Gal}(N/K)$ is real valued, and its conductor is the relative discriminant $\mathfrak{d}(N/K)$. Thus (1.4) is a potential generalization of a classical result, namely that

$$(\mathfrak{d}(N/K))_{\mathfrak{o}_K} \quad \textit{is a square.} \tag{1.6}$$

This in turn is a consequence of a famous theorem of Hecke (cf. [He1] Satz 176) on the absolute different $\mathfrak{D}_N$ of $\mathfrak{o}_N$, that

(1.7) $$(\mathfrak{D}_N)_{\mathfrak{o}_N} \quad \textit{is a square.}$$

Property (1.6) can in fact be established directly and by elementary methods (cf. [F1]) and it holds quite generally for separable extensions of quotient fields of Dedekind domains. On the other hand, the analogue to (1.7) for function fields of curves had been shown not to be universally valid (cf. [FST]). However, the case closest to that of number fields, namely that of curves over finite fields had then been left undecided. This was settled by Armitage (cf. [Arm1]) who proved the analogue of (1.7), using papers of Lamprecht and of Siegel on Gauss sums. Prior to publication, he communicated this to Serre, who then produced a new proof of (1.7) simultaneously for number fields and for function fields over finite fields which was published in Armitage's paper [Arm1] (the same proof appears also in [We1] (last theorem)). Serre's method is based on the functional equation of L-functions, and influenced the subsequent attack on (1.5). We shall describe the underlying idea.

We consider extended Artin L-functions $\tilde{L}(s,\chi) = \tilde{L}(s, N/K, \chi)$, with Euler factors at infinity (see (5.19)). These satisfy a functional equation

(1.8) $$\tilde{L}(s,\chi) = \tilde{L}(1-s,\bar{\chi})W(\chi)A(\chi)^{\frac{1}{2}-s}$$

where $\bar{\chi}$ is the conjugate complex character, $A(\chi)$ is a positive constant, and $W(\chi) = W(N/K, \chi)$ a complex constant of absolute value 1, the *Artin root number*. If χ is Abelian, then $\tilde{L}(s,\chi)$ is a Hecke L-function, and in this case one has an explicit formula for the root number. In particular if $\chi = \psi$ is non-ramified (i.e., comes from a Galois group of a non-ramified extension) then ψ may be viewed, via class field theory, as an Abelian character of $\mathrm{Cl}(\mathfrak{o}_K)$, and one has

(1.9) $$W(\psi) = \psi(\mathfrak{D}_K).$$

On the other hand if ψ is any quadratic character – i.e., any Abelian character with $\psi^2 = \varepsilon$ the identity character, then

(1.10) $$W(\psi) = 1.$$

By (1.9), (1.10), one concludes that for all quadratic characters ψ of $\mathrm{Cl}(\mathfrak{o}_K)$, $\psi(\mathfrak{D}_K) = 1$, which yields (1.7), taking into account the change in notation.

Turning now back to (1.5), one may consider χ, ψ and hence also $\chi\psi$ as characters of Galois groups for suitable N/K. In a letter to Armitage (May 1967) Serre suggested the formula

(1.11) $$\psi(\mathfrak{f}(N/K,\chi)) = W(\chi)/W(\chi\psi)$$

as a starting point for the proof of (1.5). Armitage indeed proved this formula in the case required, i.e. for χ real valued and ψ non-ramified quadratic. When, some few years later, Armitage proceeded to publish a paper (cf. [Arm 2] – see below for the background), containing his proof, Langland's work on root numbers had become known, and this included a formula which contained (1.11) as a special case (see [We3] (pp. 152 and 160–161)).

Here another development had come into play, and we will leave our main story for a moment to mention this in passing. Weil's work on characterizing modular forms via functional equations of Dirichlet series (cf. [We2]) and subsequent generalizations (cf. [We3], [J. L.]) had involved the twisting by Abelian characters and this had led to the interest of twisting formulae such as (1.11). See here also the discussion in [Se5] (§3).

Coming back now to the root number-conductor formula in Armitage's attack on the problem, one knows – and this is quite easy – that

$$W(N/K, \theta) = \pm 1, \qquad \text{if } \theta \text{ is real valued.} \tag{1.12}$$

Thus the reformulation of the conjecture (1.5), via (1.11), leads to the most remarkable of parity questions, that of the values of the Artin root numbers for real valued characters. This is also important for the existence of non-trivial real zeros of the Dedekind zeta functions. Indeed, for real valued χ the functional equation (1.8) tells us that $\tilde{L}(s, \chi)A(\chi)^{s/2}$ is either symmetric, or skew-symmetric, about $s = \frac{1}{2}$, depending on whether $W(\chi) = 1$, or $= -1$. In the latter case the function must have a pole or a zero at $s = \frac{1}{2}$. If, say, χ is a character of $\mathrm{Gal}(N/K)$, this implies in turn that the Dedekind zeta function of N vanishes at $s = \frac{1}{2}$ (cf. [Arm2]).

Using these observations, Armitage knew by March 1968 that if Dedekind zeta functions never vanished at $s = \frac{1}{2}$, then (1.5) would hold. He approached a number of experts on zeta functions, but none could give an answer, although the non-vanishing hypothesis was generally believed to be true.

In attempts to settle the two parity questions (1.5), (1.12), it became clear that the class of real valued characters was a bad one to work with. Instead one was led to concentrate on two subclasses – that of *orthogonal characters*, corresponding to matrix representations of a group Γ which factorize through $\Gamma \to O_n(\mathbb{C})$ (the orthogonal group) or equivalently through $\Gamma \to \mathrm{GL}_n(\mathbb{R})$, and that of *symplectic characters*, corresponding to representations which factorize through $\Gamma \to Sp_{2n}(\mathbb{C})$ (the symplectic group), or equivalently through $\Gamma \to \mathrm{GL}_n(\mathbb{H})$, $\mathbb{H}$ the Hamiltonian quaternions. (The irreducible symplectic characters are those of real Schur index 2.) Every real valued character is a sum of characters of these two types, and every irreducible real valued character is either orthogonal, or symplectic, but not both. Using new induction theorems, Serre [Se3] then proved (1.4) for χ orthogonal (as well as for all real valued χ coming from tame extensions N/K), and subsequently Fröhlich–Queyrut proved, that $W(\chi) = 1$, whenever χ is orthogonal (cf. [FQ]). This had been conjectured by Serre (letter to Langlands April 1970 and letter to the author January 1972). Later Deligne gave another proof of this result, via a local formula (cf. [De2] – see also [Tt2]). There is also an elegant proof by Serre of this result in the function field case, which uses the geometry, more precisely the Weil pairing on homology (or cohomology), published in [FQ]. At present this has no number field analogue.

Important examples of irreducible orthogonal representations are the surjections $\Gamma \to D_{2m}$, the *dihedral matrix group* D_{2m} ($m \geqslant 3$) being generated by two matrices

$$(1.13)\qquad \begin{cases} \begin{pmatrix} y & 0 \\ 0 & y^{-1} \end{pmatrix}, & y \text{ a primitive } m\text{-th root of } 1, \\ \begin{pmatrix} 0 & 1 \\ 1 & 0 \end{pmatrix}. \end{cases}$$

We shall call the character of such a representation a *dihedral* character. Similarly, important examples of irreducible symplectic representations are the surjections $\Gamma \to H_{4m}$, the (generalized) *quaternion matrix group* H_{4m} $(m \geqslant 2)$ being generated by two matrices

$$(1.14)\qquad \begin{cases} \begin{pmatrix} y & 0 \\ 0 & y^{-1} \end{pmatrix}, & y \text{ a primitive } 2m\text{-th root of } 1, \\ \begin{pmatrix} 0 & 1 \\ -1 & 0 \end{pmatrix}. \end{cases}$$

The character of such a representation will be called a *quaternion* character. Induction theorems essentially reduce orthogonal, and symplectic representations to these two types, respectively. [Note [1] (later the symbols D_{2m}, H_{4m} will also stand for the corresponding abstract groups).]

In his paper [Se3], Serre also gave an example of a conductor $\mathfrak{f}(N/K, \chi)$, with real valued χ (necessarily not tame, i.e., with N/K not tame, and not orthogonal), whose ideal class is not a square.

This was crucial. Armitage could now reverse his planned procedure for the proof of (1.5) and instead produce an L-function with real valued χ and $W(\chi) = -1$ (cf. [Arm2]). For further such, unpublished, computations by Serre and by Armitage and their influence on the birth of a new theory, see below. Jumping ahead of the chronological order, we mention that subsequently the author devised a standard method (cf. [F7]) to produce "at random" symplectic characters with given root number $+1$ or -1, or with given conductor class modulo ideal class squares. For the conductors the characters involved had by necessity to come from wild extensions, but for the root numbers one could achieve the same within a "tame framework" (see also [F18] (Theorem 18) and [Ge]). In a more special context, for tame extensions of $\mathbb{Q}$ with H_8 as Galois group, a very precise result of this type was obtained in [F6]. For later developments see the author's recent paper [F26], or Chapter V of this volume.

We have now come full circle. Consider again the situation in Martinet's paper [Ma2], i.e. let $N/\mathbb{Q}$ be Galois with group $\Gamma \cong H_8$. This isomorphism $\Gamma \cong H_8$, with H_8 viewed as a matrix group (1.14) (for $y = i$), defines the unique irreducible symplectic character ψ_8. Write $W(N/\mathbb{Q}) = W(N/\mathbb{Q}, \psi_8)$ for the root number. Then $W(N/\mathbb{Q}) = \pm 1$.

Serre then conceived, early in 1971, what he called the "crazy idea" that perhaps one has $W(N/\mathbb{Q}) = -1$ precisely when $\mathfrak{o}_N$ does not possess a normal integral basis. As he said subsequently "this looked to me at the time not as a serious conjecture but rather as wishful thinking – trop beau pour être vrai". He verified this for two of Martinet's three original examples. Moreover, with the original aim of getting information on multiplicities of zeros at $s = \frac{1}{2}$, Armitage had done computations on twelve more such fields. He checked Serre's idea and found in each case that it fitted. He communicated this to the author in December 1971. Stimulated by Serre's challenge and encouraged by the strong testimony of all these computations, the author then succeeded in January 1972 in obtaining a proof (cf. [F6]). To give a neat statement, introduce a module invariant $U(N/\mathbb{Q}) = \pm 1$, for tame fields N with $\mathrm{Gal}(N/\mathbb{Q}) \cong H_8$, setting $U(N/\mathbb{Q}) = +1$ precisely if $\mathfrak{o}_N$ was free over $\mathbb{Z}H_8$. Then $U(N/\mathbb{Q})$ determines $\mathfrak{o}_N$ as a module over $\mathbb{Z}H_8$, and the result of [F6] is that the analytic-arithmetic and the module theoretic invariants coincide, i.e. that

$$U(N/\mathbb{Q}) = W(N/\mathbb{Q}). \tag{1.15}$$

Moreover, as already indicated above, it was shown that both values ± 1 occur infinitely often.

This was a very astonishing result indeed, and at that stage rather mysterious – and the proof in [F6] certainly did nothing to dispel the mystery. Nevertheless it seemed obvious that there must be a general theory underlying such a connection between two seemingly unrelated topics. It was initially however far from obvious what this should be, and it took some optimism to push on with the subject. In fact the discovery of (1.15) turned out to be the starting point of a rapid development.

Going further in the same direction, a result corresponding to (1.15) was established – again by hard explicit computations – for tame Galois extensions $N/\mathbb{Q}$ with Galois group H_{12} in [Q1]. The groups $\Gamma = H_8$ and $\Gamma = H_{12}$ have two properties in common, which were relevant in these initial break throughs, and which are not shared by other quaternion groups. Firstly there are exactly two non-isomorphic locally free $\mathbb{Z}\Gamma$-modules, thus giving rise to numerical module invariants $U(N/\mathbb{Q}) = \pm 1$ as above, and secondly there is exactly one irreducible symplectic, i.e., quaternion character ψ of such a group, giving rise to a numerical arithmetic invariant $W(N/\mathbb{Q}) = W(N/\mathbb{Q}, \psi) = \pm 1$. A further advance thus depended on (a) a systematic formulation of the module theoretic problem, and (b) a systematic analysis of the data provided by root numbers of symplectic characters.

We discuss the second aspect first. It is of course still possible to distinguish "canonically" between irreducible symplectic characters, which are not conjugate over $\mathbb{Q}$, and in fact the crucial result in this context was the *Galois invariance* of root numbers $W(N/K, \chi)$ for (irreducible) symplectic characters χ, i.e. the equation

$$W(N/K, \chi) = W(N/K, \chi') \tag{1.16}$$

if χ and χ' are conjugate over $\mathbb{Q}$, and N/K is tame. (It is false in the wild case.) This was proved in [F7] (see also [F16]) for quaternion groups, but the result then extends to arbitrary groups by induction of symplectic characters (for this see note [1]). This proof was replaced by another one, which is both smoother and gives more insight into the real reasons behind the Galois invariance (cf. [F17] – there was actually an earlier version of this paper, which circulated at the time and incorporated most of the general theory developed at that stage). In the case of quaternion groups H_{4m}, m odd, Eq. (1.16) could also be checked easily by direct computations (cf. [F8], also [F9]), and it was specifically for all the groups H_{4l^r}, l an odd prime, $r \geqslant 1$, that the first explicit application was given, generalizing Queyrut's result on H_{12} (cf. [F8]). Here general methods, applicable to a range of groups, rather than explicit computations dealing with one given group Γ, come into play for the first time. The conjugacy classes of irreducible symplectic characters ψ of H_{4l^r} correspond to the values of a parameter $j = 1, \ldots, r$; ψ belongs to j, if it is lifted from a faithful character of the quotient group H_{4l^j}. All ψ with the same j have the same root number $W(N/\mathbb{Q}, \psi)$ – where now $N/\mathbb{Q}$ is Galois and tame with $\mathrm{Gal}(N/\mathbb{Q}) \cong H_{4l^r}$. We denote this by $W_j(N/\mathbb{Q})$. There are then also natural module invariants $U_j(X)$ ($j = 1, \ldots, r$), defined for every locally free $\mathbb{Z}H_{4l^r}$-module X, and one gets

(1.17) $$U_j(\mathfrak{o}_N) = \left(\frac{W_j(N/\mathbb{Q})}{l}\right).$$

We now come to the other basic task, mentioned above, which arose naturally at an early stage, that of formulating the structure problem for the Galois module $\mathfrak{o}_N$ more systematically. One needs a classification of locally free $\mathbb{Z}\Gamma$-modules X, recalling the earlier local result that if $N/\mathbb{Q}$ is tame, then $\mathfrak{o}_N$ is locally free over $\mathbb{Z}\Gamma$. Such a classification is given by the rank of X, which for $\mathfrak{o}_N$ as above is 1, and by the class $(X)_{\mathbb{Z}\Gamma}$ of X in the locally free classgroup $\mathrm{Cl}(\mathbb{Z}\Gamma)$, a finite Abelian group. The two invariants determine X to within stable isomorphism (see note [2]), and for "most" Γ even to within isomorphism, the free modules giving the identity of $\mathrm{Cl}(\mathbb{Z}\Gamma)$. In the cases $\Gamma = H_8$, $\Gamma = H_{12}$, we have $\mathrm{Cl}(\mathbb{Z}\Gamma) \cong \pm 1$, and (1.15) can be read as an equation in $\mathrm{Cl}(\mathbb{Z}H_8)$, and similarly for H_{12}.

A further motivation for the development of a general, systematic approach was a conjecture of Martinet leading to the second important milestone in the history of the subject, after (1.15). Let $\mathfrak{M}$ be a maximal order in $\mathbb{Q}\Gamma$, containing $\mathbb{Z}\Gamma$, and let $N/\mathbb{Q}$ be a Galois extension with Galois group Γ. The $\mathfrak{M}$-module $\mathfrak{o}_N\mathfrak{M}$ generated by $\mathfrak{o}_N$ is certainly locally free and Martinet conjectured that

(1.18) $$\mathfrak{o}_N\mathfrak{M} \text{ is stably free over } \mathfrak{M},$$

(i.e., $\mathfrak{o}_N\mathfrak{M} \oplus \mathfrak{M} \cong \mathfrak{M} \oplus \mathfrak{M}$). For tame extensions and arbitrary Γ this was indeed proved in [F17] (see also [F16]), thereby setting in evidence the power of the new theory. The effectiveness of the method is emphasized by the fact that the corresponding result for wild extensions is not in general true (cf. [Co2]).

The proof of (1.18) uses the theory of class groups. With $\mathfrak{M}$ as above, one has a surjection of class groups $\mathrm{Cl}(\mathbb{Z}\Gamma) \to \mathrm{Cl}(\mathfrak{M})$, whose kernel $D(\mathbb{Z}\Gamma)$, the *kernel group*, does not depend on the choice of $\mathfrak{M}$, and so (1.18) amounted to proving that in the tame case

(1.19) $$(\mathfrak{o}_N)_{\mathbb{Z}\Gamma} \in D(\mathbb{Z}\Gamma).$$

For a different, more classical interpretation of (1.19), we write $\mathbb{Q}\Gamma = \prod_i A_i$, as product of simple algebras. Let $F_i = \mathrm{cent}(A_i)$, and let C_i be the ideal classgroup of

$\mathfrak{o}_{F_i}$ in the usual sense, except that when F_i is totally real and A_i is a matrix ring over a totally definite quaternion algebra, then C_i is the ideal classgroup modulo the totally positive principal ideals. One then has, for each i, a projection c_i: $\mathrm{Cl}(\mathbb{Z}\Gamma) \to C_i$, and $D(\mathbb{Z}\Gamma) = \mathrm{Ker} \prod_i c_i$. Thus (1.19) now reads $c_i((\mathfrak{o}_N)_{\mathbb{Z}\Gamma}) = 1$ for all i.

The Eq. (1.17) also arose out of the new general approach. In fact the U_j occurring there where certain canonical surjections

$$D(\mathbb{Z}H_{4l^r}) \to \pm 1,$$

and in the paper [F8] it was already necessary to prove what is essentially (1.19) for the groups $\Gamma = H_{4l^r}$.

After the initial stages it soon became apparent that the restriction to basefield $\mathbb{Q}$ was unnecessary, and in fact undesirable, as it would have precluded the use of powerful induction and restriction techniques. One only has to observe that if $\Gamma = \mathrm{Gal}(N/\mathbb{Q})$, and we want to induce up from or restrict to a subgroup Δ, then we have to study the extension $N/K, K = N^\Delta$. So from now on we consider more generally tame Galois extensions N/K. We still however view $\mathfrak{o}_N$ primarily as a $\mathbb{Z}\Gamma$-module, $\Gamma = \mathrm{Gal}(N/K)$. As such it is still locally free, although of rank $[K:\mathbb{Q}]$, and determines an element

$$(1.20) \qquad (\mathfrak{o}_N)_{\mathbb{Z}\Gamma} = U_{N/K} \in \mathrm{Cl}(\mathbb{Z}\Gamma),$$

and we shall throughout continue to use the symbol $U_{N/K}$ in this sense.

We can still speak of a NIB. This is now a $\mathbb{Z}$-basis $\{a_i^\gamma\}$ of $\mathfrak{o}_N$, $((i = 1, \ldots, [K:\mathbb{Q}]), \gamma \in \Gamma)$, or equivalently a free $\mathbb{Z}\Gamma$-basis $\{a_i\}$. Its existence would imply $U_{N/K} = 1$, and in fact if $K \neq \mathbb{Q}$ the converse is true as well (as for modules of rank > 1 the class together with the rank determines the isomorphism class). All the results mentioned earlier (e.g. (1.15), (1.17), (1.18)) remain true in this new context and in particular

$$(1.21) \qquad U_{N/K} \in D(\mathbb{Z}\Gamma),$$

and are given in this general form in [F17] and [F18].

It is now time to say something about the theory of the author presented mainly in [F17], which lies behind these results and others, obtained during this period, about the tools which had to be developed, and the deeper connections, which arose and which are as significant and interesting, as the results directly affecting the Galois module structure of $\mathfrak{o}_N$. There are three main ingredients. (i) The first is an entirely new presentation of classgroups of group rings and more generally orders, in particular of $\mathrm{Cl}(\mathbb{Z}\Gamma)$, in terms of Galois homomorphisms from the group of virtual characters into certain other groups. Connected with this is a generalization of the notion of a group determinant – the generalized group determinants themselves being functions on characters. This provided the appropriate language for a description of $U_{N/K}$ and at the same time led to a deeper understanding of classgroups. (ii) The second ingredient was a generalization to non-Abelian characters of the classical Lagrange resolvent (already investigated in [F4] and [F13], and closely connected with the module conductors mentioned earlier (cf. [F14])). These new resolvents will actually give a description of $U_{N/K}$ in the new classgroup language, and this is their main function. (iii) The third ingredient, which brings the actual number theory of number fields into the picture is the arithmetic theory of (local and global) Galois Gauss sums. These generalize the classical Gauss sums and can be used to define globally the Artin root numbers (cf. (1.8)) and locally the corresponding Langlands constants (cf. [De1] or [Tt2]). The deepest part and core of the whole theory is then a close relation between Galois Gauss sums and generalized resolvents – this for all characters, not only symplectic ones. The Eq. (1.21) is an immediate consequence of this.

The connection between Galois Gauss sums and resolvents is however of independent interest, apart from the application to Galois module structure and in fact throws new light on the arithmetic nature of Galois Gauss sums and root numbers (cf. [FT]). Thus tame Galois Gauss sums can in turn be characterized intrinsically in terms of what may be called Galois module invariants.

Once the basic theory with its three main ingredients had been established, further progress was made by simultaneous advances in each of the three areas. The principal tool was the *method of congruences*. One established congruences for values of generalized group determinants, for generalized resolvents and for Galois Gauss sums, and every new result of this type led to new insight into Galois module structure. The big aim was to generalize the original sporadic theorems connecting it with the values of symplectic root numbers. It was suspected very early on that obstructions to $\mathfrak{o}_N$ being free over $\mathbb{Z}\Gamma$, more precisely to $U_{N/K}$ having value 1, were rooted in the existence of division algebras or of matrix rings over those, occurring in the decomposition of $\mathbb{Q}\Gamma$. A number of computations made it clear, that indeed the only relevant such algebras should be quaternion algebras. This then led soon to the conjecture, that the only obstructions for $U_{N/K} = 1$ are the values $= -1$ of root numbers of symplectic characters. This can be stated quantitatively and in more precise terms. The values of these root numbers determine an element of $D(\mathbb{Z}\Gamma)$, which (for good reasons to become obvious) we denote by $tW_{N/K}$, which was first defined by Cassou-Noguès in [CN2]. The basic conjecture, made by the author, then reads:

(1.22) $$tW_{N/K} = U_{N/K}.$$

Most of the work done at this stage (cf. [CN1], [CN2], [CN3], [CN4], [Ty1], [Ty2], [Ty3], [Ty4], [Ty5], [F12], [F13], [F15], [F23]) was in this general direction, even before the precise formulation (1.22) had appeared, and more so after. The results were either proofs of (1.22) for particular classes of Galois groups (for groups without irreducible symplectic characters this meant $U_{N/K} = 1$), or approximations to the general equation, e.g., by proving it modulo some canonical small subgroups of $D(\mathbb{Z}\Gamma)$, or by proving some general consequence.

In spite of being overtaken now, as far as their principal aim is concerned, many of these papers will retain more than historical interest, in that they contribute results, which had to be used subsequently or which are of significance in themselves. The final success came fairly recently when M. Taylor succeeded in proving the conjecture (1.22) in full generality, i.e., for all tame Galois extensions N/K of number field (cf. [Ty6], [Ty7]). If the relation between resolvents and Galois Gauss sums is the core of the theory, the beautifully simple and deep theorem (1.22) is its crowning glory. The proof is based on a refinement of methods which had been developed previously and in particular the method of congruences and their combination with an entirely new tool, the logarithm for local grouprings first introduced in [Ty6] and then applied to generalized group determinants on the one hand and to Galois Gauss sums on the other.

Taylor's theorem left only problems related to the elements $tW_{N/K}$ of classgroups in the area to which this volume is mainly devoted. New research on Galois module structure of rings of algebraic integers is now going on mainly in four directions: (i) Arithmetic and metric properties of normal bases and normal integral bases. (ii) Wild extensions. (iii) The Hermitian theory, studying Galois module structure in conjunction with the trace form. (iv) The structure problem for tame extensions N/K with $\mathfrak{o}_N$ viewed as a module, say, over $\mathfrak{o}_K\Gamma$. This last line of research leads to connections with other areas of algebraic number theory, such as (a) Stickelberger relations and their generalizations, in the sense of Galois module structure of ideal class groups and (b) the classical embedding problem. (i) is very much still in its initial stage, (ii) and (iii) are really separate topics and will only very briefly be referred to, but we shall give a sketch of some developments under heading (iv).

One should also mention recent work by Chinburg, suggesting a connection, or at least a similarity of ideas between our subject, on the one hand, and the Galois module structure of S-units and Stark's conjecture in Tate's form on the other (cf. [Chb]).

§2. The Classgroup

Here we shall introduce the first basic tool of the theory, the classgroup of an order, and in particular of an integral group ring and its "Hom description". In this section Γ is a finite group (its representation as a Galois group is irrelevant here), K is a number field, $\mathfrak{o}$ its ring of integers and $\mathfrak{A}$ is an order in $K\Gamma$, i.e., a subring of $K\Gamma$ with $1 \in \mathfrak{A}$, which is a finitely generated $\mathfrak{o}$-module and spans $K\Gamma$. For us the most important of these is the integral group ring $\mathfrak{o}\Gamma$. (We could more generally replace $K\Gamma$ by an arbitrary finite dimensional semi-simple algebra, but this will not be needed).

A *locally free* (say right) $\mathfrak{A}$-module X is a finitely generated $\mathfrak{A}$-module so that the $\mathfrak{A}_\mathfrak{p}$-module $X_\mathfrak{p}$ is free, for all prime divisors $\mathfrak{p}$ of K. Its rank, $r(X)$, is defined as the rank of the free $K\Gamma$-module $X \otimes_\mathfrak{o} K = XK$, spanned by X. This rank is finite, and is also the rank of $X_\mathfrak{p}$ over $\mathfrak{o}_\mathfrak{p}\Gamma$, for all $\mathfrak{p}$.

The Grothendieck group $\mathfrak{K}_0(\mathfrak{A})$ of locally free $\mathfrak{A}$-modules is the Abelian group with generators $[X]$, corresponding to the ($\mathfrak{A}$-isomorphism classes of) locally free $\mathfrak{A}$-modules X and with relations

$$[X \oplus Y] = [X] + [Y].$$

(This means of course that in the end we may have $[X] = [Y]$ without $X \cong Y$ - see the analogous discussion for the classgroup in note [2]). The map $\mathbb{N} \to \mathfrak{K}_0(\mathfrak{A})$, which takes n into the class $[\mathfrak{A}^n]$ of the free $\mathfrak{A}$-module $\mathfrak{A}^n$ of rank n, extends to a homomorphism $\mathbb{Z} \to \mathfrak{K}_0(\mathfrak{A})$, and we define the locally free classgroup $\mathrm{Cl}(\mathfrak{A})$ to be its cokernel. It thus appears in a defining exact sequence

(2.1) $$0 \to \mathbb{Z} \to \mathfrak{K}_0(\mathfrak{A}) \to \mathrm{Cl}(\mathfrak{A}) \to 1.$$

We shall denote the image of $[X]$ in $\mathrm{Cl}(\mathfrak{A})$ by (X) and use multiplicative notation for the classgroup.

We shall now introduce an alternative description of $\mathrm{Cl}(\mathfrak{A})$ which is absolutely fundamental in all that follows. Observe first of all that the additive group R_Γ of virtual characters of Γ is a module over the absolute Galois group Ω_K (see "notations"), where

$$\chi^\omega(\gamma) = \chi(\gamma)^\omega, \qquad \text{for} \quad \chi \in R_\Gamma, \quad \omega \in \Omega_K, \quad \gamma \in \Gamma.$$

Next let $\mathfrak{J}(\mathbb{Q}^c)$ be the idele group of $\mathbb{Q}^c$, i.e. the direct limit, and in fact the union, of the idele groups $\mathfrak{J}(E)$, as E runs through the number fields in $\mathbb{Q}^c$ – of course with respect to the embedding $\mathfrak{J}(E) \subset \mathfrak{J}(E')$ if $E \subset E'$. We shall then have to work in the group $\mathrm{Hom}_{\Omega_K}(R_\Gamma, \mathfrak{J}(\mathbb{Q}^c))$. This group is in fact not as formidable as it may look – and the same is true for other similar groups which we shall have to introduce. The values of the characters of Γ, i.e., the numbers $\chi(\gamma)$ $(\chi \in R_\Gamma, \gamma \in \Gamma)$ lie in some number field E containing K (and other given number fields which may occur), which is Galois over $\mathbb{Q}$. But then

$$\mathrm{Hom}_{\Omega_K}(R_\Gamma, \mathfrak{J}(\mathbb{Q}^c)) = \mathrm{Hom}_{\Omega_K}(R_\Gamma, \mathfrak{J}(E)) = \mathrm{Hom}_G(R_\Gamma, \mathfrak{J}(E)),$$

with $G = \mathrm{Gal}(E/K)$. Thus computations can always take place in $\mathfrak{J}(E)$. In this sense one can speak of local components $f_\mathfrak{p}$ of a homomorphism f. To be more precise, let $\mathfrak{p}$ be a prime divisor of a number field $F \subset \mathbb{Q}^c$. Then the $\mathfrak{p}$-component $\mathfrak{J}(\mathbb{Q}^c)_\mathfrak{p}$ of $\mathfrak{J}(\mathbb{Q}^c)$ is the direct limit of the groups $\mathfrak{J}(E')_\mathfrak{p} = \prod_{\mathfrak{P}|\mathfrak{p}} E'^*_\mathfrak{P}$, with E' running over the number fields with $F \subset E' \subset \mathbb{Q}^c$, or alternatively $\mathfrak{J}(\mathbb{Q}^c)_\mathfrak{p}$ is the multiplicative group $(\mathbb{Q}^c)^*_\mathfrak{p}$ of $(\mathbb{Q}^c)_\mathfrak{p} = \mathbb{Q}^c \otimes_F F_\mathfrak{p}$ embedded in $\mathfrak{J}(\mathbb{Q}^c)$. For $f \in \mathrm{Hom}(R_\Gamma, \mathfrak{J}(\mathbb{Q}^c))$, the $\mathfrak{p}$-component $f_\mathfrak{p}$ is its image in $\mathrm{Hom}(R_\Gamma, \mathfrak{J}(\mathbb{Q}^c)_\mathfrak{p})$. But in fact if $F \subset E$, E as before, we can compute the $f_\mathfrak{p}(\chi)$ in $\mathfrak{J}(E)_\mathfrak{p}$.

Our first aim is the generalization of the notion of a determinant. Let B be a commutative K-algebra (say $B = K_\mathfrak{p}$). We let Ω_K act on $\mathbb{Q}^c \otimes_K B$ via $\mathbb{Q}^c$, i.e. trivially on B. Consider a representation

(2.2) $$T: \Gamma \to \mathrm{GL}_n(\mathbb{Q}^c).$$

We extend this to a homomorphism

(2.3) $$T: B\Gamma \to \mathrm{M}_n(\mathbb{Q}^c \otimes_K B)$$

of algebras, which in turn yields a homomorphism

(2.4) $$T: (B\Gamma)^* \to \mathrm{GL}_n(\mathbb{Q}^c \otimes_K B)$$

of groups. Composing with the determinant

$$\mathrm{GL}_n(\mathbb{Q}^c \otimes_K B) \to (\mathbb{Q}^c \otimes_K B)^*,$$

we end up with a homomorphism

$$\text{(2.5)} \qquad \mathrm{Det}_\chi\colon (B\Gamma)^* \to (\mathbb{Q}^c \otimes_K B)^*,$$

where χ is the character of T, and where indeed the map Det_χ only depends on χ, not on T. If θ is a further character, then clearly

$$\text{(2.6)} \qquad \mathrm{Det}_{\chi+\theta}(b) = \mathrm{Det}_\chi(b)\,\mathrm{Det}_\theta(b).$$

Thus the map $\chi \mapsto \mathrm{Det}_\chi(b)$ (for $b \in B\Gamma^*$) can be extended to a homomorphism

$$\text{(2.7)} \qquad \mathrm{Det}(b)\colon R_\Gamma \to (\mathbb{Q}^c \otimes_K B)^*, \qquad \mathrm{Det}(b)(\chi) = \mathrm{Det}_\chi(b),$$

the (generalized) *determinant* of b.

Next we shall prove that for $b \in B\Gamma$, $\chi \in R_\Gamma$, $\omega \in \Omega_K$

$$\text{(2.8)} \qquad (\mathrm{Det}_{\chi^{\omega^{-1}}}(b))^\omega = \mathrm{Det}_\chi(b).$$

Without loss of generality we assume χ to be the character of a representation T. Then $\chi^{\omega^{-1}}$ is that of $T^{\omega^{-1}}$, where $T^{\omega^{-1}}(\gamma) = T(\gamma)^{\omega^{-1}}$. Write $b = \sum_{\gamma\in\Gamma} b_\gamma \gamma$, $b_\gamma \in B$, and so $b_\gamma^\omega = b_\gamma$. Then we have

$$\begin{aligned}(\mathrm{Det}_{\chi^{\omega^{-1}}}(b))^\omega &= (\mathrm{Det}(\textstyle\sum b_\gamma T(\gamma)^{\omega^{-1}}))^\omega = \mathrm{Det}((\textstyle\sum b_\gamma T(\gamma)^{\omega^{-1}})^\omega)\\ &= \mathrm{Det}(\textstyle\sum b_\gamma T(\gamma)) = \mathrm{Det}_\chi(b).\end{aligned}$$

(Remark: More generally defining in the same way $\mathrm{Det}(b)$, now for b in $(\mathbb{Q}^c \otimes_K B)\Gamma^*$, with Ω_K acting via $\mathbb{Q}^c$, we get $(\mathrm{Det}_{\chi^{\omega^{-1}}}(b))^\omega = \mathrm{Det}_\chi(b^\omega)$). We conclude that the map $b \mapsto \mathrm{Det}(b)$ is a homomorphism

$$\text{(2.9)} \qquad \mathrm{Det}\colon (B\Gamma)^* \to \mathrm{Hom}_{\Omega_K}(R_\Gamma, (\mathbb{Q}^c \otimes_K B)^*).$$

Now take $B = K_\mathfrak{p}$, and in $K_\mathfrak{p}\Gamma$ consider the order $\mathfrak{A}_\mathfrak{p}$. If $\mathfrak{p}$ is infinite then of course by our convention in this case $\mathfrak{A}_\mathfrak{p} = K_\mathfrak{p}\Gamma$. We write $\mathbb{Q}^c \otimes_K K_\mathfrak{p} = (\mathbb{Q}^c)_\mathfrak{p}$. We get the group $\mathrm{Hom}_{\Omega_K}(R_\Gamma, (\mathbb{Q}^c)_\mathfrak{p}^*)$. Note again that for E a large enough number field this is the same as $\mathrm{Hom}_{\Omega_K}(R_\Gamma, (E_\mathfrak{p})^*)$, with $E_\mathfrak{p} = E \otimes_K K_\mathfrak{p}$. We denote by $\mathfrak{U}_\mathfrak{p}(\mathbb{Q}^c)$ the group of units of the ring of integers in $(\mathbb{Q}^c)_\mathfrak{p}$ (the integral closure of $\mathfrak{o}_{K,\mathfrak{p}}$). Restricting the map Det from $(K_\mathfrak{p}\Gamma)^*$ to $\mathfrak{A}_\mathfrak{p}^*$, we obtain a homomorphism

$$\text{(2.10)} \qquad \mathrm{Det}\colon \mathfrak{A}_\mathfrak{p}^* \to \mathrm{Hom}_{\Omega_K}(R_\Gamma, \mathfrak{U}_\mathfrak{p}(\mathbb{Q}^c)),$$

as $\mathrm{Det}_\chi(b)$ is clearly a unit for $b \in \mathfrak{A}_\mathfrak{p}^*$. Let

$$\text{(2.11)} \qquad \mathfrak{U}(\mathfrak{A}) = \prod_\mathfrak{p} \mathfrak{A}_\mathfrak{p}^* \qquad \text{(product over all prime divisors)}$$

be the group of unit ideles of $\mathfrak{A}$. Similarly write

$$\mathfrak{U}(\mathbb{Q}^c) = \prod_{\mathfrak{p}} \mathfrak{U}_{\mathfrak{p}}(\mathbb{Q}^c).$$

If L is a number field, denote by $\mathfrak{U}(L)$ the group of unit ideles, i.e., of ideles which are units at all finite prime divisors. Then indeed $\mathfrak{U}(\mathbb{Q}^c)$ is the limit (or union) of the $\mathfrak{U}(L)$ ($L \subset \mathbb{Q}^c$). Going over to the product, we get from (2.10) a homomorphism

$$\text{(2.12)} \qquad \mathrm{Det}\colon \mathfrak{U}\mathfrak{A} \to \mathrm{Hom}_{\Omega_K}(R_\Gamma, \mathfrak{U}(\mathbb{Q}^c)) \subset \mathrm{Hom}_{\Omega_K}(R_\Gamma, \mathfrak{J}(\mathbb{Q}^c)),$$

whose image we denote by $\mathrm{Det}(\mathfrak{U}\mathfrak{A})$.

Now we can come to the "Hom-description" of $\mathrm{Cl}(\mathfrak{A})$. We shall state this in an explicit form which gives a rule for computing the class of a rank one locally free module.

Theorem 1. (i) *Let X be a locally free rank one $\mathfrak{A}$-module. Choose a free generator v of $X \otimes_{\mathfrak{o}} K$ over $K\Gamma$ and for each prime divisor $\mathfrak{p}$ in K choose a free generator $x_{\mathfrak{p}}$ of $X_{\mathfrak{p}}$ over $\mathfrak{A}_{\mathfrak{p}}$. Then both v and $x_{\mathfrak{p}}$ are free generators of $V_{\mathfrak{p}}$ over $K_{\mathfrak{p}}\Gamma$, and so*

$$x_{\mathfrak{p}} = v\lambda_{\mathfrak{p}}, \qquad \lambda_{\mathfrak{p}} \in (K_{\mathfrak{p}}\Gamma)^*.$$

Let for each $\mathfrak{p}$, each χ,

$$f_{\mathfrak{p}}(\chi) = f(\chi)_{\mathfrak{p}} = \mathrm{Det}_\chi(\lambda_{\mathfrak{p}}), \qquad f_{\mathfrak{p}} \in \mathrm{Hom}_{\Omega_K}(R_\Gamma, (\mathbb{Q}^c)_{\mathfrak{p}}^*).$$

Then

$$f \in \mathrm{Hom}_{\Omega_K}(R_\Gamma, \mathfrak{J}(\mathbb{Q}^c))$$

and its class $[f]$ modulo $\mathrm{Hom}_{\Omega_K}(R_\Gamma, \mathbb{Q}^{c*})\,\mathrm{Det}(\mathfrak{U}\mathfrak{A})$ *only depends on the isomorphism class of X.*

(ii) *There is a unique isomorphism*

$$\text{(2.13)} \qquad \mathrm{Cl}(\mathfrak{A}) \cong \mathrm{Hom}_{\Omega_K}(R_\Gamma, \mathfrak{J}(\mathbb{Q}^c))/[\mathrm{Hom}_{\Omega_K}(R_\Gamma, \mathbb{Q}^{c*})\,\mathrm{Det}(\mathfrak{U}\mathfrak{A})]$$

so that for every locally free rank one module X, the class (X) maps onto the corresponding class $[f]$ as constructed above. □

The description of the classgroup by homomorphisms, given here, is enforced on one by the way in which the class of a ring of integers as a Galois module is computed – this will be seen in §4 (cf. Theorem 4). Apart from this, this description can also be seen to have tremendous advantages when one considers the functorial properties of classgroups or when one wants to compute classgroups.

We shall prove Theorem 1 (i) here, and give the proof of (ii), as based on some simple facts of $\mathfrak{K}$-theory, in II §1. It is however already clear from (i) that the group on the right hand side of (2.13) yields good invariants for modules X and that the knowledge of the class of $[f]$ as in (i) contains good information. Thus for example it is clear from the description of $[f]$ that the invariant of a free module is the identity element. Moreover the explicit description given can easily be extended to modules of higher rank and then takes direct sums of modules into products of their classes (see Remark 1 below). Thus one could reasonably accept (2.13) simply as definition of $\mathrm{Cl}(\mathfrak{A})$ and disregard the $\mathfrak{K}$-theoretic point of view altogether.

Proof of (*i*). The existence of an element $\lambda_{\mathfrak{p}}$, as stated, is immediate. Moreover if $X \cong Y$ then we can choose generators of $Y \otimes_{\mathfrak{o}} K$ and of $Y_{\mathfrak{p}}$ so as to get the same $\lambda_{\mathfrak{p}}$. The element f which we have obtained with this choice lies in $\mathrm{Hom}_{\Omega_K}(R_\Gamma, \prod_{\mathfrak{p}} (\mathbb{Q}^c)^*_{\mathfrak{p}})$. If we change the generator v of V to another one, that will be of form vb, $b \in (K\Gamma)^*$; similarly we may replace $x_{\mathfrak{p}}$ by $x_{\mathfrak{p}}u_{\mathfrak{p}}$, $u_{\mathfrak{p}} \in \mathfrak{A}^*_{\mathfrak{p}}$. For the element g which replaces f we then have

$$g_{\mathfrak{p}}(\chi) = \mathrm{Det}_\chi(b_{\mathfrak{p}}^{-1}\lambda_{\mathfrak{p}}u_{\mathfrak{p}}) = \mathrm{Det}(b_{\mathfrak{p}}^{-1})(\chi)f_{\mathfrak{p}}(\chi)\,\mathrm{Det}(u_{\mathfrak{p}})(\chi),$$

i.e.

$$g = \mathrm{Det}(b^{-1})f\,\mathrm{Det}(u),$$

$$b \in (K\Gamma)^*, \qquad u \in \mathfrak{U}\mathfrak{A}$$

whence indeed the class of f modulo $\mathrm{Hom}_{\Omega_K}(R_\Gamma, \mathbb{Q}^{c*})\,\mathrm{Det}(\mathfrak{U}\mathfrak{A})$ is unique. Finally note that for almost all $\mathfrak{p}$, we may take $x_{\mathfrak{p}} = v$ and thus $\lambda_{\mathfrak{p}} = 1$. Therefore indeed the values of $f(\chi)_{\mathfrak{p}}$ are units for almost all $\mathfrak{p}$, i.e., $f \in \mathrm{Hom}_{\Omega_K}(R_\Gamma, \mathfrak{J}(\mathbb{Q}^c))$. □

Remark 1. One can also extend a representation T as in (2.2) to homomorphisms

(2.3′) $$T\colon \mathrm{M}_q(B\Gamma) \to \mathrm{M}_{nq}(\mathbb{Q}^c \otimes_K B),$$

(2.4′) $$T\colon \mathrm{GL}_q(B\Gamma) \to \mathrm{GL}_{nq}(\mathbb{Q}^c \otimes_K B),$$

and get an extension of (2.5):

(2.5′) $$\mathrm{Det}_\chi\colon \mathrm{GL}_q(B\Gamma) \to (\mathbb{Q}^c \otimes_K B)^*.$$

This has the same nice properties as before, and the further property that if $b \in \mathrm{GL}_q(B\Gamma)$, $c \in \mathrm{GL}_m(B\Gamma)$, i.e.

$$\begin{pmatrix} b & 0 \\ 0 & c \end{pmatrix} \in \mathrm{GL}_{m+q}(B\Gamma),$$

then

(2.14) $$\mathrm{Det}_\chi \begin{pmatrix} b & 0 \\ 0 & c \end{pmatrix} = \mathrm{Det}_\chi(b)\,\mathrm{Det}_\chi(c).$$

If now X is a locally free module of rank q, say, we choose bases $\{v_i\}$, $\{x_{\mathfrak{p},i}\}$ and get $\lambda_{\mathfrak{p}} \in \mathrm{GL}_q(K_{\mathfrak{p}}\Gamma)$. Now one can proceed as above, to get an invariant in the group on the right hand side of (2.13), and by (2.14) we get in fact a homomorphism from $\mathfrak{K}_0(\mathfrak{A})$.

Remark 2. Instead of going via the localizations we could have constructed $[f]$ in the theorem directly by taking $B = Ad(K)$ as the adele ring of K, and observing that for any number field E, $Ad(E)^* = \mathfrak{J}(E)$.

We shall also need a formula for restriction of scalars. Here we shall see for the first time how convenient the Hom-language is for the description of maps on classgroups. Observe that if A, B are $\Omega_{\mathbb{Q}}$-modules then $\Omega_{\mathbb{Q}}$ acts by conjugation on $\mathrm{Hom}_{\Omega_K}(A, B)$: $f^{\omega}(a) = f(a^{\omega^{-1}})^{\omega}$. Thus one gets the usual trace map $\mathrm{Hom}_{\Omega_K} \to \mathrm{Hom}_{\Omega_{\mathbb{Q}}}$. In our case where the group operation is multiplicative we shall write this as "*norm*" map,

$$(2.15) \qquad \mathcal{N}_{K/\mathbb{Q}}\colon \mathrm{Hom}_{\Omega_K}(R_\Gamma, G) \to \mathrm{Hom}_{\Omega_{\mathbb{Q}}}(R_\Gamma, G)$$

(G a multiplicative $\Omega_{\mathbb{Q}}$-module). It is defined by

$$(2.16) \qquad \mathcal{N}_{K/\mathbb{Q}} f(\chi) = \prod_{\sigma} f(\chi^{\sigma^{-1}})^{\sigma}$$

where $\{\sigma\}$ is a right transversal of Ω_K in $\Omega_{\mathbb{Q}}$, i.e., a complete set of representatives in $\Omega_{\mathbb{Q}}$ for the embeddings $K \to \mathbb{Q}^c$. The definition is independent of the choice of $\{\sigma\}$.

Theorem 2. *There is a commutative diagram*

$$\begin{array}{ccc} \mathrm{Cl}(\mathfrak{o}_K\Gamma) \cong & \mathrm{Hom}_{\Omega_K}(R_\Gamma, \mathfrak{J}(\mathbb{Q}^c))/[\mathrm{Hom}_{\Omega_K}(R_\Gamma, \mathbb{Q}^{c*}) \cdot \mathrm{Det}(\mathfrak{U}\mathfrak{o}_K\Gamma)] \\ \downarrow & \downarrow \mathcal{N}_{K/\mathbb{Q}} \\ \mathrm{Cl}(\mathbb{Z}\Gamma) \cong & \mathrm{Hom}_{\Omega_{\mathbb{Q}}}(R_\Gamma, \mathfrak{J}(\mathbb{Q}^c))/[\mathrm{Hom}_{\Omega_{\mathbb{Q}}}(R_\Gamma, \mathbb{Q}^{c*}) \cdot \mathrm{Det}(\mathfrak{U}\mathbb{Z}\Gamma)], \end{array}$$

where the rows are the isomorphisms (2.13), *the right hand column is induced by* $\mathcal{N}_{K/\mathbb{Q}}$ (*and indeed* $\mathcal{N}_{K/\mathbb{Q}}$ *maps* $\mathrm{Det}(\mathfrak{U}\mathfrak{o}_K\Gamma)$) *into* $\mathrm{Det}(\mathfrak{U}\mathbb{Z}\Gamma)$) *and the left hand column is* $(X)_{\mathfrak{o}_K\Gamma} \mapsto (X)_{\mathbb{Z}\Gamma}$, *for any locally free* $\mathfrak{o}_K\Gamma$*-module.* □

As $\mathfrak{o}_K\Gamma$ is free over $\mathbb{Z}\Gamma$, a locally free (or a free) $\mathfrak{o}_K\Gamma$-module is indeed a locally free (a free) $\mathbb{Z}\Gamma$-module.

For the background to this theorem see II §3, in particular Theorem 13.

We now consider $\mathbb{Q}\Gamma$. A character χ will be called *symplectic* if it corresponds to a symplectic representation, i.e., a representation T which factorizes through $\Gamma \to Sp_{2m}(\mathbb{Q}^c)$, Sp_{2m} the symplectic group. An equivalent property is that the composite map $\Gamma \to \mathrm{GL}_{2m}(\mathbb{C})$ factorizes through $\mathrm{GL}_m(\mathbb{H})$. We write R_Γ^s for the subgroup of R_Γ generated by symplectic characters. Any actual character lying in R_Γ^s is then indeed symplectic. Accordingly we shall call the virtual characters in R_Γ^s also symplectic. Note here also that R_Γ^s is $\Omega_{\mathbb{Q}}$-stable.

Now let G be an $\Omega_{\mathbb{Q}}$-submodule of $\mathfrak{J}(\mathbb{Q}^c)$, e.g., $\mathfrak{U}(\mathbb{Q}^c)$. As the $\chi \in R_\Gamma^s$ are real valued, we conclude that for $f \in \mathrm{Hom}_{\Omega_{\mathbb{Q}}}(R_\Gamma, G)$, the values $f(\chi)_{\mathfrak{p}}$ will be real, for all $\chi \in R_\Gamma^s$ and all infinite prime divisors $\mathfrak{p}$, whether real or complex. Those f for which moreover $f(\chi)_{\mathfrak{p}} > 0$, for all $\chi \in R_\Gamma^s$ and all infinite prime divisors $\mathfrak{p}$, will form a subgroup which we shall denote by $\mathrm{Hom}^+_{\Omega_{\mathbb{Q}}}(R_\Gamma, G)$. Now R_Γ^s is generated by the irreducible symplectic characters and the characters $\chi = \theta + \bar{\theta}$, $\bar{\theta}$ the conjugate complex of θ. For the latter type, and for all infinite $\mathfrak{p}$, we get

$f(\chi)_{\mathfrak{p}} = f(\theta)_{\mathfrak{p}}\overline{f(\theta)_{\mathfrak{p}}} > 0$, whenever $f \in \mathrm{Hom}_{\Omega_{\mathbb{Q}}}(R_\Gamma, G)$. Thus for $\mathrm{Hom}^+_{\Omega_{\mathbb{Q}}}(R_\Gamma, G)$ it will suffice to postulate the condition $f(\chi)_{\mathfrak{p}} > 0$ for all irreducible symplectic χ.

For the evaluation of groups $\mathrm{Hom}^+_{\Omega_{\mathbb{Q}}}(R_\Gamma, G)$, with e.g., $G = \mathfrak{J}(\mathbb{Q}^c)$ or $\mathfrak{U}(\mathbb{Q}^c)$ etc., we may again replace $\mathbb{Q}^c$ by some sufficiently big number field E. Here the property is crucial that an idele a of E is positive at all infinite prime divisors of E, whether real or not, if and only if it fulfils the same condition in any extension E' of E. The usual definition of "totally positive", in terms of real prime divisors only, does not satisfy such a criterion.

2.1. Proposition. *If $\mathfrak{A}$ is an order in $\mathbb{Q}\Gamma$ then*

$$\mathrm{Det}(\mathfrak{U}\mathfrak{A}) \subset \mathrm{Hom}^+_{\Omega_{\mathbb{Q}}}(R_\Gamma, \mathfrak{U}(\mathbb{Q}^c)).$$

Proof. The inclusion relation is certainly true when the $^+$ is omitted (see (2.10)). Let then T be a symplectic representation of Γ. Localizing at infinity we have $\mathfrak{A}_\infty = \mathbb{R}\Gamma$, and so we get T: $(\mathbb{R}\Gamma)^* \to \mathrm{GL}_{2m}(\mathbb{C})$. This factorizes through $(\mathbb{R}\Gamma)^* \to \mathrm{GL}_m(\mathbb{H})$ and it is well known that the image of $\mathrm{GL}_m(\mathbb{H}) \to \mathrm{GL}_{2m}(\mathbb{C}) \xrightarrow{\mathrm{Det}} \mathbb{C}^*$ is real and positive. In other words for $a \in (\mathbb{R}\Gamma)^*$, $\mathrm{Det}_\chi(a) > 0$, χ being the character of T. □

Taking in particular $\mathfrak{A} = \mathbb{Z}\Gamma$ we get an exact sequence

$$(2.17)\quad 1 \to D(\mathbb{Z}\Gamma) \to \mathrm{Cl}(\mathbb{Z}\Gamma) \to \mathrm{Hom}_{\Omega_{\mathbb{Q}}}(R_\Gamma, \mathfrak{J}(\mathbb{Q}^c))/[\mathrm{Hom}_{\Omega_{\mathbb{Q}}}(R_\Gamma, \mathbb{Q}^{c*})\,\mathrm{Hom}^+_{\Omega_{\mathbb{Q}}}(R_\Gamma, \mathfrak{U}(\mathbb{Q}^c))] \to 1,$$

where the so called *kernel group* $D(\mathbb{Z}\Gamma)$ is defined as the kernel of the right hand surjection from $\mathrm{Cl}(\mathbb{Z}\Gamma)$. For $D(\mathbb{Z}\Gamma)$ we immediately get from Theorem 1 the isomorphism

$$(2.18)\quad D(\mathbb{Z}\Gamma) \cong \mathrm{Hom}^+_{\Omega_{\mathbb{Q}}}(R_\Gamma, \mathfrak{U}(\mathbb{Q}^c))/[\mathrm{Hom}^+_{\Omega_{\mathbb{Q}}}(R_\Gamma, \mathfrak{Y}(\mathbb{Q}^c)) \cdot \mathrm{Det}(\mathfrak{U}\mathbb{Z}\Gamma)],$$

where $\mathfrak{Y}(\mathbb{Q}^c)$ is the group of global units in $\mathbb{Q}^c$, i.e., of units of the integral closure of $\mathbb{Z}$ in $\mathbb{Q}^c$. Equivalently $\mathfrak{Y}(\mathbb{Q}^c)$ is the direct limit (or union) of the groups $\mathfrak{Y}(E)$ of global units in number fields E.

Interpretation 1. If $\mathfrak{M}$ is a maximal order in $\mathbb{Q}\Gamma$ then indeed

$$(2.19)\qquad \mathrm{Det}(\mathfrak{U}\mathfrak{M}) = \mathrm{Hom}^+_{\Omega_{\mathbb{Q}}}(R_\Gamma, \mathfrak{U}(\mathbb{Q}^c)).$$

This is really a local result. Indeed if $\mathfrak{U}_p(\mathbb{Q}^c)$ denotes the group of units of integers in $(\mathbb{Q}^c)_p$, or equivalently the unit ideles of $\mathbb{Q}^c$ which have components 1 at all prime divisors not above p, we have

2.2. Proposition. *For all finite p*

$$\mathrm{Det}(\mathfrak{M}_p^*) = \mathrm{Hom}_{\Omega_{\mathbb{Q}}}(R_\Gamma, \mathfrak{U}_p(\mathbb{Q}^c)).$$

Also

$$\mathrm{Det}(\mathbb{R}\Gamma^*) = \mathrm{Hom}^+_{\Omega_{\mathbb{Q}}}(R_\Gamma, (\mathbb{Q}^c \otimes_{\mathbb{Q}} \mathbb{R})^*). \qquad \square$$

(For a proof see II.§1).

From (2.19) and (2.17) we then get

$$(2.20) \qquad D(\mathbb{Z}\Gamma) = \mathrm{Ker}[\mathrm{Cl}(\mathbb{Z}\Gamma) \to \mathrm{Cl}(\mathfrak{M})].$$

The map on the right is extension of scalars:

$$(X)_{\mathbb{Z}\Gamma} \mapsto (X\mathfrak{M})_{\mathfrak{M}} = (X \otimes_{\mathbb{Z}\Gamma} \mathfrak{M})_{\mathfrak{M}}$$

where $\mathfrak{M} \supset \mathbb{Z}\Gamma$. For rank one modules this is also immediately clear from the description in Theorem 1 (i).

Interpretation 2. Let C_i be the ideal classgroups defined in §1, following Eq. (1.19). Thus if $\mathbb{Q}\Gamma = \prod_i A_i$, where A_i is a simple algebra with centre F_i, then C_i is the group of ideal classes of $\mathfrak{o}_{F_i}$, either modulo principal ideals, or modulo totally positive principal ideals. The A_i correspond to the $\Omega_{\mathbb{Q}}$-orbits of irreducible characters of Γ. For each i, choose a representative χ_i out of the corresponding orbit. If $f \in \mathrm{Hom}_{\Omega_{\mathbb{Q}}}(R_\Gamma, \mathfrak{J}(\mathbb{Q}^c))$, then $f(\chi_i) \in \mathfrak{J}(F_i)$, as $F_i = \mathbb{Q}(\chi_i)$ is the field of values of χ_i on Γ and thus Ω_{F_i} is precisely the stabilizer of χ_i in $\Omega_{\mathbb{Q}}$. Thus the evaluation map

$$f \mapsto \{f(\chi_i)\}_i$$

yields an isomorphism

$$\mathrm{Hom}_{\Omega_{\mathbb{Q}}}(R_\Gamma, \mathfrak{J}(\mathbb{Q}^c)) \cong \prod_i \mathfrak{J}(F_i).$$

By checking on the effect of this isomorphism on subgroups of the Hom group we end up with an isomorphism

$$(2.21) \qquad D(\mathbb{Z}\Gamma) \cong \mathrm{Ker}\left[\mathrm{Cl}(\mathbb{Z}\Gamma) \to \prod_i C_i\right].$$

Example 1 (cf. [F17], Appendix). Let S be a commutative ring and X a right $S\Gamma$-module. Then $\hat{X} = \mathrm{Hom}_S(X, S)$ has again the structure of a right $S\Gamma$-module. Defining the action of $\hat{X}$ on X in terms of a pairing

$$\langle\ ,\ \rangle\colon \hat{X} \times X \to S$$

we have $\langle y\gamma, x\rangle = \langle y, x\gamma^{-1}\rangle$, $y \in \hat{X}$, $x \in X$, $\gamma \in \Gamma$. Moreover $\hat{X}$ may be identified with $\mathrm{Hom}_{S\Gamma}(X, S\Gamma)$, via a pairing

$$[\ ,\ ]\colon \hat{X} \times X \to S\Gamma$$

where $[y, x] = \sum_\gamma \langle y, x\gamma^{-1}\rangle\gamma$. Then $[y\gamma, x] = \gamma^{-1}[y, x]$. The action of $S\Gamma$ on

$\hat{X}$ can also be described in terms of the involution on $S\Gamma$ given by $\overline{\sum_\gamma s_\gamma \gamma} = \sum s_\gamma \gamma^{-1}$ ($s_\gamma \in S$). We have

$$[ya, x] = \bar{a}[y, x] \qquad \text{for} \quad a \in S\Gamma.$$

Now let $\mathfrak{o}_K$ be again the ring of integers in a number field K. We have an involution $^-$ on R_Γ, where $\bar{\chi}(\gamma) = \chi(\gamma^{-1})$. In terms of the field $\mathbb{C}$, $\bar{\chi}$ is the complex conjugate of χ. In terms of representation theory, $\bar{\chi}$ is the *contragredient* of χ – i.e., if χ corresponds to a $\mathbb{Q}^c\Gamma$-module X, then $\bar{\chi}$ corresponds to $\hat{X}$. The map $f \mapsto \bar{f}$ with $\bar{f}(\chi) = f(\bar{\chi})$ then sets up an automorphism of $\mathrm{Hom}_{\Omega_K}(R_\Gamma, A)$ for any Ω_K-module A. Next observe that if X is a locally free $\mathfrak{o}_K\Gamma$-module, then so is $\hat{X}$ and we wish to describe this new module in terms of the action of $^-$ on the Hom group. Let X be then of rank one, so that $\hat{X}$ is also of rank one. We shall show with reference to Theorem 1 (i)

(2.22) If X gives rise to the class $[f]$, then $\hat{X}$ gives rise to $[\bar{f}^{-1}]$.

Indeed, we may take X as a right fractional ideal of $\mathfrak{o}_K\Gamma$ in $K\Gamma$, so that in the description of Theorem 1(i), we may take $v = 1$ in $V = K\Gamma$. Also now $x_\mathfrak{p} = \lambda_\mathfrak{p} \in (K_\mathfrak{p}\Gamma)^*$. For the dual module we have

$$(\hat{X})_\mathfrak{p} = \mathrm{Hom}_{\mathfrak{o}_{K,\mathfrak{p}}\Gamma}(X_\mathfrak{p}, \mathfrak{o}_{K,\mathfrak{p}}\Gamma) = [a \in K_\mathfrak{p}\Gamma, \bar{a}X_\mathfrak{p} \subset \mathfrak{o}_{K,\mathfrak{p}}\Gamma].$$

Thus $(\hat{X})_\mathfrak{p}$ is free on $\bar{x}_\mathfrak{p}^{-1}$. If now $f(\chi)_\mathfrak{p} = \mathrm{Det}_\chi(x_\mathfrak{p})$ then $\mathrm{Det}_\chi(\bar{x}_\mathfrak{p}^{-1}) = \mathrm{Det}_{\bar{\chi}}(x_\mathfrak{p})^{-1} = f(\bar{\chi})_\mathfrak{p}^{-1} = \bar{f}^{-1}(\chi)_\mathfrak{p}$ as we had to show.

Example 2 (cf. [SW2], [U5]). The classical examples for locally free rank one $\mathbb{Z}\Gamma$-modules, which are not necessarily free are the Swan-modules $S(r)$, where r is a (say positive) integer prime to the order of Γ; $S(r)$ is defined as the ideal of $\mathbb{Z}\Gamma$ generated by r and by $\sum = \sum_{\gamma \in \Gamma} \gamma$, the sum of the group elements. We shall show that in $\mathrm{Hom}_{\Omega_\mathbb{Q}}(R_\Gamma, \mathfrak{J}(\mathbb{Q}^c))$ the module $S(r)$ is represented by a map f, where

$$(2.23) \quad \begin{cases} f(\chi)_p = 1 & \text{for any prime divisor not dividing order } (\Gamma), \\ f(\chi)_p = r^{\langle \varepsilon, \chi \rangle} & \text{if } p \mid \text{order } (\Gamma). \end{cases}$$

Here ε is the identity representation, $\langle\ ,\ \rangle$ is the usual inner product in R_Γ, i.e., $\langle \varepsilon, \chi \rangle$ is the multiplicity of ε in χ. To prove (2.23) we replace $S(r)$ by the isomorphic module $S'(r) = (r^{-1}e_1 + e_0)S(r)$ where e_0 is the idempotent $(\text{order } \Gamma)^{-1}\sum$, and $e_1 = 1 - e_0$. Note that this module is generated by

$$(r^{-1}e_1 + e_0)\textstyle\sum = \sum \qquad \text{and} \qquad (r^{-1}e_1 + e_0)r = e_1 + re_0.$$

If $p \nmid \text{order } (\Gamma)$ (p finite) then $\mathbb{Z}_p\Gamma$ contains e_1, e_0 and $S'(r)_p = \mathbb{Z}_p\Gamma$. Thus $f(\chi)_p = 1$. The same argument applies at the infinite prime. If on the other hand $p \mid \text{order } (\Gamma)$ then $\sum = r^{-1}(e_1 + re_0)\sum \in (e_1 + re_0)\mathbb{Z}_p\Gamma$, i.e., $S'(r)_p$ is free on $e_1 + re_0$, and hence $f(\chi)_p = \mathrm{Det}_\chi(e_1 + re_0)$. If χ is irreducible, $\chi \neq \varepsilon$ then $\mathrm{Det}_\chi(e_1 + re_0) = 1 = r^{\langle \varepsilon, \chi \rangle}$. If $\chi = \varepsilon$ then $\mathrm{Det}_\chi(e_1 + re_0) = r = r^{\langle \varepsilon, \chi \rangle}$. Thus in general $f(\chi)_p = r^{\langle \varepsilon, \chi \rangle}$, as we had to show.

§3. Ramification and Module Structure

We shall recall here the basic facts on ramification which will be needed. For details and proofs see the standard texts – e.g. [Se1] or [F5] – however we will give here a proof of Noether's theorem.

Let E/F be an extension of non-Archimedean local fields, with $[E:\mathbb{Q}_p] < \infty$. We shall view the group $\mathfrak{I}(F)$ of fractional ideals of $\mathfrak{o}_F$ as embedded in $\mathfrak{I}(E)$. For the respective maximal ideals we then have the equation $\mathfrak{p}_F = \mathfrak{p}_E^e$, where $e = e(E/F)$ is the *ramification index*. Writing $\bar{E} = \mathfrak{o}_E/\mathfrak{p}_E$, $\bar{F} = \mathfrak{o}_F/\mathfrak{p}_F$ for the residue classfields, the *residue class degree* is

$$f = f(E/F) = [\bar{E}:\bar{F}].$$

Then $ef = [E:F]$. Moreover if $F \supset G \supset \mathbb{Q}_p$, then

$$e(E/G) = e(E/F)e(F/G),$$

$$f(E/G) = f(E/F)f(F/G).$$

If $e(E/F) = 1$ then E/F is *non-ramified*, if $(e(E/F), p) = 1$ then E/F is *tame*.

Every extension E/F has a unique maximal non-ramified subextension E'/F. If in particular E/F is Galois with Galoisgroup Δ, then we have an exact sequence

$$(3.1) \qquad 1 \to \Delta_0 \to \Delta \to \bar{\Delta} \to 1$$

where $\bar{\Delta} = \mathrm{Gal}(\bar{E}/\bar{F})$ and $\Delta \to \bar{\Delta}$ is given by "restriction" of operators to $\bar{E}$. Δ_0 is called the *inertia group* of E/F. Moreover E' is the fixed field E^{Δ_0}. The order of Δ_0 is e. Thus E/F is tame if, and only if, Δ_0 is of order prime to p. Moreover in this case Δ_0 is cyclic, E' contains the primitive e-th roots of unity and $E = E'(\pi^{1/e})$, where π is an element of E' with $(\pi) = \mathfrak{p}_{E'}$.

With Δ as above, suppose Δ is a subgroup of a group Γ. We then also consider the Galois algebra

$$(3.2) \qquad A = \prod E^{\gamma} \qquad \text{(direct sum of fields)},$$

where the product extends over a right transversal $\{\gamma\}$ of Δ in Γ, with the E^{γ} isomorphic copies of E. The group Γ acts in the obvious way, i.e. we have isomorphisms of $F\Gamma$-modules

$$(3.3) \qquad A \cong \mathrm{Hom}_{F\Delta}(F\Gamma, E) \qquad (\cong E \otimes_{F\Delta} F\Gamma).$$

Write $\mathfrak{o}_A = \prod \mathfrak{o}_E^{\gamma}$. This is the integral closure of $\mathfrak{o}_F$ in A. In the next theorem let $\mathbb{Q}_p \subset G \subset F \subset E$, as above, but with E/F at first not necessarily Galois. Let $t_{E/F}$ denote the trace.

Theorem 3. [Note (3)]. *The two conditions*

(i) *E/F is tame,*

(ii) $t_{E/F}\mathfrak{o}_E = \mathfrak{o}_F$,

are equivalent.

Suppose moreover that E/F is Galois with Galois group Δ, and let A, Γ be given as above ((3.2), (3.3)). *Then conditions* (i), (ii) *are also equivalent with each of*

(iii) $\mathfrak{o}_E$ *is free over* $\mathfrak{o}_G\Delta$,

(iv) $\mathfrak{o}_A$ *is free over* $\mathfrak{o}_G\Gamma$.

Remark. We see that if (iii) or (iv) hold for some field G as above, then they hold for all such G.

Proof. For $a \in \mathfrak{o}_E$, let $t_1(a)$ be the trace of the endomorphism of the $\bar{F}$-module $\mathfrak{o}_E/\mathfrak{p}_F$, induced by multiplication by a. Then

$$t_1(a) = \overline{t_{E/F}(a)} \tag{3.4}$$

where $\bar{\ }$ denotes the residue class map mod $\mathfrak{p}_E$ or mod $\mathfrak{p}_F$. But, as $\bar{F}$-module, $\mathfrak{o}_E/\mathfrak{p}_F$ is the product of e copies of $\bar{E}$, and hence

$$t_1(a) = e t_{\bar{E}/\bar{F}}(\bar{a}).$$

Hence by (3.4)

$$\overline{t_{E/F}(a)} = e t_{\bar{E}/\bar{F}}(\bar{a}). \tag{3.5}$$

But $t_{E/F}\colon \mathfrak{o}_E \to \mathfrak{o}_F$ is surjective if, and only if, $\bar{t}_{E/F}\colon \mathfrak{o}_E \to \bar{F}$ is surjective, and, by (3.5), this is the case precisely when $e \not\equiv 0 \pmod{p}$. We have now established the equivalence of (i) and (ii).

Now suppose E/F is Galois with group Δ. If (ii) holds, then in particular there is an element $b \in \mathfrak{o}_E$ with $t_{E/F}b = 1$, i.e. $\sum_\delta b^\delta = 1$. Let X be a free $\mathfrak{o}_G\Delta$-module, and g: $X \to \mathfrak{o}_E$ a surjective homomorphism of $\mathfrak{o}_G\Delta$-modules. We shall prove that $\mathfrak{o}_E$ is projective over $\mathfrak{o}_G\Delta$, by constructing a homomorphism f: $\mathfrak{o}_E \to X$ of $\mathfrak{o}_G\Delta$-modules, with $g \circ f = 1_{\mathfrak{o}_E}$. As $\mathfrak{o}_E$ is free over $\mathfrak{o}_G$ there certainly exists a homomorphism h: $\mathfrak{o}_E \to X$ of $\mathfrak{o}_G$-modules, with $g \circ h = 1_{\mathfrak{o}_E}$. Put $f(x) = \sum_\delta h(x^{\delta^{-1}}b)^\delta$. Then f commutes with Δ-action and

$$\begin{aligned} g \circ f(x) &= \sum g(h(x^{\delta^{-1}}b)^\delta) = \sum (g(h(x^{\delta^{-1}}b)))^\delta \\ &= \sum (x^{\delta^{-1}}b)^\delta = x. \end{aligned}$$

Thus indeed $\mathfrak{o}_E$ is projective over $\mathfrak{o}_G\Delta$, and as $E = \mathfrak{o}_E \otimes_{\mathfrak{o}_G} G$ is certainly free over $G\Delta$ it follows from a theorem of Swan (cf. [Sw1] Corollary 6.4, or [SE] Theorem 2.21), that $\mathfrak{o}_E$ is free over $\mathfrak{o}_G\Delta$. Thus (ii) ⇒ (iii).

Next (iii) ⇒ (iv), as

$$\mathfrak{o}_A \cong \mathfrak{o}_E \otimes_{\mathfrak{o}_G\Delta} \mathfrak{o}_G\Gamma.$$

If (iv) holds, then, on representing 1 in terms of a free basis of $\mathfrak{o}_A$ over $\mathfrak{o}_G\Gamma$, we find an element $a \in \mathfrak{o}_A$ with $t_{A/F}a = 1$. This implies that for some γ, the E^γ-component of a (in the decomposition (3.2)) has trace a unit. By replacing a by $a^{\gamma^{-1}}$ we conclude that $t_{E/F}(a^{\gamma^{-1}})$ is a unit, i.e., we get (ii). This then completes the proof of Theorem 3. □

Now we look at a Galois extension N/K of number fields with Galois group Γ. Given a non-zero prime ideal $\mathfrak{p}$ of $\mathfrak{o}_K$, and a prime ideal $\mathfrak{P}$ of $\mathfrak{o}_N$ above it, the ramification index $e(N_{\mathfrak{P}}/K_{\mathfrak{p}}) = e_{\mathfrak{p}}$ only depends on $\mathfrak{p}$. The completion $N_{\mathfrak{P}}$ is a local extension field of $K_{\mathfrak{p}}$ with Galois group $\Gamma_{\mathfrak{P}} = \Delta \subset \Gamma$; if we put $K_{\mathfrak{p}} = F$, $N_{\mathfrak{P}} = E$, then $N_{\mathfrak{p}}$ ($= N \otimes_K K_{\mathfrak{p}}$ by definition) plays the role of A in the preceding theorem. Moreover if L is a subfield of K, let $L_{\mathfrak{p}}$ be its completion at the prime below $\mathfrak{p}$. We then have

Corollary 1. *The following conditions are equivalent*:
(i) $N_{\mathfrak{P}}/K_{\mathfrak{p}}$ *is tame, i.e.,* N/K *is tame at* $\mathfrak{p}$.
(ii) $\mathfrak{o}_{N,\mathfrak{P}}$ *is a free* $\mathfrak{o}_{L_{\mathfrak{p}}}$ Δ*-module.*
(iii) $\mathfrak{o}_{N_{\mathfrak{p}}}$ *is a free* $\mathfrak{o}_{L_{\mathfrak{p}}}$ Γ*-module.* □

Collecting all primes together, we now get

Corollary 2. *The following conditions are equivalent*:
(i) N/K *is tame.*
(ii) $\mathfrak{o}_N$ *is locally free over* $\mathfrak{o}_L\Gamma$.
(iii) $t_{N/K}\mathfrak{o}_N = \mathfrak{o}_K$. □

§4. Resolvents

The topic of this section is the second basic ingredient of the theory, the (generalized) resolvent. To begin with, E/F is a Galois extension of fields of finite degree with Galois group Γ, and R_Γ is the ring of virtual characters over $E^c = F^c$. Although this is not really necessary, we shall assume the fields to be of characteristic zero, so as to be able to keep to ordinary group representations.

Let B be a commutative F-algebra. Then $E \otimes_F B$ is free on one generator over $B\Gamma$, Γ acting via E. If $a \in E \otimes_F B$ we define its resolvent element by $\sum a^\gamma \gamma^{-1} \in (E \otimes_F B)\Gamma$.

4.1. Proposition. *If* a *is a free generator of* $E \otimes_F B$ *over* $B\Gamma$ *then* $\sum a^\gamma \gamma^{-1} \in ((E \otimes_F B)\Gamma)^*$.

Remark. The converse is true as well.

Proof. The determinant of the basis $\{a^\gamma\}$ satisfies

$$\mathrm{Det}(a^{\delta\sigma^{-1}})_{\delta,\sigma} \in (E \otimes_F B)^*. \tag{4.1}$$

This is classically true for a basis $\{a^\gamma\}$ of E over F, which can also be viewed as a basis of $E \otimes_F B$ over B. Transforming the basis means multiplying the basis determinant by an invertible element. Thus (4.1) will always hold.

We wish to solve

$$\left(\sum_{\gamma\in\Gamma} a^{\gamma}\gamma^{-1}\right)\left(\sum_{\delta\in\Gamma} c_{\delta}\delta\right) = 1, \qquad c_{\delta}\in E\otimes_F B, \tag{4.2}$$

i.e.,

$$\sum_{\delta\in\Gamma} c_{\delta}a^{\delta\sigma^{-1}} = \begin{cases} 1, & \sigma = 1, \\ 0, & \sigma\in\Gamma, \quad \sigma\neq 1, \end{cases} \tag{4.3}$$

and by (4.1) this is possible. □

Now define the *resolvent* of a by

$$(a\,|\,\chi) = \mathrm{Det}_{\chi}(\textstyle\sum a^{\gamma}\gamma^{-1}). \tag{4.4}$$

In other words if $T\colon \Gamma\to \mathrm{GL}_n(F^c) = \mathrm{GL}_n(E^c)$ is a representation with character χ, then

$$(a\,|\,\chi) = \mathrm{Det}(\textstyle\sum a^{\gamma}T(\gamma)^{-1}).$$

From Proposition 4.1 and by (2.8), we now get the

Corollary. *If a is a free generator of $E\otimes_F B$ over $B\Gamma$, then*

$$(a\,|\,\chi)\in(F^c\otimes_F B)^*,$$

and so the map $\chi\mapsto(a\,|\,\chi)$ lies in $\mathrm{Hom}_{\Omega_E}(R_{\Gamma},(F^c\otimes_F B)^*)$. □

In the particular case when E/F is Abelian and χ is an Abelian character, i.e., a homomorphism $\Gamma\to F^{c*}$ then of course $(a\,|\,\chi)$ is the Lagrange resolvent. The non-Abelian generalization of this most classical of notions is mentioned in passing by Noether (cf. [No]) and was first formally treated in a paper in 1966 (cf. [F4]). The definition there is formally different from that stated here, but equivalent to it. The new concept was then developed and applied in [F8], [F13] and [F15]. Discussions with S. Ullom were helpful at this stage. The role of resolvents as one cornerstone of the theory, as we see it today, received its final formulation in the basic paper [F17].

The principal application is based on the next proposition.

4.2. Proposition. *If $a\in E\otimes_F B$, $\lambda\in B\Gamma$, and if a^{λ} is the image of a under the operator λ, then in $(E\otimes_F B)\Gamma$*

$$\sum_{\gamma} a^{\lambda\gamma}\gamma^{-1} = (\textstyle\sum a^{\gamma}\gamma^{-1})\lambda.$$

Proof. Write

$$\lambda = \sum_{\sigma\in\Gamma} c_{\sigma}\sigma, \qquad c_{\sigma}\in B.$$

Then

$$a^{\lambda} = \sum_{\sigma} c_{\sigma}a^{\sigma},$$

and therefore

$$\sum_{\gamma} a^{\lambda\gamma}\gamma^{-1} = \sum_{\sigma,\gamma} c_\sigma a^{\sigma\gamma}\gamma^{-1}\sigma^{-1}\sigma = \sum_{\gamma} a^{\gamma}\gamma^{-1} \cdot \sum_{\sigma} c_\sigma \sigma,$$

as we had to show. □

Corollary. *If $\lambda \in (B\Gamma)^*$ then, together with a, also a^λ is a free generator of $E \otimes_F B$ over $B\Gamma$, and*

$$(4.5) \qquad (a^\lambda \mid \chi) = (a \mid \chi)\, \mathrm{Det}_\chi(\lambda).$$

The first assertion is obvious, the second one is an immediate consequence of the proposition. □

Now suppose F is the quotient field of a Dedekind domain $\mathfrak{o}$ and $B = F \otimes_{\mathfrak{o}} \mathfrak{o}_{\mathfrak{p}}$, where $\mathfrak{o}_{\mathfrak{p}}$ is the completion of $\mathfrak{o}$ at some prime ideal $\mathfrak{p}$. Write $\mathfrak{o}_E$ for the integral closure of $\mathfrak{o}$ in E and $\mathfrak{o}_{E,\mathfrak{p}} = \mathfrak{o}_E \otimes_{\mathfrak{o}} \mathfrak{o}_{\mathfrak{p}}$.

4.3. Proposition. *Suppose that $\mathfrak{p}$ is non-ramified in E and that a is a free generator of $\mathfrak{o}_{E,\mathfrak{p}}$ over $\mathfrak{o}_{\mathfrak{p}}\Gamma$. Then*

$$\sum a^\gamma \gamma^{-1} \in (\mathfrak{o}_{E,\mathfrak{p}}\Gamma)^*.$$

Proof. Clearly $\sum a^\gamma\gamma^{-1} \in \mathfrak{o}_{E,\mathfrak{p}}\Gamma$. We have to show that the solution $\sum_\delta c_\delta \delta$ of (4.2) also lies in $\mathfrak{o}_{E,\mathfrak{p}}\Gamma$, i.e., that the c_δ lie in $\mathfrak{o}_{E,\mathfrak{p}}$. This however is now immediate as the determinant $\mathrm{Det}(a^{\delta\sigma^{-1}})$ (which has entries in $\mathfrak{o}_{E,\mathfrak{p}}$) actually lies in $\mathfrak{o}^*_{E,\mathfrak{p}}$. □

From now on we consider again a Galois extension N/K of number fields with Galois group Γ. We assume moreover that N/K is tame. By Corollary 2 to Theorem 3, $\mathfrak{o}_N$ is locally free over $\mathfrak{o}_K\Gamma$, and thus defines a class

$$(\mathfrak{o}_N)_{\mathfrak{o}_K\Gamma} \in \mathrm{Cl}(\mathfrak{o}_K\Gamma).$$

Using resolvents, we shall now find an element $f \in \mathrm{Hom}_{\Omega_K}(R_\Gamma, \mathfrak{J}(\mathbb{Q}^c))$ whose class $[f]$, in the sense of Theorem 1 (i) (or (ii)), is the image of $(\mathfrak{o}_N)_{\mathfrak{o}_K\Gamma}$. Such an f will be called a *representative* of $(\mathfrak{o}_N)_{\mathfrak{o}_K\Gamma}$. It will determine the class $(\mathfrak{o}_N)_{\mathfrak{o}_K\Gamma}$. Analogously we shall speak later of a representative of $(\mathfrak{o}_N)_{\mathbb{Z}\Gamma}$. Recall that $\mathfrak{o}_{N,\mathfrak{p}} = \mathfrak{o}_N \otimes_{\mathfrak{o}_K} \mathfrak{o}_{K,\mathfrak{p}}$ is free of rank one over $\mathfrak{o}_{K,\mathfrak{p}}\Gamma$, for all prime divisors $\mathfrak{p}$ of K.

Theorem 4. *Let b be a free generator of N over $K\Gamma$ and, for each prime divisor $\mathfrak{p}$ of K, let $a_\mathfrak{p}$ be a free generator of $\mathfrak{o}_{N,\mathfrak{p}}$ over $\mathfrak{o}_{K,\mathfrak{p}}\Gamma$. For $\chi \in R_\Gamma$, define $(a \mid \chi) \in \prod (\mathbb{Q}^c_\mathfrak{p})^*$ by*

$$(a \mid \chi)_\mathfrak{p} = (a_\mathfrak{p} \mid \chi).$$

Then

$$(a \mid \chi) \in \mathfrak{J}(\mathbb{Q}^c)$$

and the map

$$\chi \mapsto (a \mid \chi) \cdot (b \mid \chi)^{-1}$$

is a representative of $(\mathfrak{o}_N)_{\mathfrak{o}_K\Gamma}$.

This theorem is the main reason for introducing resolvents into the subject. It also justifies the use of the Hom-language for classgroups, introduced in Theorem 1.

Remark. We can view a itself as an adele, with local components $a_\mathfrak{p}$, and indeed as a free generator of $Ad(\mathfrak{o}_N)$ over $Ad(\mathfrak{o}_K)\Gamma$. Then $(a \mid \chi)$ is just its resolvent – taking now $B = Ad(K)$.

Proof. Refer back to Theorem 1 (i). Take $X = \mathfrak{o}_N$, $V = N$, $v = b$, $x_\mathfrak{p} = a_\mathfrak{p}$. Then we have

$$a_\mathfrak{p} = b^{\lambda_\mathfrak{p}}, \tag{4.6}$$

and $(\mathfrak{o}_N)_{\mathfrak{o}_K\Gamma}$ is represented by f, where

$$f_\mathfrak{p}(\chi) = \mathrm{Det}_\chi(\lambda_\mathfrak{p}). \tag{4.7}$$

But by (4.5), with the appropriate change in notation,

$$\mathrm{Det}_\chi(\lambda_\mathfrak{p}) = \frac{(a_\mathfrak{p} \mid \chi)}{(b \mid \chi)} = \frac{(a \mid \chi)_\mathfrak{p}}{(b \mid \chi)}. \tag{4.8}$$

Now the elements of $\prod_\mathfrak{p} \mathbb{Q}_\mathfrak{p}^{c*}$ with local components $\mathrm{Det}_\chi(\lambda_\mathfrak{p})$, and $(b \mid \chi)_\mathfrak{p}$ respectively, are ideles, hence so is $(a \mid \chi)$ and so by (4.7), (4.8) $\chi \mapsto (a \mid \chi)(b \mid \chi)^{-1}$ indeed represents $(\mathfrak{o}_N)_{\mathfrak{o}_K\Gamma}$. □

The object, in which we are really interested, is the class of $\mathfrak{o}_N$ over $\mathbb{Z}\Gamma$. We shall use throughout the notation

$$(\mathfrak{o}_N)_{\mathbb{Z}\Gamma} = U_{N/K}. \tag{4.9}$$

Here we shall apply Theorem 2. As there, let $\{\sigma\}$ be a right transversal of Ω_K in $\Omega_\mathbb{Q}$. We define

$$\mathcal{N}_{K/\mathbb{Q}}(b \mid \chi) = \prod_\sigma (b \mid \chi^{\sigma^{-1}})^\sigma, \tag{4.10}$$

and the same with a replacing b. Note that this definition of $\mathcal{N}_{K/\mathbb{Q}}$ does still depend on the choice of $\{\sigma\}$ (but only modulo roots of unity, as we shall presently see). Keeping $\{\sigma\}$ fixed, and observing that $\mathcal{N}_{K/\mathbb{Q}}(a \mid \chi)\mathcal{N}_{K/\mathbb{Q}}(b \mid \chi)^{-1} = \mathcal{N}_{K/\mathbb{Q}} f(\chi)$, where $f(\chi) = (a \mid \chi)(b \mid \chi)^{-1}$, we have the

Corollary. $U_{N/K}$ *is represented in* $\mathrm{Hom}_{\Omega_\mathbb{Q}}(R_\Gamma, \mathfrak{J}(\mathbb{Q}^c))$ *by*

$$\chi \mapsto \mathcal{N}_{K/\mathbb{Q}}(a \mid \chi)\mathcal{N}_{K/\mathbb{Q}}(b \mid \chi)^{-1}.$$ □

We briefly return to resolvents in the general situation we had considered at the beginning of §4, i.e., $\Gamma = \mathrm{Gal}(E/F)$, F an arbitrary field (of characteristic zero), B a commutative F-algebra, and a a free generator of $E \otimes_F B$ over $B\Gamma$. We moreover suppose that F is an extension of finite degree of a field F_0, and that $B = F \otimes_{F_0} B_0$, B_0 a commutative F_0-algebra. We let Ω_{F_0} act on $F^c \otimes_{F_0} B_0$ via F^c. The restriction of Det_χ to Γ is an Abelian character of Γ, denoted by $\det_\chi$, which via the surjection $\Omega_F \to \Gamma$ also becomes an Abelian character of Ω_F.

4.4. Proposition. (i) *If* $\omega \in \Omega_F$ *then*

$$(a \mid \chi^{\omega^{-1}})^\omega = (a \mid \chi) \det_\chi(\omega).$$

(ii) *If both* $\{\delta\}$ *and* $\{\sigma\}$ *are right transversals of* Ω_F *in* Ω_{F_0}, *then*

$$\prod_\delta (a \mid \chi^{\delta^{-1}})^\delta = \prod_\sigma (a \mid \chi^{\sigma^{-1}})^\sigma \det_\chi(\gamma) \qquad (\gamma \in \Gamma \text{ independent of } \chi).$$

Proof. We shall prove (i), assuming χ to be the character of a representation T. Then

$$\begin{aligned}(a \mid \chi^{\omega^{-1}})^\omega &= \left(\mathrm{Det}\left(\sum_\gamma a^\gamma T^{\omega^{-1}}(\gamma)\right)\right)^\omega \\ &= \mathrm{Det}\left(\sum_\gamma a^{\gamma\omega} T(\gamma)^{-1}\right).\end{aligned}$$

Define T on Ω_F via its surjection onto Γ. Then we get

$$\begin{aligned}(a \mid \chi^{\omega^{-1}})^\omega &= \mathrm{Det}\left(T(\omega) \cdot \sum_\gamma a^{\gamma\omega} T(\gamma\omega)^{-1}\right) \\ &= \mathrm{Det}\, T(\omega) \cdot \mathrm{Det} \sum_\gamma a^\gamma T(\gamma)^{-1} \\ &= \det_\chi(\omega) \cdot (a \mid \chi),\end{aligned}$$

as required.

For (ii) we use (i). The factors in each product, corresponding to a particular coset $\Omega_F \sigma$ are $(a \mid \chi^{\sigma^{-1}})^\sigma$ and $(a \mid \chi^{\sigma^{-1}\omega^{-1}})^{\omega\sigma}$, respectively, with some $\omega \in \Omega_F$ (depending on σ). They thus differ by $(\det_{\chi^{\sigma^{-1}}}(\omega))^\sigma = \det_\chi(\omega)$. The result is now immediate. □

§5. *L*-Functions and Galois Gauss Sums

In this section the third basic concept for our theory will be introduced, and the fundamental theorem, which underlies the connection between the Galois module structure of rings of integers and the functional equation will be stated. The introductory discussion of L-functions is more leisurely than is strictly needed, but this should help give a proper understanding of the ramifications of our subject.

Once one has given the basic definitions of L-functions, Abelian Gauss sums and conductors of characters of class field theory, via their local decomposition, one can proceed in one of two ways. Either one goes straight via the global theory to the functional equation of the Artin L-function, which yields directly the global Galois Gauss sum and the global root number. Or one continues entirely locally to get the Langlands constants and the local Galois Gauss sums. The two approaches complement each other; both have their merits and really represent different aspects of the same theory. In the present section we shall choose the first, essentially global point of view, as this puts the surprising relation to Galois module structure into its sharpest relief. Subsequently in Chap. III, the proofs will however be reduced to genuine local theorems, and then the central object will be the local Galois Gauss sum, allied to the local, i.e., Langlands constants.

The numerical invariants, such as Galois Gauss sums and root numbers arise in the first place as complex numbers which happen to be algebraic. From this point of view it becomes necessary to consider $\mathbb{Q}^c$ not merely as *an* algebraic closure of $\mathbb{Q}$, but as *the* algebraic closure inside $\mathbb{C}$. There is thus always a distinguished infinite prime divisor, and every algebraic number a possesses one distinguished absolute value, denoted by $|a|$. The complex conjugate $\bar{a}$ is uniquely defined, and so is the *positive* square root of a real algebraic number, giving us the formula $|a| = (a\bar{a})^{1/2}$. All this will be used, see e.g. (5.7), (5.7b), (5.8) and also the choice of the standard positive imaginary square root i of -1 in (5.1).

In the present section it will suffice to view characters as attached to fixed finite Galois groups. Later in Chap. III we shall have to adopt a more sophisticated point of view.

Let F be a local field, here either Archimedean or non-Archimedean. We consider continuous homomorphisms $\theta\colon F^* \to \mathbb{C}^*$ of finite order, also called *multiplicative characters* of F. Example: The identity character $\varepsilon = \varepsilon_F : F^* \to 1$. If $F = \mathbb{C}$ this is the only one, if $F = \mathbb{R}$ there is a further one, "sign", defined by $\operatorname{sign}(a) = a/|a|$ $(= \pm 1)$. We define the *root number* $W(\theta) \in \mathbb{C}$ by

$$\text{(5.1)} \qquad \begin{cases} W(\varepsilon) = 1, \\ W(\operatorname{sign}) = -i, \end{cases}$$

and the *L-functions* of the complex variable s by

$$\text{(5.2)} \qquad \begin{cases} \tilde{L}(s, \varepsilon) = \pi^{-(s+1/2)}\Gamma(s/2)\Gamma((s+1)/2), & F = \mathbb{C}, \\ \left.\begin{array}{l} \tilde{L}(s, \varepsilon) = \pi^{-s/2}\Gamma(s/2) \\ \tilde{L}(s, \operatorname{sign}) = \pi^{-(s+1)/2}\Gamma((s+1)/2) \end{array}\right\}, & F = \mathbb{R}. \end{cases}$$

(No confusion should arise from the fact that here the symbol Γ denotes the Gamma function and π the circle constant).

From now on F will be non-Archimedean, of finite degree over $\mathbb{Q}_p$. We shall write

$$\mathfrak{U}(F) = \mathfrak{U}^{(0)}(F) = \mathfrak{o}_F^*,$$

$$\mathfrak{U}^{(n)}(F) = 1 + \mathfrak{p}_F^n, \qquad n > 0.$$

(Here $\mathfrak{p}_F^n$ is to be viewed as a subset of F.) This is a decreasing sequence of subgroups. If now θ is a multiplicative character of F, then for some $m \geqslant 0$, $\operatorname{Ker}\theta \supset \mathfrak{U}^{(m)}(F)$. Choosing m minimal we define the *conductor* of θ by

(5.3) $$\mathfrak{f}(\theta) = \mathfrak{p}_F^m.$$

If $m = 0$, θ is *non-ramified*. We can then view θ as a function on $F^*/\mathfrak{U}(F)$, i.e., on the group $\mathfrak{J}(F)$ of fractional ideals of F.

If $m > 0$, θ is *ramified*. In this case we shall also associate with θ a function of $\mathfrak{J}(F)$, again denoted by θ, simply setting $\theta(\mathfrak{a}) = 0$. The L-function of θ is now

(5.4) $$\tilde{L}(s, \theta) = (1 - \theta(\mathfrak{p}_F)\mathbf{N}_F\mathfrak{p}_F^{-s})^{-1}.$$

Thus if θ is ramified then $\tilde{L}(s, \theta) = 1$. We shall always use the term "ramified", as it was defined here, i.e. as the negation of "non-ramified".

Next we define the *Gauss sum* $\tau(\theta)$. For this we need the *standard additive character* $\psi_F: F$ (additive) $\to \mathbb{C}^*$. This is the composition of the trace $t_{F/\mathbb{Q}_p}$ with a homomorphism $\psi_p: \mathbb{Q}_p$ (additive) $\to \mathbb{C}^*$, given by

$$\psi_p(\mathbb{Z}_p) = 1, \qquad \psi_p\left(\frac{1}{p^r}\right) = e^{2\pi i/p^r}.$$

Now assume θ is ramified. Choose $c \in F^*$ with $(c) = \mathfrak{f}(\theta)\mathfrak{D}_F$, where $\mathfrak{D}_F$ is the different of $F/\mathbb{Q}_p$. Then

(5.5) $$\tau(\theta) = \sum_u \theta(uc^{-1})\psi_F(uc^{-1}),$$

with u running through a complete set of representatives of $\mathfrak{U}(F) \bmod 1 + \mathfrak{f}(\theta)$. It is clear on a moments consideration that $\tau(\theta)$ is independent of (a) the choice of representatives u and (b) the choice of c within the stated condition. If θ is non-ramified, we have

(5.6) $$\tau(\theta) = \theta(\mathfrak{D}_F)^{-1}.$$

One shows that for any θ

(5.7) $$|\tau(\theta)| = \mathbf{N}_F\mathfrak{f}(\theta)^{1/2}$$

where $\mathbf{N}_F\mathfrak{f}(\theta)$ is the absolute norm of $\mathfrak{f}(\theta)$ as an ideal of $\mathfrak{o}_F$ and $^{1/2}$ is the positive square root.

We illustrate the proof in the case when $\mathfrak{f}(\theta) = \mathfrak{p}_F$, which together with the trivial case $\mathfrak{f}(\theta) = \mathfrak{o}_F$, is the only one we need. Indeed

$$\begin{aligned}\tau(\theta)\tau(\theta^{-1}) &= \sum_{u,v} \theta(uv^{-1})\psi_F((u+v)c^{-1}) \\ &= \sum_{v,w} \theta(w)\psi_F(v(w+1)c^{-1}) \qquad (\text{put } w = uv^{-1}).\end{aligned}$$

If $w \not\equiv -1 \pmod{\mathfrak{p}_F}$ then $v \mapsto \psi_F(v(w+1)c^{-1})$ is non-trivial on $\mathfrak{o}_F/\mathfrak{p}_F$, so

$$\sum_{x \in \mathfrak{o}_F \bmod \mathfrak{p}_F} \psi_F(x(w+1)c^{-1}) = 0,$$

i.e.,

$$\sum_{v \in \mathfrak{o}_F^* \bmod \mathfrak{p}_F} \psi_F(v(w+1)c^{-1}) = -1.$$

On the other hand for $w \equiv -1 \pmod{\mathfrak{p}_F}$ the sum over v is $\mathbf{N}_F\mathfrak{p}_F - 1$, as $\psi_F(v(w+1)c^{-1}) = 1$. Thus

$$\tau(\theta)\tau(\theta^{-1}) = \theta(-1)\mathbf{N}_F\mathfrak{p}_F - \sum_w \theta(w) = \theta(-1)\mathbf{N}_F\mathfrak{p}_F. \tag{5.7a}$$

But

$$\overline{\tau(\theta)} = \sum_u \theta^{-1}(uc^{-1})\overline{\psi_F(uc^{-1})} = \sum_u \theta^{-1}(uc^{-1})\psi_F(-uc^{-1}),$$

i.e.,

$$\overline{\tau(\theta)} = \theta(-1)\tau(\theta^{-1}). \tag{5.7b}$$

The two Eqs. (5.7a) and (5.7b) give (5.7). For general $\mathfrak{f}(\theta)$ the computations are a bit longer, but essentially the same.

The *root number* $W(\theta)$ is now defined by

$$W(\theta) = \tau(\bar{\theta})/\mathbf{N}_F\mathfrak{f}(\theta)^{1/2} = \tau(\bar{\theta})/|\tau(\bar{\theta})|, \tag{5.8}$$

with $\bar{\theta} = \theta^{-1}$ the complex conjugate or inverse of θ.

Now we turn to "class field theoretic" characters of a number field K. There are various ways in which one can define these – the important property at this stage is that they have local components. Here we shall define them as continuous homomorphisms $\mathfrak{J}(K) \to \mathbb{C}^*$, which are of finite order and map $K^* \to 1$ (*idele class characters of finite order*). Equivalently we can view such a character as a continuous homomorphism $\theta\colon \mathfrak{J}(K) \to \mathbb{Q}^{c*}$ (with discrete topology) and with $\theta(K^*) = 1$. The composite $K_\mathfrak{p}^* \to \mathfrak{J}(K) \xrightarrow{\theta} \mathbb{C}^*$ is the *local component* $\theta_\mathfrak{p}$ of θ at $\mathfrak{p}$. This $\theta_\mathfrak{p}$ is a multiplicative character of $K_\mathfrak{p}^*$. We then define the conductor of θ by

$$\mathfrak{f}(\theta) = \prod_\mathfrak{p} \mathfrak{f}(\theta_\mathfrak{p}), \tag{5.9}$$

(product over the finite prime divisors). This is a finite product, as $\theta_\mathfrak{p}$ is non-ramified at almost all $\mathfrak{p}$. The global Gauss sum is

$$\tau(\theta) = \prod_\mathfrak{p} \tau(\theta_\mathfrak{p}) \tag{5.10}$$

(product over all finite prime divisors). This is again a finite product. We also write

$$W_\infty(\theta) = \prod_{\mathfrak{p}|\infty} W(\theta_\mathfrak{p}) \tag{5.11}$$

(product over the infinite prime divisors). Now we define the extended global L-function of θ by

$$\tilde{L}(s,\theta) = \prod_\mathfrak{p} \tilde{L}(s,\theta_\mathfrak{p}) \tag{5.12}$$

(product over all prime divisors). This is an infinite product, which certainly converges for (real part of s) > 1. The central theorem is (cf. [He2], [He3], [Tt1], [Lg]).

Hecke's theorem. *$\tilde{L}(s, \theta)$ can be extended to a meromorphic function in the whole complex plane, and it satisfies a functional equation*

$$\tilde{L}(s, \theta) = W(\theta)A(\theta)^{\frac{1}{2}-s}\tilde{L}(1-s, \bar{\theta})$$

where $A(\theta) > 0$ and $|W(\theta)| = 1$. □

From Tate's proof of the functional equation (cf. [Tt1], [Lg]) one has

$$(5.13)\qquad \begin{cases} W(\theta) = \prod_{\mathfrak{p}} W(\theta_{\mathfrak{p}}) & \text{(product over all prime divisors)}, \\ A(\theta) = \mathbf{N}_K(\mathfrak{D}_K \mathfrak{f}(\theta)), & \end{cases}$$

where $\mathfrak{D}_K$ is the absolute different of K. Alternatively, using (5.8)–(5.11) one gets

$$(5.14)\qquad W(\theta) = \tau(\bar{\theta})W_{\infty}(\theta)/\mathbf{N}_K\mathfrak{f}(\theta)^{1/2}.$$

We now turn to characters (not necessarily Abelian characters as hitherto in this section) of Galois groups. Let first E/F be a Galois extension of number fields or of local fields, Archimedean or non-Archimedean, with Galois group Γ. If χ is an Abelian character of Γ, view it also as a character of Γ^{ab}, and let $A_{E/F}$ be the Artin map of F^* (local case) or of $\mathfrak{J}(F)$ (global case) into Γ^{ab}. There is then a character θ_χ (a multiplicative character of F in the local case, an idele class character in the global case) so that $\chi(A_{E/F}a) = \theta_\chi(a)$. (Strictly speaking the map $\chi \mapsto \theta_\chi$ depends on E/F, but in fact by the functorial properties of the Artin symbol this dependence is purely formal, and need not be indicated in our notation). The general aim is then to use this map $\chi \mapsto \theta_\chi$, together with additivity, and inductivity, to define appropriate functions on characters of Galois groups, analogous to those defined above on the characters θ. To formalize this, consider a function g with values in some multiplicative Abelian group G, defined on pairs (L, θ), where L runs through the intermediate fields between F and E, and θ is a multiplicative character of L (local case), or an idele class character of L (global case). A function g' with values in G, defined on pairs $(E/L, \chi)$, with L as above and with $\chi \in R_\Delta$, $\Delta = \mathrm{Gal}(E/L)$, will be called an *extension* of g, if the following conditions hold.

$$(5.15)\qquad \begin{cases} \text{(a)}\quad g'(E/L, \chi) = g(L, \theta_\chi), & \text{if } \chi \text{ is Abelian}, \\ \text{(b)}\quad g'(E/L, \chi + \chi') = g'(E/L, \chi)g'(E/L, \chi'), & \\ \text{(c)}\quad g'(E/L, \chi) = g'(E/M, \mathrm{ind}_\Delta^\Sigma(\chi)), & \text{if}\quad \deg(\chi) = 0, \end{cases}$$

where in (c) $L \supset M \supset F$, $\Delta \subset \Sigma = \mathrm{Gal}(E/M)$, $\mathrm{ind}_\Delta^\Sigma$ the induction map. Properties (a)–(c) will determine g' uniquely (cf. [Tt2]). We shall then usually employ the same symbol for the extension as for the original function g. Property (c) is *inductivity in degree zero*. If the equation in (c) holds without restriction on the degree we shall say that g is *fully inductive*.

Remark. Actually the functions we are mainly interested in are also invariant under inflation (from characters of quotient groups), and one could thus dispense with the top field E i.e., go over to characters of Ω_F. This will indeed be done later – at this stage we wish however to keep formal definitions and postulates to a minimum, and this alternative approach will not be needed here.

Now consider first an Archimedean local field F, i.e. $F = \mathbb{R}$, $E = \mathbb{R}$ or $\mathbb{C}$, or $F = E = \mathbb{C}$. As in this case every character is sum of Abelian characters, there is a unique function satisfying (a), (b) for $g = W$, or $g = \tilde{L}(s,\ )$, and one verifies then easily that W is inductive in degree zero, and $\tilde{L}$ is fully inductive. One can actually give a neat formula. If χ is a character of $\mathrm{Gal}(E/F)$ with underlying module V, then $V = V_+ \oplus V_-$, and accordingly $\chi = \chi_+ + \chi_-$, where the generator of Γ acts on V_+ via the eigenvalue $+1$, and on V_- via the eigenvalue -1 (this being missing if $\Gamma = 1$). Now one has

$$(5.16)\quad \begin{cases} W(E/F,\chi) = i^{-\deg(\chi_-)}, \\ \tilde{L}(s,E/F,\chi) = \begin{cases} (\pi^{-(s+\frac{1}{2})}\Gamma(s/2)\Gamma((s+1)/2))^{\deg(\chi)} & \text{for } F = \mathbb{C}, \\ (\pi^{-s/2}\Gamma(s/2))^{\deg(\chi_+)}(\pi^{-(s+1)/2}\Gamma((s+1)/2))^{\deg(\chi_-)} & \text{for } F = \mathbb{R}. \end{cases} \end{cases}$$

(See the remark after (5.2).)

Next we consider a non-Archimedean local field F. Here we only need the L-functions at this stage. We shall first define the notion of the *non-ramified part* $\mathrm{n}\chi$ of a character χ, which will be of importance throughout. Here $\chi \mapsto \mathrm{n}\chi$ is an endomorphism of the additive group R_Γ, depending on E/F. It is determined by its value on the irreducible characters χ, as follows (with Γ_0 the inertia group of E/F, see §3 (3.1)). Here $\mathrm{Ker}\,\chi$ is the kernel of the corresponding representation.

$$(5.17)\quad \begin{cases} \mathrm{Ker}\,\chi \supset \Gamma_0, & \text{i.e. } \chi \text{ non-ramified} \Rightarrow \chi = \mathrm{n}\chi, \\ \mathrm{Ker}\,\chi \not\supset \Gamma_0, & \text{i.e., } \chi \text{ ramified} \Rightarrow \mathrm{n}\chi = 0. \end{cases}$$

Alternatively, if χ is the character of a representation with underlying $\mathbb{Q}^c\Gamma$-module V, then $\mathrm{n}\chi$ is that of the representation of Γ on the fixed submodule V^{Γ_0} of Γ_0. As Γ/Γ_0 is Abelian, $\mathrm{n}\chi$ is a $\mathbb{Z}$-linear combination of Abelian characters χ_i, i.e.,

$$\mathrm{n}\chi = \sum m_i\chi_i \qquad (m_i \in \mathbb{Z}).$$

We now define the L-function of χ by

$$(5.18)\qquad \tilde{L}(s,E/F,\chi) = \prod_i \tilde{L}(s,\theta_{\chi_i})^{m_i}.$$

We thus have the equation

$$\tilde{L}(s,E/F,\chi) = \tilde{L}(s,E/F,\mathrm{n}\chi)$$

for all $\chi \in R_\Gamma$, and analogously with F replaced by any field between F and E. This L-function will satisfy (5.15), and in fact it is fully inductive. This follows quite easily from the properties of the map $\chi \mapsto \mathrm{n}\chi$, to be established later (see Proposition III, 1.3). To get the usual form of the local L-function, assume E/F to be non-ramified, $\chi = \mathrm{n}\chi$ and $\theta_i = \theta_{\chi_i}$ as above. Then $A_{E/F}(\mathfrak{p}_F) = \sigma_{E/F}$ is the Frobenius and

$$\tilde{L}(s, E/F, \chi) = \prod_i \tilde{L}(s, \theta_i)^{m_i} = \prod_i (1 - \theta_i(\mathfrak{p}_F)\mathbf{N}_F\mathfrak{p}_F^{-s})^{-m_i}$$

$$= \mathrm{Det}_\chi(1 - \sigma_{E/F}\mathbf{N}_F\mathfrak{p}_F^{-s})^{-1}.$$

Now we return to number fields, and we use again the symbol N/K for a Galois extension, and Γ for $\mathrm{Gal}(N/K)$. If $\mathfrak{p}$ is a prime divisor of K, $\mathfrak{P}$ a prime divisor of N above $\mathfrak{p}$ then $N_{\mathfrak{P}}/K_{\mathfrak{p}}$ is a Galois extension of local fields, whose Galois group $\Gamma_{\mathfrak{p}}$ embeds into Γ as the stabilizer of $\mathfrak{P}$ and this embedding is uniquely determined by $\mathfrak{p}$, to within conjugacy. The local component $\chi_{\mathfrak{p}}$ of a character χ of Γ is simply the restriction of χ to $\Gamma_{\mathfrak{p}}$, and if χ is Abelian then $\theta_{\chi,\mathfrak{p}} = \theta_{\chi_{\mathfrak{p}}}$. The *Artin L-function* of χ is

$$\tilde{L}(s, N/K, \chi) = \prod_{\mathfrak{p}} \tilde{L}(s, N_{\mathfrak{P}}/K_{\mathfrak{p}}, \chi_{\mathfrak{p}}) \tag{5.19}$$

(product over all prime divisors $\mathfrak{p}$ of K) where for each $\mathfrak{p}$ one chooses some $\mathfrak{P}$ above it. Again $\tilde{L}(s, N/K, \chi)$ satisfies (5.15) for varying $(N/K, \chi)$, and again it is fully inductive. All this follows from the corresponding properties of the local L-functions. Moreover if we put

$$W_\infty(N/K, \chi) = \prod_{\mathfrak{p}|\infty} W(N_{\mathfrak{P}}/K_{\mathfrak{p}}, \chi_{\mathfrak{p}}) \tag{5.20}$$

(product over the infinite prime divisors of K), where the right hand sides were defined e.g., in (5.16), then W_∞ is an extension of the function defined in (5.11). (It satisfies (5.15) for all layers but is not always fully inductive (cf. [F13]).)

By Brauer's induction theorem (cf. [Br] or [Se2] (§10)) we have a representation

$$\chi = \sum m_i \,\mathrm{ind}_{\Delta_i}^{\Gamma}(\phi_i) \tag{5.21}$$

for every character χ of Γ, where the Δ_i are subgroups of Γ, the ϕ_i Abelian character of Δ_i. By (5.15), with $M_i = N^{\Delta_i}$,

$$\tilde{L}(s, N/K, \chi) = \prod_i \tilde{L}_{M_i}(s, \theta_{\phi_i})^{m_i}.$$

By Hecke's theorem we now have (cf. [Art2], [Br]).

Theorem. *The function $\tilde{L}(s, N/K, \chi)$ can be extended to a meromorphic function in the whole complex plane, and it satisfies a functional equation*

$$\tilde{L}(s, N/K, \chi) = W(N/K, \chi)A(N/K, \chi)^{\frac{1}{2}-s}\tilde{L}(1 - s, N/K, \bar{\chi})$$

where $\bar{\chi}$ is the contragredient (complex conjugate) of χ,

$$|W(N/K, \chi)| = 1, \qquad A(N/K, \chi) > 0. \qquad \square$$

As $\tilde{L}$ cannot satisfy two distinct functional equations, the constants $W(N/K, \chi)$ (the *Artin root numbers*) and $A(N/K, \chi)$ are uniquely determined, in spite of the non-uniqueness of (5.21). We conclude therefore from the corresponding properties of $\tilde{L}(s, N/K, \chi)$ that

For varying $(N/K, \chi)$, $W(N/K, \chi)$ *and* $A(N/K, \chi)$ *satisfy* (5.15) *and are in fact fully inductive.*

The underlying functions on idele class characters are of course those in Hecke's theorem. We now write purely formally

$$(5.22)\qquad \begin{cases} A(N/K, \chi) = \mathbf{N}_K(\mathfrak{D}_K)^{\deg(\chi)} \cdot \mathbf{N}_K\mathfrak{f}(N/K, \chi), \\ W(N/K, \chi) = \tau(N/K, \bar{\chi}) W_\infty(N/K, \chi)/\mathbf{N}_K\mathfrak{f}(N/K, \chi)^{1/2}. \end{cases}$$

Here $A(N/K, \chi)$, $W(N/K, \chi)$ are defined in the above theorem, $\mathfrak{D}_K$ is the different of K, the first equation defines formally the positive number $\mathbf{N}_K\mathfrak{f}(N/K, \chi)$, and the second equation the complex number $\tau(N/K, \chi)$, the (global) Galois Gauss sum. The map $(K, \theta) \mapsto \mathbf{N}_K(\mathfrak{D}_K)$ has the extension $(N/K, \chi) \mapsto \mathbf{N}_K(\mathfrak{D}_K)^{\deg(\chi)}$, satisfying (5.15). As $A(N/K, \chi)$ extends $A(\theta)$ it follows from (5.13) and (5.22) that $(N/K, \chi) \mapsto \mathbf{N}_K\mathfrak{f}(N/K, \chi)$ extends $(K, \theta) \mapsto \mathbf{N}_K\mathfrak{f}(\theta)$, i.e., satisfies (5.15). Thus finally it follows from (5.14) and (5.22), that $(N/K, \chi) \mapsto \tau(N/K, \chi)$ extends $(K, \theta) \to \tau(\theta)$. In other words, the Galois Gauss sum does satisfy (5.15). It is however in general not fully inductive. As $\tau(\theta) \in \mathbb{Q}^{c*}$, it follows that the $\tau(N/K, \chi)$ are *non-zero* algebraic numbers. Moreover $|W(N/K, \chi)| = |W_\infty(N/K, \chi)| = 1$, hence by (5.22)

$$(5.23)\qquad |\tau(N/K, \chi)| = \mathbf{N}_K\mathfrak{f}(N/K, \chi)^{1/2}.$$

The global Galois Gauss sum τ and the infinite factor W_∞ were (in a slightly different form) first introduced in [Ha], in the same global context as here. See also [F13], in particular for a detailed discussion of their inductive properties.

Now we can state the basic theorem of the whole subject. Suppose here that N/K is tame. Let $\mathcal{N}_{K/\mathbb{Q}}(a \mid \chi)$ be again as defined in Theorem 4, and in (4.10), i.e., for each prime divisor $\mathfrak{p}$ of K,

$$\mathfrak{o}_{N,\mathfrak{p}} = a_\mathfrak{p}\mathfrak{o}_{K,\mathfrak{p}}\Gamma, \qquad (a \mid \chi)_\mathfrak{p} = (a_\mathfrak{p} \mid \chi) \qquad \text{and} \qquad \mathcal{N}_{K/\mathbb{Q}}(a \mid \chi) = \prod_\sigma (a \mid \chi^{\sigma^{-1}})^\sigma,$$

where $\{\sigma\}$ is a right transversal of Ω_K in $\Omega_\mathbb{Q}$. Put, for all $\chi \in R_\Gamma$,

$$(5.24)\qquad u(\chi) = \mathcal{N}_{K/\mathbb{Q}}(a \mid \chi) \cdot \tau(N/K, \chi)^{-1} W'(N/K, \chi),$$

where $\chi \mapsto W'(N/K, \chi)$ is the homomorphism $R_\Gamma \to \mathbb{Q}^{c*}$, given on irreducible characters χ by

$$(5.25)\qquad W'(N/K, \chi) = \begin{cases} 1, & \chi \text{ not symplectic}, \\ W(N/K, \chi), & \chi \text{ symplectic}. \end{cases}$$

Although this is not yet essential at this stage it is worth noting that in fact the possible values of $W'(N/K, \chi)$ are ± 1 (recall (1.12) and see Proposition 6.1 to follow).

It is trivial that u is a homomorphism

$$R_\Gamma \to \mathfrak{J}(\mathbb{Q}^c).$$

But in fact we have a much stronger result.

Theorem 5 (cf. [F17]). $u \in \mathrm{Hom}^+_{\Omega_\mathbb{Q}}(R_\Gamma, \mathfrak{U}(\mathbb{Q}^c))$. □

Remark. This theorem implies an interpretation of the Galois Gauss sums, which arise in the first place out of the functional equation of the Artin L-function, in terms of the module structure of rings of integers. It is this, at first sight entirely unsuspected, relation which "makes" the theory, i.e., gives it both its power and its depth. Also, we have here a phenomenon which turns up in a similar manner in various aspects of algebraic number theory: Some arithmetic invariant, e.g. given in terms of representations of Galois groups, such as special values of functions, or Galois Gauss sums etc., which in the first place is defined to have complex values, actually has algebraic ones and these in turn appear then as parameters for algebraic constructions. Two classical examples are (i) the basic cyclotomic unit in quadratic fields, coming out of the class number formula, (ii) the classical Gauss sums, appearing in the evaluation of $s = 1$ of Dirichlet L-functions.

Essentially the last theorem contains three separate assertions, which we shall restate, and also prove, separately.

Theorem 5A. *$u \in \mathrm{Hom}^+(R_\Gamma, \mathfrak{J}(\mathbb{Q}^c))$, i.e., $u(\chi)_\mathfrak{p} > 0$ for all infinite prime divisors whether real or not, whenever $\chi \in R^s_\Gamma$ (i.e. χ is symplectic).* □

Theorem 5B. $u \in \mathrm{Hom}_{\Omega_\mathbb{Q}}(R_\Gamma, \mathfrak{J}(\mathbb{Q}^c))$. □

Theorem 5C. $u \in \mathrm{Hom}(R_\Gamma, \mathfrak{U}(\mathbb{Q}^c))$. □

Putting it differently, the Theorem asserts that the norm resolvent and the adjusted Galois Gauss sum have closely related behaviour with respect to the action of the absolute Galois group $\Omega_\mathbb{Q}$, and with respect to their signatures and their values at finite primes.

Combining Theorems 4 and 5 we now get, with u as in Theorem 5,

Theorem 6 (cf. [F17] [Notes 4 and 5]). *$U_{N/K}$ is represented by $u \in \mathrm{Hom}^+_{\Omega_\mathbb{Q}}(R_\Gamma, \mathfrak{U}(\mathbb{Q}^c))$, and in particular*

$$U_{N/K} \in D(\mathbb{Z}\Gamma).$$

Moreover we can assume that

$u(\chi)_p = 1$ *at all prime divisors of* $\mathbb{Q}$ *other than the finite primes* l *dividing* (order Γ),

$$u(\chi)_l = \mathcal{N}_{K/\mathbb{Q}}(a_l \mid \chi)\tau(N/K, \chi)_l^{-1} W'(N/K, \chi)_l^{-1} \qquad \textit{if } l \mid (\text{order } \Gamma),$$

where a_l is an element of N_l with $\mathfrak{o}_{N,l} = a_l(\mathfrak{o}_{K,l}\Gamma)$.

Proof. By Theorems 4 and 5, $U_{N/K}$ is represented by f, where

$$f(\chi) = u(\chi)(\tau(N/K,\chi)W'(N/K,\chi)\mathcal{N}_{K/\mathbb{Q}}(b\,|\,\chi)^{-1})$$

$$= u(\chi)h(\chi) \qquad \text{say.}$$

As f, $u \in \mathrm{Hom}_{\Omega_{\mathbb{Q}}}(R_\Gamma, \mathfrak{J}(\mathbb{Q}^c))$, and $h \in \mathrm{Hom}(R_\Gamma, \mathbb{Q}^{c*})$, we get $h \in \mathrm{Hom}_{\Omega_{\mathbb{Q}}}(R_\Gamma, \mathbb{Q}^{c*})$. Thus also $fh^{-1} = u$ represents $U_{N/K}$. By (2.18) $U_{N/K} \in D(\mathbb{Z}\Gamma)$.

The formula for $u(\chi)_l$ is obtained by observing, that for any prime divisor $\mathfrak{p}$ of K above l, the component $a_{l,\mathfrak{p}} = a_{\mathfrak{p}}$ satisfies the conditions of Theorem 4. On the other hand, if p is a prime not dividing (order Γ) or the infinite prime, then $\mathbb{Z}_p\Gamma = \mathfrak{M}_p$ is a maximal order (p finite), or "$\mathbb{Z}_p\Gamma$" $= \mathbb{R}\Gamma$. Thus by Proposition 2.2., $u_p \in \mathrm{Det}(\mathbb{Z}_p\Gamma^*)$; hence we can replace the original u by $u \prod_p u_p^{-1}$ (product over all prime divisors other than those dividing (order Γ)). □

Remark. One can find an element $c \in N$ so that for each $l\,|\,$(order Γ), $c_l \mathfrak{o}_{K,l}\Gamma = \mathfrak{o}_N$, and then replace $\mathcal{N}_{K/\mathbb{Q}}(a_l\,|\,\chi)$ by $\mathcal{N}_{K/\mathbb{Q}}(c\,|\,\chi)_l$.

Corollary. *If $\mathfrak{M}$ is a maximal order in $\mathbb{Q}\Gamma$, containing $\mathbb{Z}\Gamma$ then*

$$(\mathfrak{o}_N\mathfrak{M}) = 1,$$

i.e., $\mathfrak{o}_N\mathfrak{M}$ *is stably free.* □

In the case $K = \mathbb{Q}$ this was conjectured by Martinet.

The Corollary follows from (2.20).

The proofs will occupy a considerable part of Chap. III.

We now come back to conductors, first for local non-Archimedean fields. There is an extension $\mathfrak{f}(E/F,\chi)$, in the sense of (5.15), of the local conductor defined in (5.3). This is given by an explicit formula [see however note 6], involving the sequence of ramification groups (cf. [Art1], [Se1], [Se4]). We shall not give this here, except in the case when E/F is tame. Then we have indeed

(5.26) $$\mathfrak{f}(E/F,\chi) = \mathfrak{p}_F^{\deg(\chi)-\deg(\mathrm{n}\chi)},$$

with $\mathrm{n}\chi$ as defined in (5.17). In the tame case the proof of inductivity in degree zero will be seen to be straightforward (cf. III (2.5), (2.6)).

Now let N/K be a Galois extension of number fields. One defines the global Artin conductor as the product

(5.27) $$\mathfrak{f}(N/K,\chi) = \prod_{\mathfrak{p}} \mathfrak{f}(N_{\mathfrak{P}}/K_{\mathfrak{p}}, \chi_{\mathfrak{p}})$$

(product over all finite prime divisors). From the uniqueness properties for (5.15) it now follows, that the number $\mathbf{N}_K\mathfrak{f}(N/K,\chi)$, occurring in (5.22), is indeed the norm of $\mathfrak{f}(N/K,\chi)$.

The next theorem is in a sense a weak analogue to Theorem 5.

Theorem 7 (cf. [F17]). *Let N/K be tame. With $(a\,|\,\chi)$ as in Theorem 4 we have the equation in* $\mathfrak{J}(\mathbb{Q}^c)$

$$((a\,|\,\chi)(a\,|\,\bar{\chi})) = \mathfrak{f}(N/K, \chi).$$

This will be proved in III, §7. □

To understand the significance of the last theorem we shall deduce here a theorem which may be considered as an analogue to Theorem 6.

Theorem 8. *Let N/K be tame. With $(a\,|\,\chi)$ as in Theorem 4, the map*

$$\chi \mapsto (a\,|\,\chi)(a\,|\,\bar{\chi})$$

lies in $\mathrm{Hom}_{\Omega_K}(R_\Gamma, \mathfrak{J}(\mathbb{Q}^c))$, *and represents the class* $((\mathfrak{o}_N)(\hat{\mathfrak{o}}_N)^{-1})_{\mathfrak{o}_K\Gamma}$, *where* $\hat{\mathfrak{o}}_N \cong \mathrm{Hom}_{\mathfrak{o}_K}(\mathfrak{o}_N, \mathfrak{o}_K)$. □

Remark 1. Of course $\hat{\mathfrak{o}}_N \cong \mathfrak{D}_{N/K}^{-1}$, where $\mathfrak{D}_{N/K}$ is the relative different.

Remark 2. Combining Theorems 7 and 8, we see that if $\mathfrak{f}(N/K, \chi)$ is not a principal ideal in the field $K(\chi)$ of values of χ over K, then $(\mathfrak{o}_N)_{\mathfrak{o}_K\Gamma} \neq 1$. For otherwise also $(\hat{\mathfrak{o}}_N)_{\mathfrak{o}_K\Gamma} = 1$, hence certainly $\mathfrak{f}(N/K, \chi)$ is principal.

Proof of Theorem 8. By Theorem 4 and §2 Example 1 (in particular (2.22)), the class $((\mathfrak{o}_N)(\hat{\mathfrak{o}}_N)^{-1})_{\mathfrak{o}_K\Gamma}$ is represented by

$$\chi \mapsto (a\,|\,\chi)(a\,|\,\bar{\chi})(b\,|\,\chi)^{-1}(b\,|\,\bar{\chi})^{-1},$$

with a and b as in Theorem 4. Theorem 8 now follows by showing that (cf. [F13] Proposition 5.1)

$$(b\,|\,\chi)(b\,|\,\bar{\chi}) = \mathrm{Det}_\chi\Big(\sum_\gamma t_{N/K}(b \cdot b^\gamma)\gamma\Big), \tag{5.28}$$

i.e., $\chi \mapsto (b\,|\,\chi)(b\,|\,\bar{\chi})$ lies in $\mathrm{Det}(K\Gamma^*)$. Indeed $(b\,|\,\bar{\chi}) = \mathrm{Det}_{\bar{\chi}}(\sum b^\gamma\gamma^{-1}) = \mathrm{Det}_\chi(\sum b^\gamma\gamma)$. But

$$\sum_\gamma b^\gamma\gamma \cdot \sum_\gamma b^\gamma\gamma^{-1} = \sum_{\sigma,\gamma} b^\sigma b^\gamma \sigma\gamma^{-1} = \sum_{\delta,\gamma} b^\gamma b^{\delta\gamma}\delta = \sum_\delta t_{N/K}(b \cdot b^\delta)\delta,$$

and this immediately gives (5.28). □

Final remark. We can now illustrate what we meant in §1 by the "method of congruences". Let N/K be tame, Galois with Galois group Γ. We consider functions g on R_Γ of three types: (i) $g \in \mathrm{Det}\,\mathfrak{U}\mathbb{Z}\Gamma$, (ii) $g(\chi) = \tau(N/K, \chi)$ and (iii) $g(\chi) = \mathscr{N}_{K/\mathbb{Q}}(a\,|\,\chi)$, with a as in (5.24). In its simplest and most naive form, the method consists in proving statements of the type $g(\theta) \equiv 1 \pmod{\mathfrak{a}}$, for some given $\theta \in R_\Gamma$ and some integral ideal $\mathfrak{a}$ in a number field E. For suitable θ and $\mathfrak{a}$, this gives a restriction on the map u of (5.24), i.e., on $U_{N/K}$.

§6. Symplectic Root Numbers and the Class $U_{N/K}$

We now come to what may rightly be called the ultimate result of the theory of Galois module structure for tame extensions (Theorem 9). It describes the class $U_{N/K}$ in terms of the Artin root numbers for symplectic characters. A result of this nature had been suspected from the beginning, even before it was formally conjectured. A major step in directing effort towards such a goal was the proof of the Galois invariance of Artin root numbers early on – see Proposition 6.2.

Let throughout Γ be a finite group. Let S_Γ be the subgroup, generated by the irreducible symplectic characters, of the group R_Γ of virtual characters of Γ. S_Γ is a direct component of the $\Omega_{\mathbb{Q}}$-module R_Γ. We shall then define a homomorphism

(6.1) $$t'_\Gamma : \mathrm{Hom}_{\Omega_{\mathbb{Q}}}(S_\Gamma, \pm 1) \to \mathrm{Hom}^+_{\Omega_{\mathbb{Q}}}(R_\Gamma, \mathfrak{U}(\mathbb{Q}^c)).$$

As noted before we may replace $\mathfrak{U}(\mathbb{Q}^c)$ by $\mathfrak{U}(E)$ for a sufficiently large number field E, Galois over $\mathbb{Q}$. We shall define t'_Γ by prescribing the values $(t'_\Gamma f)(\chi)_{\mathfrak{p}}$, for any $f \in \mathrm{Hom}_{\Omega_{\mathbb{Q}}}(S_\Gamma, \pm 1)$, any irreducible character χ and any prime divisor $\mathfrak{p}$ of E, and extending to arbitrary χ by linearity. The rule is

(6.2) $$(t'_\Gamma f)(\chi)_{\mathfrak{p}} = \begin{cases} f(\chi) \quad (= \pm 1 \in \mathbb{Q}_p^*) & \text{if } \chi \text{ symplectic, } \mathfrak{p} \text{ finite above } p, \\ 1 & \text{in all other cases.} \end{cases}$$

It is clear that the particular choice of E is irrelevant.

Composing (6.1) with the surjection

$$\mathrm{Hom}^+_{\Omega_{\mathbb{Q}}}(R_\Gamma, \mathfrak{U}(\mathbb{Q}^c)) \to D(\mathbb{Z}\Gamma)$$

(see (2.18)) we now get a homomorphism (cf. [CN2])

(6.3) $$t = t_\Gamma : \mathrm{Hom}_{\Omega_{\mathbb{Q}}}(S_\Gamma, \pm 1) \to D(\mathbb{Z}\Gamma).$$

The group $\mathrm{Hom}_{\Omega_{\mathbb{Q}}}(S_\Gamma, \pm 1)$ is clearly a vector space over the field $\mathbb{F}_2$ of 2 elements, with a basis, which is in bijection with the $\Omega_{\mathbb{Q}}$-orbits of irreducible symplectic characters of Γ, i.e., of irreducible real valued characters of Γ of Schur index 2. We shall give a second description of the group, which puts into evidence its behaviour under a "change of Γ" – something which is not clear from the original definition.

Let R^s_Γ be again the subgroup of R_Γ generated by the symplectic characters. Every actual character which lies in R^s_Γ is in fact symplectic, and we may naturally call R^s_Γ the group of virtual symplectic characters. If $\chi \in R_\Gamma$, denote by $\bar{\chi}$ its contragredient (or complex conjugate) and write $\mathrm{Tr}(\chi) = \chi + \bar{\chi}$. Then Tr is an $\Omega_{\mathbb{Q}}$-homomorphism $R_\Gamma \to R^s_\Gamma$, and the map $S_\Gamma \to R^s_\Gamma \to R^s_\Gamma/\mathrm{Tr}(R_\Gamma)$ yields an isomorphism $S_\Gamma/2S_\Gamma \cong R^s_\Gamma/\mathrm{Tr}(R_\Gamma)$. Therefore

(6.4) $$\mathrm{Hom}_{\Omega_{\mathbb{Q}}}(S_\Gamma, \pm 1) \cong \mathrm{Hom}_{\Omega_{\mathbb{Q}}}(R^s_\Gamma/\mathrm{Tr}(R_\Gamma), \pm 1).$$

Now let N/K be a Galois extension of number fields with Galois group Γ.

6.1. Proposition. *If χ is a real valued character of Γ (in particular if χ is symplectic) then $W(N/K,\chi) = \pm 1$. For any χ, $W(N/K,\mathrm{Tr}(\chi)) = 1$.*

Indeed, almost immediately from the functional equation of the Artin L-function, one gets $W(N/K,\bar{\chi}) = \overline{W(N/K,\chi)}$. Hence if $\bar{\chi} = \chi$, then $W(N/K,\chi)$ is real; as it is of absolute value 1 it must be $= \pm 1$. Moreover for any χ, $W(N/K,\mathrm{Tr}(\chi)) = W(N/K,\chi), \overline{W(N/K,\chi)} = 1$. □

What is crucial in the sequel is the next proposition, asserting the Galois invariance of $W_{N/K}$ (cf. [F7] Theorem 5, where this was proved for quaternion characters – the general result follows then by induction. A more conceptual proof soon became available and will be given later in this volume).

Proposition 6.2. *If N/K is tame then the map $\chi \mapsto W(N/K,\chi)$ $(\chi \in S_\Gamma)$, defines an element*

$$W_{N/K} \in \mathrm{Hom}_{\Omega_{\mathbb{Q}}}(S_\Gamma, \pm 1).$$ □

This will be proved in Chap. III, §3 (Theorem 21). Note that

$$W_{N/K}(\chi) = W'(N/K,\chi), \qquad \text{for} \quad \chi \in S_\Gamma \tag{6.5}$$

where $W'(N/K,\chi)$ was defined in (5.25).

Proposition 6.2 "had" to be proved, if one wanted to find an "invariant" interpretation for the symplectic root numbers. In view of Proposition 6.1, the result simply means that

$$W(N/K,\chi) = W(N/K,\chi^{\omega})$$

for tame symplectic χ and for $\omega \in \Omega_{\mathbb{Q}}$. This is not true for wild extensions (cf. [F7] §9).

In the above notation we can now state the main theorem,

Theorem 9. *If N/K is tame then*

$$tW_{N/K} = U_{N/K}$$

(cf. [Ty7]). □

Corollary 1. *If Γ has no irreducible symplectic character then $U_{N/K} = 1$, and in fact $\mathfrak{o}_N$ is a free $\mathbb{Z}\Gamma$-module.*

The second statement follows from the fact that for such groups Γ, $(X)_{\mathbb{Z}\Gamma} = 1$ implies that X is free over $\mathbb{Z}\Gamma$ (see note [2]). Examples: 1) All Abelian groups Γ (see [Ty4] for a direct proof). 2) All odd order groups. 3) All dihedral groups. □

Corollary 2. $U_{N/K}^2 = 1$.

For, clearly $W_{N/K}^2 = 1$. □

One can say more:

$\mathfrak{o}_N \oplus \mathfrak{o}_N$ *is always a free* $\mathbb{Z}\Gamma$*-module.*

For, its rank is > 1 (see note [2]). □

Now recall the definition of $(\hat{\mathfrak{o}}_N) = \hat{U}_{N/K}$ (Example 1 in §2).

Corollary 3. $U_{N/K}\hat{U}_{N/K} = 1$, *i.e.,* $\mathfrak{o}_N$ *is stably self dual.*

Under somewhat stronger hypotheses this was proved already in [Ty3]. Another general proof was given by S. Chase (see note [2] in II). Both proofs use the theory of Swan-modules (see Example 2 in §2).

To get the last Corollary, use (2.22) and the fact that $\bar{W}_{N/K} = W_{N/K}$, whence $tW_{N/K} \cdot t\hat{W}_{N/K} = 1$. □

We shall briefly indicate the basic strategy in the proof of Theorem 9. It will be reduced to two other theorems, one dealing with the arithmetic properties of the invariants involved, in particular the Galois Gauss sums, and the other a fixed point theorem for group determinants. We shall first state the latter in a global variant of the form given in [Ty6], [Ty7].

We shall begin by introducing the appropriate notations. Let F be a number field, Galois over $\mathbb{Q}$. In the statement of the next theorem and in its proof we shall be concerned with the action of the Galois group Gal$(F/\mathbb{Q})$ on determinants of group rings of Γ. (To avoid notational confusion between the two roles of finite groups, we shall exceptionally use italics for certain Galois groups. Specifically we shall write

$$G = \mathrm{Gal}(F/\mathbb{Q}).)$$

Let then l be a fixed prime number. For any prime divisor $\mathfrak{l}$ of F above l, let $F_{\mathfrak{l},1}$ be the maximal tame extension of $\mathbb{Q}_l$ in $F_{\mathfrak{l}}$, and write $\mathfrak{o}_{\mathfrak{l},1}$ for its ring of integers. Then put

$$\mathfrak{O}_{F,l} = \prod_{\mathfrak{l}|l} \mathfrak{o}_{\mathfrak{l},1}, \tag{6.6}$$

for the product of these. $\mathbb{Z}_l$ embeds diagonally in $\mathfrak{O}_{F,l}$, and $(\mathbb{Z}_l\Gamma)^*$ diagonally in $(\mathfrak{O}_{F,l}\Gamma)^*$ and is elementwise fixed under G-action, G acting via $\mathfrak{O}_{F,l}$.

Next observe that G also acts on $\mathrm{Hom}_{\Omega_F}(R_\Gamma, \mathfrak{U}(\mathbb{Q}^c))$ by conjugation. Indeed if $f \in \mathrm{Hom}_{\Omega_F}(R_\Gamma, \mathfrak{U}(\mathbb{Q}^c))$, and $\omega \in \Omega_{\mathbb{Q}}$ maps onto $g \in G = \Omega_{\mathbb{Q}}/\Omega_F$, then f^g is uniquely defined by $f^g(\chi) = f(\chi^{\omega^{-1}})^\omega$ for all $\chi \in R_\Gamma$. Now the map Det: $(\mathfrak{O}_{F,l}\Gamma)^* \to \mathrm{Hom}_{\Omega_F}(R_\Gamma, \mathfrak{U}(\mathbb{Q}^c))$ preserves G-action, as can be seen by a slight generalization of the proof of (2.8). Therefore $\mathrm{Det}(\mathfrak{O}_{F,l}\Gamma^*)$ is a G-submodule of the Hom group above. Thus

$$(\mathrm{Det}(\mathfrak{O}_{F,l}\Gamma^*))^G \subset \mathrm{Hom}_{\Omega_F}(R_\Gamma, \mathfrak{U}(\mathbb{Q}^c))^G = \mathrm{Hom}_{\Omega_{\mathbb{Q}}}(R_\Gamma, \mathfrak{U}(\mathbb{Q}^c)),$$

while evidently

$$\mathrm{Det}(\mathbb{Z}_l\Gamma^*) \subset (\mathrm{Det}(\mathfrak{O}_{F,l}\Gamma^*))^G.$$

The crucial point is that we get the opposite inclusion, i.e., we have (cf. [Ty6], [Ty7]).

Theorem 10. $(\mathrm{Det}(\mathfrak{O}_{F,l}\Gamma^*))^G = \mathrm{Det}(\mathbb{Z}_l\Gamma^*)$. □

The proof will be completed in II §6. The main tool is the logarithm for group rings, introduced in [Ty6], see II §5.

Now consider again a tame extension N/K of number fields with Galois group Γ. For every prime number l, dividing the order of Γ, choose a free generator a_l of $\mathfrak{o}_{N,l}$ over $\mathfrak{o}_{K,l}\Gamma$. Define $v\colon R_\Gamma \to \mathfrak{U}(\mathbb{Q}^c)$ by its local components, as follows:

$$(6.7)\qquad \begin{cases} v(\chi)_l = \mathcal{N}_{K/\mathbb{Q}}(a_l \mid \chi)\tau(N/K,\chi)_l^{-1}, & \text{if } l \mid \text{order } (\Gamma), \\ v(\chi)_p = 1, & \text{if } p \text{ is any other prime divisor in } \mathbb{Q}. \end{cases}$$

Compare v with the map u, appearing in Theorem 6 (with the same local free generators at all l). By Proposition 6.1, $(u(\chi)v(\chi)^{-1})_p = \pm 1$ for *all* p, and so indeed $v(\chi) \in \mathfrak{U}(\mathbb{Q}^c)$.

Let $\mu(\mathbb{Q}^c)$ be the group of roots of unity in $\mathbb{Q}^c$. We now have (cf. [Ty7])

Theorem 11. *There exists an element $y^* \in \mathrm{Hom}^+_{\Omega_\mathbb{Q}}(R_\Gamma, \mu(\mathbb{Q}^c))$, and a Galois extension F of $\mathbb{Q}$ so that for each prime divisor l of order (Γ),*

$$y_l^* v_l \in \mathrm{Det}(\mathfrak{O}_{F,l}\Gamma^*). \qquad \square$$

The proof of this Theorem is essentially local, and its main tools are the logarithm for group rings (cf. [Ty6]) and the method of congruences, already referred to earlier. The proof will occupy most of Chap. IV.

What we shall do here, is to show that Theorems 10 and 11 imply Theorem 9.

Observe first that by its definition, the class $tW_{N/K}$ is represented by the map

$$\chi \mapsto t'W_{N/K}(\chi) = W'(N/K,\chi) \cdot W'(N/K,\chi)_\infty^{-1},$$

with W' defined in (5.25), and the subscript ∞ denoting its idelic component at infinity. (Do not confuse with the root number $W_\infty(N/K,\chi)$ at infinity, defined in (5.20)!) Choosing the same local free generators a_l in Theorem 6 as here, it now follows by Theorem 6, that $U_{N/K} \cdot tW_{N/K} = U_{N/K} \cdot tW_{N/K}^{-1}$ is represented by $u^{(1)}$, where

$$\begin{aligned} u^{(1)}(\chi)_l &= v(\chi)_l \qquad \text{if } l \text{ divides order } (\Gamma), \\ u^{(1)}(\chi)_p &= W'(N/K,\chi)_p \quad (=\pm 1) \qquad \text{for any other finite } p, \\ u^{(1)}(\chi)_\infty &= 1. \end{aligned}$$

Now the map y^* of Theorem 11 lies in $\mathrm{Hom}^+_{\Omega_\mathbb{Q}}(R_\Gamma, \mathfrak{Y}(\mathbb{Q}^c))$, and so by (2.18), $U_{N/K} \cdot tW_{N/K}^{-1}$ is also represented by $y^* u^{(1)}$.

Next define $u^{(2)}$ by

$$u^{(2)}(\chi)_l = 1, \qquad \text{if } l \text{ divides order } (\Gamma),$$

$$u^{(2)}(\chi)_p = y^*(\chi)_p u^{(1)}(\chi)_p, \qquad \text{for any other finite } p,$$
$$u^{(2)}(\chi)_\infty = y^*(\chi)_\infty.$$

For any finite $p, p \nmid$ order (Γ),

$$u_p^{(2)} \in \mathrm{Hom}_{\Omega_\mathbb{Q}}(R_\Gamma, \mathfrak{U}_p(\mathbb{Q}^c)) = \mathrm{Det}(\mathbb{Z}_p\Gamma^*),$$

as for these p, $\mathbb{Z}_p\Gamma$ is a maximal order. (See Proposition 2.2.) Also $u_\infty^{(2)} \in \mathrm{Det}(\mathbb{R}\Gamma^*)$. Thus $u^{(2)} \in \mathrm{Det}(\mathfrak{U}\mathbb{Z}\Gamma)$. Therefore $y^* u^{(1)} u^{(2)-1}$ represents $U_{N/K} \cdot tW_{N/K}^{-1}$. But $y^* u^{(1)} u^{(2)-1} = \prod_l y_l^* v_l$ (product over all l dividing order (Γ)). Thus finally

$$(6.8) \qquad \begin{cases} \text{(i)} \quad y_l^* v_l \in \mathrm{Hom}_{\Omega_\mathbb{Q}}(R_\Gamma, \mathfrak{U}_l(\mathbb{Q}^c)), & \text{all } l \mid \text{order } (\Gamma), \\ \text{(ii)} \quad \prod_l y_l^* v_l \text{ represents } U_{N/K} \cdot tW_{N/K}^{-1}. \end{cases}$$

By Theorem 11, $y_l^* v_l \in \mathrm{Det}(\mathfrak{O}_{F,l}\Gamma^*)$, hence by (6.8) (i), $y_l^* v_l \in \mathrm{Det}(\mathfrak{O}_{F,l}\Gamma^*)^G$, and thus, by Theorem 10, $y_l^* v_l \in \mathrm{Det}(\mathbb{Z}_l\Gamma^*)$. Thus $\prod_l y_l^* v_l \in \mathrm{Det}(\mathfrak{U}\mathbb{Z}\Gamma)$. Therefore, by (6.8) (ii), finally $U_{N/K} \cdot tW_{N/K}^{-1} = 1$ as we had to show. □

§7. Some Problems and Examples

Assuming Theorem 9, we shall discuss in this section some of the problems, which arise as a consequence of it, and some of the answers to these in a special case. We shall later return to this topic. (See II §4 and V.)

Theorem 9 reduces the determination of $U_{N/K}$ to that of the image of $W_{N/K}$ under the map t. In many of the questions which now arise one will – instead of considering a fixed extension N/K – consider all tame Galois extensions over a given K "with a given Galois group". This means a change of point of view, in that we now fix a group Γ and consider its representations as Galois group over K. A precise formulation will be given later on. In the present section we shall still get by without this.

The first question is the representation theoretic one: which groups Γ have irreducible symplectic characters? More precisely one wants to know – for a given Γ – the number of $\Omega_\mathbb{Q}$-orbits of such characters, i.e., the $\mathbb{F}_2$-dimension of $\mathrm{Hom}_{\Omega_\mathbb{Q}}(S_\Gamma, \pm 1)$, i.e. the number of simple components of $\mathbb{Q}\Gamma$, which are matrix rings over quaternion division algebras. For generalized quaternion groups, this is of course known (see also II §4). We shall however not discuss these questions for general Γ. In the present section we shall repeatedly use quaternion 2-groups for illustration, i.e. groups

$$\Gamma = H_{2^n} = gp[\sigma, \omega \mid \sigma^{2^{n-2}} = \omega^2,\ \omega^4 = 1,\ \omega^{-1}\sigma\omega = \sigma^{-1}]$$

(with $n \geqslant 3$). These possess exactly one $\Omega_\mathbb{Q}$-orbit of irreducible symplectic characters, i.e., for these

$$(7.1) \qquad \mathrm{Hom}_{\Omega_\mathbb{Q}}(S_\Gamma, \pm 1) \cong \pm 1.$$

The next problem is to "find t_Γ", e.g., by determining $\operatorname{Im} t_\Gamma$. For this purpose one may need a description of $D(\mathbb{Z}\Gamma)$ or of its 2-primary part $D(\mathbb{Z}\Gamma)_2$, which can be used for computations. To do this for our example, let ψ be an irreducible symplectic character of H_{2^n}, and let δ be the regular character of the quotient group $H_{2^n}^{ab}$ (the Klein Vier group) lifted to H_{2^n}, i.e., the sum of all Abelian characters of H_{2^n}. The field $\mathbb{Q}(\psi)$ of values of ψ is the maximal real subfield $\mathbb{Q}^+(2^{n-1})$ of the field of 2^{n-1}-st roots of unity. Let $\mathfrak{L}$ be the unique maximal ideal in its ring of integers, lying above 2. Let ρ be the character mod $\mathfrak{L}^2$ of order 2, and let ν be the rational residue class character mod 8, with $\nu(-1) = 1$. Then (cf. [F23] (3.4) – which restates the result of [FKW], see also [F17] (Lemma 14.2)) we have

7.1. Proposition. *With $\Gamma = H_{2^n}$, and ψ, δ, ρ, ν as above, the map*

$$S\colon \operatorname{Hom}^+_{\Omega_\mathbb{Q}}(R_\Gamma, \mathfrak{U}(\mathbb{Q}^c)) \to \pm 1,$$

given by

$$S(f) = \nu(f(\delta)_2) \cdot \rho(f(\psi)_2),$$

sets up an isomorphism

$$D(\mathbb{Z}\Gamma) \cong \pm 1,$$

(*via the isomorphism* (2.18)). □

We can now see how in our example the behaviour of t_Γ can immediately be deduced. Let g be the non-trivial element of $\operatorname{Hom}_{\Omega_\mathbb{Q}}(S_\Gamma, \pm 1)$, i.e., $g(\psi) = -1$. Then $t_\Gamma g$ is represented by $t'_\Gamma g$ (cf. (6.1), (6.2)), where $t'_\Gamma g(\delta) = 1$, $t'_\Gamma g(\psi)_2 = -1$. Hence $S t_\Gamma g = \rho(-1)$. If $n > 3$, i.e., $\mathbb{Q}(\psi) \neq \mathbb{Q}$, then $-1 \equiv 1 \pmod{\mathfrak{L}^2}$, whence $\rho(-1) = 1$. If $\Gamma = H_8$ then $\rho(-1) = -1$. Thus we have (cf. [F23] Proposition 1)

7.2. Proposition. *For $\Gamma = H_8$, t_Γ is an isomorphism. For $\Gamma = H_{2^n}, n > 3$, t_Γ is null.* □

Note here, that in all cases the two groups $\operatorname{Hom}_{\Omega_\mathbb{Q}}(S_\Gamma, \pm 1)$ and $D(\mathbb{Z}\Gamma)$ are actually isomorphic – a phenomenon which recurs elsewhere.

The next problem which arises is that of an explicit description of $W_{N/K}$, i.e., of the symplectic root numbers, in terms of other, easily computed, invariants. This is important to get specific information for given fields, as well as to obtain existence and density theorems. The first example which was computed, dealt with the case $K = \mathbb{Q}$ and $\Gamma = H_8$ (cf. [F6]), and in fact the proof of the main result of that paper (quoted in (1.15)) was based on an explicit description of all $N/\mathbb{Q}$ with $\operatorname{Gal}(N/\mathbb{Q}) \cong H_8$ and a computation of the $W(N/\mathbb{Q}, \psi)$ for the (unique) irreducible symplectic character ψ – for the remainder of this section to be denoted by $W(N/\mathbb{Q})$. We also write $W_\infty(N/\mathbb{Q}) = +1$ or -1, according to whether N is real or imaginary. This is actually the root number, at infinity, $W_\infty(N/\mathbb{Q}, \psi)$ defined in (5.16). Let $L_N = L$ be the maximal Abelian subfield of N (where always $\operatorname{Gal}(N/\mathbb{Q}) \cong H_8$). Then $L = \mathbb{Q}(\sqrt{d_1}, \sqrt{d_2})$ is of degree 4. For any number field F, denote by $\Delta(F)$ the product of the prime numbers dividing its discriminant. Now we get (cf. [F6])

7.3. Proposition. (i) *Let* $L = \mathbb{Q}(\sqrt{d_1}, \sqrt{d_2})$ *be of degree* 4. *Then* $L = L_N$, *for some* N, *with* $\mathrm{Gal}(N/\mathbb{Q}) \cong H_8$, *if, and only if,*

$$\left(\frac{d_1, d_2}{p}\right)\left(\frac{-1, d_1}{p}\right)\left(\frac{-1, d_2}{p}\right) = 1, \qquad \text{for all } p, \tag{7.2}$$

(*where* $(,/p)$ *denotes the Hilbert symbol*).

(ii) *If* $\mathrm{Gal}(N/\mathbb{Q}) \cong H_8$ *and* N *is tame then*

$$W(N/\mathbb{Q}) = W_\infty(N/\mathbb{Q})\left(\frac{2}{\Delta(L)}\right)\left(\frac{-1}{\Delta(N)}\right), \qquad L = L_N,$$

(*where* $(-)$ *denotes the Jacobi symbol.*)

(iii) *Suppose that* $L = \mathbb{Q}(\sqrt{d_1}, \sqrt{d_2})$ *satisfies* (7.2) *and that furthermore all prime divisors of* $\Delta(L)$ *are* $\equiv 1 \pmod 4$. *Then, for all* N *with* $L = L_N$, $W(N/\mathbb{Q}) = c_L$ *only depends on* L. *Each value* ± 1 *is taken for infinitely many* L.

(iv) *Suppose* $L = \mathbb{Q}(\sqrt{d_1}, \sqrt{d_2})$ *satisfies* (7.2), *but* $d_1 d_2$ (*is odd and*) *has prime divisors* $\equiv -1 \pmod 4$. *Then the number* m *of distinct prime divisors of* $\Delta(L)$ *is* $\geqslant 3$. *For each product* d' *of distinct odd primes not dividing* $\Delta(L)$ (*including* $d' = 1$) *there exist exactly* 2^{m-2} *fields* N, *with* $\mathrm{Gal}(N/\mathbb{Q}) \cong H_8$, $L = L_N$, *and* $\Delta(N) = \Delta(L)d'$. *For exactly half of these* $W(N/\mathbb{Q})$ *takes the values* $+1$, (*or* -1 *respectively*!). □

Remark. The constant c_L appearing in (iii) has an explicit description in terms of the divisors of $\Delta(L)$ (cf. [F6] Theorem 6).

It follows from the last proposition that $W(N/\mathbb{Q}) = +1$ or -1 both occur infinitely often, moreover with additional arithmetic boundary conditions imposed, e.g., with given quadratic subfield $\mathbb{Q}(\sqrt{d})$ – where we must have $d > 0$ and $d \equiv -1 \pmod 8$ –, with given ramification or non-ramification of finitely many prescribed primes, or with given $W_\infty(N/\mathbb{Q})$.

The distinction between the two cases occurring in the last proposition, under (ii) and (iii), for the possible variation of $W(N/\mathbb{Q})$ goes over to fields with Galois group H_{2^n}, $n > 3$ – one now has to take L_N the unique subfield with $\mathrm{Gal}(L_N/\mathbb{Q}) \cong D_{2^{n-1}}$, the dihedral group (see Theorem 3 and its Corollary 1 in [F12]). The computation of the root numbers in the general case will follow in Chap. V, §2.

Notes to Chapter I

[1] Induction theorems for real valued characters χ assert the existence of a presentation

$$\chi = \sum n_i \,\mathrm{ind}_{\Delta_i}^{\Gamma}(\phi_i),$$

where the n_i are integers, the Δ_i are subgroups belonging to some family $\mathfrak{F}$ of groups, and the ϕ_i are characters of Δ_i of prescribed types. In the classical result of this nature, the theorem of Berman-Witt (cf. [Se1] Theorem 27), the ϕ_i are again to be real valued. $\mathfrak{F}$ consists of groups which are semidirect products $\Sigma \rtimes \Pi$, where (i) Π is a p group, (ii) Σ is a cyclic group of order prime to p, (iii) Π acts on Σ by $\pi^{-1}\sigma\pi = \sigma^{f(\pi)}$, with $f\colon \Pi \to \pm 1$ a homomorphism of groups.

This theorem was not strong enough for what was needed in the arithmetic theory, and in fact better theorems were obtained. The Δ_i are as above, but on the ϕ_i further strong restrictions can be imposed. For orthogonal characters χ the ϕ_i are of three types:

(i) Real valued Abelian characters, i.e., homomorphisms into ± 1,

(ii) Dihedral characters, i.e., characters of dihedral representations (cf. (1.13)),

(iii) Bicyclic characters $\phi = \psi + \psi^{-1}$, where ψ is Abelian, ψ^{-1} its multiplicative inverse.

For symplectic characters the ϕ_i are of two types:

(i) Quaternion characters, i.e., characters of quaternion representations (cf. (1.14)),

(ii) Bicyclic characters.

These results go back to Serre, who first gave the proof of the orthogonal induction theorem in [Se3] – see also [De2] and [Ri2] for other proofs. For a detailed survey of this aspect of representation theory see [Ma3].

In the applications one uses these theorems to reduce arithmetic problems associated with orthogonal or symplectic representations to special types of field extensions. Here those with dihedral or (generalized) quaternion Galois groups play the important role. A detailed study of these extensions is contained in [F7] and from a different angle in [DM]. One knows that those with quaternion Galois group often present an exceptional picture and do not easily fit into the general theory of the embedding problem.

[2] Let X, Y be locally free $\mathbb{Z}\Gamma$-modules. Then their ranks and their classes in $\mathrm{Cl}(\mathbb{Z}\Gamma)$ coincide if, and only if, they are stably isomorphic, i.e., $X \oplus \mathbb{Z}\Gamma \cong Y \oplus \mathbb{Z}\Gamma$. Whenever the ranks are > 1 then this implies actually $X \cong Y$. Moreover for many groups Γ stable isomorphism of locally free modules always implies isomorphism i.e., we have "cancellation". This is the case whenever none of the simple components of $\mathbb{Q}\Gamma$ are totally definite quaternion division algebras over a totally real field, e.g., for Γ of odd order or Γ Abelian. We also have cancellation for $\Gamma = H_8$ and H_{16} but not for $\Gamma = H_{32}$ (see [Sw3]).

Specifically, $(X) = 1$ implies that X is stably free, i.e., $X \oplus \mathbb{Z}\Gamma$ is free over $\mathbb{Z}\Gamma$. If we have cancellation then of course $(X) = 1$ implies that X is free.

[3] Let Δ_1 be the p-Sylow group of the inertia group Δ_0 (it is unique, hence normal). Δ_1 is the first ramification group. The extension E/F is tame, precisely if $\Delta_1 = 1$. One then has a strengthening of Noether's theorem: $\mathfrak{o}_E$ is relatively projective with respect to a subgroup Σ of the Galois group if, and only if, $\Sigma \supset \Delta_1$ (cf. [F3], [Mi]).

[4] There is an alternative formulation of the theory as it has been outlined so far in §2–5, involving variants of the basic concepts. This second approach is outlined both in [F17] and in [F18]. The flavour is now more ideal theoretic, rather than idele theoretic.

(a) *Classgroup*. Let again $\mathfrak{o}$ be the ring of integers in a number field K. Fix a finite set S of primes, including those dividing order (Γ). One considers functions $\mathfrak{b}$ from the set of irreducible characters of Γ to fractional ideals such that

(i) $\mathfrak{b}(\chi)$ is a (non-zero) fractional ideal of the ring of integers in $K(\chi)$, the field of values of χ on Γ, over K.

(ii) The numerator and denominator of $\mathfrak{b}(\chi)$ are both prime to all $p \in S$.

(iii) For $\omega \in \mathrm{Gal}(K(\chi)/K)$, $\mathfrak{b}(\chi^\omega) = \mathfrak{b}(\chi)^\omega$.

These maps form a group $I(\mathfrak{o}, \Gamma)$, and in this group we consider the subgroup $G(\mathfrak{o}, \Gamma)$ of functions $\mathfrak{b}$, where $\mathfrak{b}(\chi) = (b(\chi))$, b being a function from the set of irreducible characters of Γ to $\mathbb{Q}^{c*}$, with the following properties:

(ia) for all $\omega \in \Omega_K$, $b(\chi^\omega) = b(\chi)^\omega$ (or equivalently this equation for $\omega \in \mathrm{Gal}(K(\chi)/K)$, and of course $b(\chi) \in K(\chi)^*$),

(iia) $b(\chi)$ is a unit at all primes above those in S,

(iiia) for all $p \in S$, $b(\chi)_p \in \mathrm{Det}_\chi(\mathfrak{o}_p \Gamma^*)$.

One then has

I. $$\mathrm{Cl}(\mathfrak{o}\Gamma) \cong I(\mathfrak{o}, \Gamma)/G(\mathfrak{o}, \Gamma). \qquad \square$$

Remark. One can of course extend the $\mathfrak{b}$ and the b by additivity to the whole of R_Γ, and we shall assume that we have done so.

(b) *Resolvents*. Let N/K be a tame Galois extension of number fields with Galois group Γ. The resolvents $(a \mid \chi)$, $a \in N$ lie in a one dimensional $K(\chi)$-subspace V_χ of $\mathbb{Q}^c$ (in fact of $N(\chi)$). Denote by $(\mathfrak{o} : \chi)$ the $\mathfrak{o}_{K(\chi)}$-module generated by the $(a \mid \chi)$ with $a \in \mathfrak{o}_N$. This is a rank one $\mathfrak{o}_{K(\chi)}$-submodule of V_χ. If $bK\Gamma = N$ then $(\mathfrak{o} : \chi)(b \mid \chi)^{-1}$ is a fractional ideal in $K(\chi)$.

II. *Let* $bK\Gamma = N$ *and* $b\mathfrak{o}_p\Gamma = \mathfrak{o}_{N,p}$, *for all* $p \in S$. *Then, with* $\mathfrak{b}(\chi) = (\mathfrak{o} : \chi)(b \mid \chi)^{-1}$ *we have* $\mathfrak{b} \in I(\mathfrak{o}, \Gamma)$, *and the class of* $\mathfrak{b}$ *in* $I(\mathfrak{o}, \Gamma)/G(\mathfrak{o}, \Gamma)$ *corresponds to* $(\mathfrak{o}_N)_{\mathfrak{o}\Gamma}$ *under the isomorphism* I. $\square$

Next define $\mathcal{N}_{K/\mathbb{Q}}(\mathfrak{o} : \chi)$ as the $\mathfrak{o}_{\mathbb{Q}(\chi)}$-module generated by the $\mathcal{N}_{K/\mathbb{Q}}(a \mid \chi)$, with $a \in \mathfrak{o}_N$ – or alternatively define the norm of $(\mathfrak{o} : \chi)$ directly.

III. *With* b *as in* II, *and writing*

$$\mathfrak{b}_1(\chi) = \mathcal{N}_{K/\mathbb{Q}}(\mathfrak{o} : \chi) \cdot \mathcal{N}_{K/\mathbb{Q}}(b \mid \chi)^{-1},$$

we have $\mathfrak{b}_1 \in I(\mathbb{Z}, \Gamma)$ *and the class of* $\mathfrak{b}_1$ *in* $I(\mathbb{Z}, \Gamma)/G(\mathbb{Z}, \Gamma)$ *corresponds to* $U_{N/K}$ *under the isomorphism* I. $\square$

(c) *Galois Gauss sums*. Here we have:

IV. $\mathcal{N}_{K/\mathbb{Q}}(\mathfrak{o} : \chi) = \mathfrak{o}_{\mathbb{Q}(\chi)}\tau(\chi)$. $\square$

This is a particularly neat form to express the basic connection between Galois Gauss sums and resolvents, except for the signature property.

[5] Note that the representative u, as given in Theorem 6, is unique not just modulo the denominator in the formula (2.18) for $D(\mathbb{Z}\Gamma)$, but actually already modulo $\mathrm{Det}(\mathfrak{U}\mathbb{Z}\Gamma)$. We thus get a unique class in the group $\mathrm{Hom}^+_{\Omega_\mathbb{Q}}(R_\Gamma, \mathfrak{U}(\mathbb{Q}^c))/\mathrm{Det}(\mathfrak{U}\mathbb{Z}\Gamma)$. One of the outstanding problems is the interpretation of this invariant, associated with N/K. For one attack on this, using Hermitian structure, see [F24] (Chap. VI, §2). Anticipating the theory, developed in II §1, we can remark that the above group is a subgroup of the Grothendieck group of "locally freely presented finite $\mathfrak{o}\Gamma$-modules". In fact it is the kernel in this group of the ring extension map to maximal orders. For the local analogue see II (2.3).

[6] A further description of the conductor, let us say, in the local case and irrespective of ramification (and analogously in the global case) reduces it to the discriminant. Let E/F then be a local Galois extension with Galois group Γ. Then

(a) $$\mathfrak{f}(E/F, \chi) = \mathfrak{f}(E/F, \chi^\omega),$$

for all $\chi \in R_\Gamma$, all $\omega \in \Omega_\mathbb{Q}$,

(b) $$\mathfrak{f}(E/F, \chi + \chi') = \mathfrak{f}(E/F, \chi)\mathfrak{f}(E/F, \chi'),$$

for all $\chi, \chi' \in R_\Gamma$,

(c) $$\mathfrak{f}(E/F, \mathrm{ind}_\Delta^\Gamma \varepsilon_\Delta) = \mathfrak{d}(E^\Delta/F) \qquad \text{(the relative discriminant)}$$

for any subgroup Δ of Γ, with fixed field E^Δ and identity character ε_Δ. ($\mathrm{ind}_\Delta^\Gamma$ is induction $\Delta \to \Gamma$.) These three properties determine $\mathfrak{f}(E/F, \chi)$ as a homomorphism $R_\Gamma \to \mathfrak{I}(F^c)$. One still has of course to show that the values lie in $\mathfrak{I}(F)$.

II. Classgroups and Determinants

§1. Hom-Description

Our aim here is to show how the Hom-description of the class group in Theorem 1 is deduced from well known results of algebraic $\mathfrak{K}$-theory. For some background see note [1]. There are three basic steps involved. The first leads to a $\mathfrak{K}$-theoretic interpretation of the class group, in terms of an exact sequence. We state the relevant definitions and results, but will not prove them – for a detailed discussion see e.g., [F24] (Chap. 1). The second step is the translation into the language of reduced norms, using classical results on these, and the final step then takes us via the determinants defined in I §2 to the Hom-description. In passing we shall also obtain in this way a proof of Proposition I.2.2. The notation is that of I §2. In particular K is a number field, $\mathfrak{o}$ its ring of algebraic integers, Γ a finite group and $\mathfrak{A}$ an $\mathfrak{o}$-order in $K\Gamma$. With appropriate changes (specifically replacing R_Γ by a certain Grothendieck group) the contents of this section apply equally well to any arbitrary semisimple finite dimensional algebra in place of $K\Gamma$ – but we shall avoid such unnecessary generality.

Our first observation is that the sequence I (2.1) used to define $\mathrm{Cl}(\mathfrak{A})$ splits. Indeed let $r_{\mathfrak{A}}$ be the rank map $\mathfrak{K}_0\mathfrak{A} \to \mathbb{Z}$ (see I §2). Then, if $\mathfrak{A}^n$ denotes the product of n copies of $\mathfrak{A}$, the map

$$(X) \mapsto [X] - [\mathfrak{A}^n], \qquad n = \mathrm{rank}(X) = r_{\mathfrak{A}}(X),$$

defines an isomorphism

$$\mathrm{Cl}(\mathfrak{A}) \cong \mathrm{Ker}\, r_{\mathfrak{A}}.$$

Indeed this follows from the fact that the composite map

$$\mathbb{Z} \xrightarrow{n \mapsto [\mathfrak{A}^n]} \mathfrak{K}_0(\mathfrak{A}) \xrightarrow{r_{\mathfrak{A}}} \mathbb{Z}$$

is the identity map. We thus have an exact sequence

$$(1.1) \qquad 1 \to \mathrm{Cl}(\mathfrak{A}) \to \mathfrak{K}_0(\mathfrak{A}) \xrightarrow{r_{\mathfrak{A}}} \mathbb{Z} \to 0$$

(we write the operation in $\mathrm{Cl}(\mathfrak{A})$ as multiplication).

Recall now the definition of $\mathfrak{K}_1(S)$ for an arbitrary ring S,

$$\mathfrak{K}_1(S) = \varinjlim \mathrm{GL}_n(S)^{ab},$$

with the direct limit defined via the standard embeddings $\mathrm{GL}_n(S) \to \mathrm{GL}_{n+q}(S)$,

$$P \mapsto \begin{pmatrix} P & 0 \\ 0 & I_q \end{pmatrix}, \qquad I_q \text{ the identity matrix.}$$

We then have an exact sequence

$$(1.2) \qquad \mathfrak{K}_1(\mathfrak{A}) \to \mathfrak{K}_1(K\Gamma) \to \mathfrak{K}_0 T(\mathfrak{A}) \to \mathfrak{K}_0(\mathfrak{A}) \xrightarrow{r_{\mathfrak{A}}} \mathbb{Z} \to 0,$$

where the map on the extreme left comes from the embedding $\mathfrak{A} \subset K\Gamma$, and where we shall now define $\mathfrak{K}_0 T(\mathfrak{A})$ and the maps involving it.

We consider "*locally freely presented* $\mathfrak{A}$*-torsion modules*" M, with $\mathfrak{A}$, say, acting on the right, i.e., $\mathfrak{A}$-modules of form $M = X/Y$, where X and Y are locally free $\mathfrak{A}$-modules spanning the same $K\Gamma$-module $XK = YK$. As $\mathfrak{o}$-modules these are finite. We define $\mathfrak{K}_0 T(\mathfrak{A})$ as the Grothendieck group of such modules M, modulo exact sequences. Thus $\mathfrak{K}_0 T(\mathfrak{A})$ is the Abelian group, generated by symbols $[M]$, and with relations

$$[M] = [M'] + [M''],$$

for every exact sequence $0 \to M' \to M \to M'' \to 0$ of $\mathfrak{A}$-modules. If X, Y are locally free $\mathfrak{A}$-modules, with $XK = YK$, then there is a locally free $\mathfrak{A}$-module Z, with $XK = ZK$ and $Z \subset X \cap Y$. We can then define uniquely an element $[X/Y] \in \mathfrak{K}_0 T(\mathfrak{A})$ by $[X/Y] = [X/Z] - [Y/Z]$. The map $\mathfrak{K}_0 T(\mathfrak{A}) \to \mathfrak{K}_0(\mathfrak{A})$ is given by $[X/Y] \mapsto [Y] - [X]$. On the other hand, we view $\mathrm{GL}_n(K\Gamma)$ as the automorphism group of $(K\Gamma)^n$, acting on the left. For $a \in \mathrm{GL}_n(K\Gamma)$ we then have two free $\mathfrak{A}$-modules $\mathfrak{A}^n$ and $a\mathfrak{A}^n$ spanning $(K\Gamma)^n$, and we get an element $[\mathfrak{A}^n/a\mathfrak{A}^n] \in \mathfrak{K}_0 T(\mathfrak{A})$. This only depends on the image of a under the composite map $\mathrm{GL}_n(K\Gamma) \to \mathrm{GL}_n(K\Gamma)^{ab} \to \mathfrak{K}_1(K\Gamma)$ and thus defines a homomorphism $\mathfrak{K}_1(K\Gamma) \to \mathfrak{K}_0 T(\mathfrak{A})$, and this is the one in (1.2).

We assume the exactness of sequence (1.2). In fact (1.2) is the analogue to a well known exact sequence of $\mathfrak{K}$-theory, the latter involving projective modules, rather than locally free ones, and the bigger category of finite $\mathfrak{A}$-modules of finite projective dimension. Our sequence can be easily derived from the latter one (cf. [F24], I).

From (1.1) and (1.2) we now get yet a further interpretation of $\mathrm{Cl}(\mathfrak{A})$, namely

$$(1.3) \qquad \mathrm{Cl}(\mathfrak{A}) \cong \mathrm{Cok}[\mathfrak{K}_1(K\Gamma) \to \mathfrak{K}_0 T(\mathfrak{A})].$$

The map $\mathfrak{K}_0 T(\mathfrak{A}) \to \mathrm{Cl}(\mathfrak{A})$ is given by $[X/Y] \mapsto (X)^{-1}(Y)$, as one verifies.

Now we go over to the completions at a finite prime divisor $\mathfrak{p}$ of K. We get a Grothendieck group $\mathfrak{K}_0 T(\mathfrak{A}_\mathfrak{p})$ of finite $\mathfrak{A}_\mathfrak{p}$-modules, quotients of free $\mathfrak{A}_\mathfrak{p}$-modules, and a local analogue to the exact sequence (1.2). As however the rank map $\mathfrak{K}_0(\mathfrak{A}_\mathfrak{p}) \to \mathbb{Z}$ is an isomorphism, we really end up with an exact sequence

$$(1.4) \qquad \mathfrak{K}_1(\mathfrak{A}_\mathfrak{p}) \to \mathfrak{K}_1(K_\mathfrak{p}\Gamma) \to \mathfrak{K}_0 T(\mathfrak{A}_\mathfrak{p}) \to 0.$$

On the other hand, one easily shows (see e.g., [F24] I) that

$$\mathfrak{K}_0 T(\mathfrak{A}) \cong \coprod \mathfrak{K}_0 T(\mathfrak{A}_\mathfrak{p})$$

(restricted product over all finite prime divisors), or using (1.4),

$$\mathfrak{K}_0 T(\mathfrak{A}) \cong \coprod (\mathfrak{K}_1(K_\mathfrak{p}\Gamma)/\mathfrak{K}_1(K_\mathfrak{p}\Gamma; \mathfrak{A}_\mathfrak{p})), \tag{1.5}$$

where we abbreviate $\mathfrak{K}_1(K_\mathfrak{p}\Gamma; \mathfrak{A}_\mathfrak{p}) = \operatorname{Im}[\mathfrak{K}_1(\mathfrak{A}_\mathfrak{p}) \to \mathfrak{K}_1(K_\mathfrak{p}\Gamma)]$. In the last isomorphism we may and will take the product over *all* prime divisors $\mathfrak{p}$ of K, as for the infinite $\mathfrak{p}$ we have $\mathfrak{A}_\mathfrak{p} = K_\mathfrak{p}\Gamma$, by definition.

Notation. If we have a family $\{H_i\}$ of Abelian groups, with i running over some index set, and for almost all i (i.e., with at most finitely many exceptions), a subgroup H_i' of H_i is given, the *product* $\prod_i (H_i | H_i')$ *of the* H_i, *restricted with respect to the subgroups* H_i', is defined as the subgroup of $\prod_i H_i$, of elements h, whose component h_i at i lies in H_i', for almost all i. Let then

$$\text{(1.6)} \quad \begin{cases} G = G_\mathfrak{A} = \prod\limits_\mathfrak{p} (\mathfrak{K}_1(K_\mathfrak{p}\Gamma) | \mathfrak{K}_1(K_\mathfrak{p}\Gamma; \mathfrak{A}_\mathfrak{p})), \\ G' = G'_\mathfrak{A} = \prod\limits_\mathfrak{p} (\mathfrak{K}_1(K_\mathfrak{p}\Gamma; \mathfrak{A}_\mathfrak{p})), \\ G'' = G''_\mathfrak{A} = \operatorname{Im}[\mathfrak{K}_1(K\Gamma) \to G] \qquad \text{(diagonal embedding).} \end{cases}$$

Then, by (1.5),

$$\mathfrak{K}_0 T(\mathfrak{A}) \cong G/G',$$

and hence, by (1.3),

$$\operatorname{Cl}(\mathfrak{A}) \cong G/(G' \cdot G''). \tag{1.7}$$

Going back to the definitions of the various maps, we can describe (1.7) quite explicitly. We shall do this in terms of the surjection $G \to \operatorname{Cl}(\mathfrak{A})$. Let $g \in G/(G' \cdot G'')$. Then g is the class of an element g^* of G with the following property. For some fixed n, independent of $\mathfrak{p}$, and for every $\mathfrak{p}$, the local component $g_\mathfrak{p}^*$ (in terms of the local decomposition (1.6)) is the image of an element $a_\mathfrak{p} \in \operatorname{GL}_n(K_\mathfrak{p}\Gamma)$, under the map from the linear group into $\mathfrak{K}_1(K_\mathfrak{p}\Gamma)$. Inside the free $K\Gamma$-module $(K\Gamma)^n$, spanned by $\mathfrak{A}^n$, the local equations $X_\mathfrak{p} = a_\mathfrak{p}\mathfrak{A}_\mathfrak{p}^n$, for all finite $\mathfrak{p}$, will determine a locally free $\mathfrak{A}$-module X and, under (1.7)

$$(X) \leftrightarrow g. \tag{1.8}$$

We shall now reformulate (1.7) in terms of reduced norms. If S is a finite dimensional semisimple algebra over a field F, we define the reduced norm

$$\operatorname{nrd}\colon S^* \to \operatorname{cent}(S)^*$$

via the simple components. By going over to matrix rings $\operatorname{M}_n(S)$ and abelianizing, it gives rise to a system of homomorphisms $\mathrm{GL}_n(S)^{ab} \to \mathrm{cent}(S)^*$, compatible with the standard maps $\mathrm{GL}_n(S)^{ab} \to \mathrm{GL}_{n+q}(S)^{ab}$, and hence a homomorphism (using the same symbol as before)

$$\mathrm{nrd}\colon \mathfrak{K}_1(S) \to \mathrm{cent}(S)^*. \tag{1.9}$$

One knows (cf. [SE], Prop. 8.5):

1.1. Lemma. *For $S = K\Gamma$, or $= K_\mathfrak{p}\Gamma$ (all $\mathfrak{p}$), or $= \mathfrak{A}_\mathfrak{p}$ (all finite $\mathfrak{p}$) the map $S^* \to \mathfrak{K}_1(S)$ is surjective.* □

Next, by Wang's theorem (cf. [Wg]), the reduced norm map $S^{*ab} \to \mathrm{cent}(S)^*$ is injective for $S = K\Gamma$, $S = K_\mathfrak{p}\Gamma$. Thus by Lemma 1.1 we get

1.2. Lemma. *For $S = K\Gamma$ and $S = K_\mathfrak{p}\Gamma$ the map* nrd *of* (1.9) *yields an isomorphism*

$$\mathfrak{K}_1(S) \cong \mathrm{nrd}(S^*)$$ □

Now we turn to the image of nrd, describing it in terms of a simple component A of $K\Gamma$.

1.3. Lemma. $\mathrm{nrd}(A_\mathfrak{p}^*) = \mathrm{cent}(A_\mathfrak{p})^*$ *except when $\mathfrak{p}$ is a real prime and $A_\mathfrak{p} = M_n(\mathbb{H})$, $\mathbb{H}$ the quaternion algebra. In this case* $\mathrm{nrd}(A_\mathfrak{p}^*)$ *is the group of positive reals.* □

This follows easily from the arithmetic of local division algebras. For the global image we consider the idele group $\mathfrak{J}(\mathrm{cent}(A))$ of the number field $\mathrm{cent}(A)$. We define purely formally

$$\mathrm{nrd}(\mathfrak{J}(A)) = [x \in \mathfrak{J}(\mathrm{cent}(A)) \text{ with } x_\mathfrak{p} \in \mathrm{nrd}(A_\mathfrak{p}^*) \text{ for all } \mathfrak{p}]. \tag{1.10}$$

Alternatively we can define $\mathfrak{J}(A)$ as the idele group of the algebra A, and $\mathrm{nrd}(\mathfrak{J}(A))$ as its image under the extension of nrd. By 1.3, $\mathrm{nrd}(\mathfrak{J}(A))$ differs from $\mathfrak{J}(\mathrm{cent}(A))$ at the real prime divisors at most.

1.4. Lemma.

(i) $$\mathrm{nrd}(A^*) = \mathrm{cent}(A)^* \cap \mathrm{nrd}(\mathfrak{J}(A)),$$

(ii) $$\mathfrak{J}(\mathrm{cent}(A)) = \mathrm{cent}(A)^* \cdot \mathrm{nrd}(\mathfrak{J}(A)).$$

The first assertion is the Hasse-Schilling norm theorem (cf. [HS]), the second one comes from the fact that a number field contains elements of arbitrary signature at the real primes. □

Finally, from the arithmetic of local division algebras we also have

1.5. Lemma. *If $\mathfrak{M}_\mathfrak{p}$ is a maximal order in $K_\mathfrak{p}\Gamma$ then* $\mathrm{nrd}(\mathfrak{M}_\mathfrak{p}^*) = \mathfrak{O}_\mathfrak{p}^*$, *where $\mathfrak{O}_\mathfrak{p}$ is the integral closure of $\mathfrak{o}_\mathfrak{p}$ in* $\mathrm{cent}(K_\mathfrak{p}\Gamma)$. □

To come back now to (1.7), observe first that in (1.6), $G_{\mathfrak{A}} = G_{\mathfrak{M}}$, $\mathfrak{M}$ a maximal order. Indeed, for almost all $\mathfrak{p}$, $\mathfrak{A}_\mathfrak{p} = \mathfrak{M}_\mathfrak{p}$, and the definition of a relatively restricted product is unaffected by finitely many changes. By Lemmas 1.2, 1.3 and 1.5, the maps (1.9) for $S = K_\mathfrak{p}\Gamma$, all $\mathfrak{p}$, give rise to an isomorphism of $G_{\mathfrak{A}}$ with the product of the $\mathrm{nrd}(K_\mathfrak{p}\Gamma^*)$ restricted with respect to the $\mathfrak{O}_\mathfrak{p}^*$, in other words,

$$(1.11) \qquad G_{\mathfrak{A}} \cong \mathrm{nrd}(\mathfrak{J}(K\Gamma)) \subset \mathfrak{J}(\mathrm{cent}(K\Gamma)).$$

Here of course

$$\mathfrak{J}(\mathrm{cent}(K\Gamma)) = \prod \mathfrak{J}(\mathrm{cent}(A)),$$
$$\mathrm{nrd}(\mathfrak{J}(K\Gamma)) = \prod \mathrm{nrd}(\mathfrak{J}(A)),$$

with the product running over the simple components A of $K\Gamma$. The isomorphism (1.11) will, by Lemma 1.1, yield isomorphisms

$$G'_{\mathfrak{A}} \cong \prod_{\mathfrak{p}} \mathrm{nrd}(\mathfrak{A}_\mathfrak{p}^*) = \mathrm{nrd}\left(\prod_{\mathfrak{p}} \mathfrak{A}_\mathfrak{p}^*\right) = \mathrm{nrd}(\mathfrak{U}\mathfrak{A}),$$

and

$$G'' \cong \mathrm{nrd}(K\Gamma)^*.$$

Here, as in I §2 we write $\mathfrak{U}\mathfrak{A} = \prod \mathfrak{A}_\mathfrak{p}^*$ (product over all $\mathfrak{p}$). Thus the reduced norm induces by (1.7) an isomorphism

$$(1.12) \qquad \mathrm{Cl}(\mathfrak{A}) \cong \mathrm{nrd}(\mathfrak{J}(K\Gamma))/(\mathrm{nrd}(K\Gamma^*) \cdot \mathrm{nrd}(\mathfrak{U}\mathfrak{A})).$$

By Lemma 1.4 however, we can rewrite this as

$$(1.13) \qquad \mathrm{Cl}(\mathfrak{A}) \cong \mathfrak{J}(\mathrm{cent}\, K\Gamma)/((\mathrm{cent}\, K\Gamma)^* \, \mathrm{nrd}(\mathfrak{U}\mathfrak{A})).$$

For, by that Lemma, the map from the right hand side of (1.12) to that of (1.13) is an isomorphism. Going back to the explicit description leading up to (1.8) we now have, under (1.13),

$$(1.14) \quad (X) \leftrightarrow \text{class of } \mathrm{nrd}(a), \qquad \text{where} \qquad \mathrm{nrd}(a)_\mathfrak{p} = \mathrm{nrd}(a_\mathfrak{p}), \quad \text{for all } \mathfrak{p}.$$

The isomorphism (1.13) is that in [F10] (Theorem 2). We now come to the final step. For this we need the notion of a determinant (see I §2, (2.3)–(2.9)), but extended to matrices over group rings (cf. (2.3)′–(2.5)′).

Let Ir_Γ be the set of (absolutely) irreducible characters of Γ. This is an Ω_K-set, and a free generating set of the Abelian group R_Γ. We thus have an isomorphism

$$(1.15) \qquad \mathrm{Map}_{\Omega_K}(\mathrm{Ir}_\Gamma, \;\;) \cong \mathrm{Hom}_{\Omega_K}(R_\Gamma, \;\;),$$

where Map_{Ω_K} is the group of Ω_K-set maps from Ir_Γ into some Ω_K-module. On the other hand we have a bijection of sets

$$(1.16) \qquad \mathrm{Ir}_\Gamma \cong \mathrm{Alg}_K(\mathrm{cent}(K\Gamma), \mathbb{Q}^c); \qquad \chi \mapsto g_\chi,$$

where on the right we have the set of homomorphisms $g_\chi\colon \mathrm{cent}(K\Gamma) \to \mathbb{Q}^c$ of K-algebras. Here g_χ is given by the restriction to $\mathrm{cent}(K\Gamma)$ of any representation $T\colon K\Gamma \to \mathrm{M}_m(\mathbb{Q}^c)$ with character χ. We shall also denote again by g_χ the unique extension to a homomorphism $\mathrm{cent}(B\Gamma) \to \mathbb{Q}^c \otimes_K B$, with B a commutative K-algebra, e.g. $K_\mathfrak{p}$. Ω_K will act on $\mathrm{Alg}_K(\mathrm{cent}(K\Gamma), \mathbb{Q}^c)$ by composition, i.e., we have $g^\omega(x) = g(x)^\omega$, for $x \in \mathrm{cent}(K\Gamma)$, and $\omega \in \Omega_K$. The map (1.16) is one of Ω_K-sets.

1.6. Lemma. *Let $B = K$ or $= K_\mathfrak{p}$. With $b \in \mathrm{cent}(B\Gamma)^*$ associate $f_b\colon \mathrm{Ir}_\Gamma \to \mathbb{Q}^c \otimes_K B$, where*

$$f_b(\chi) = g_\chi(b).$$

Via (1.15) *this map $b \mapsto f_b$ becomes an isomorphism*

$$\mathrm{cent}(B\Gamma)^* \cong \mathrm{Hom}_{\Omega_K}(R_\Gamma, (\mathbb{Q}^c \otimes_K B)^*).$$

Moreover the diagram

$$\begin{array}{ccc} & \mathrm{GL}_n(B\Gamma) & \\ {}^{\mathrm{nrd}}\swarrow & & \searrow^{\mathrm{Det}} \\ \mathrm{cent}(B\Gamma)^* & \cong & \mathrm{Hom}_{\Omega_K}(R_\Gamma, (\mathbb{Q}^c \otimes_K B)^*) \end{array}$$

commutes.

Proof. Clearly $f_b \in \mathrm{Map}_{\Omega_K}(\mathrm{Ir}_\Gamma, (\mathbb{Q}^c \otimes_K B)^*)$, and $b \mapsto f_b$ is an injective homomorphism. Conversely if $f \in \mathrm{Map}_{\Omega_K}(\mathrm{Ir}_\Gamma, (\mathbb{Q}^c \otimes_K B)^*)$, and A is a simple component of $\mathrm{cent}(K\Gamma)$, then there exists a unique element $c = c_A \in \mathrm{cent}(A)^*$ so that $g_\chi(c) = f(\chi)$ for all $\chi \in \mathrm{Ir}_\Gamma$ which correspond to A. (These form a single Ω_K-orbit). Now let $b = \sum c_A$; then $f = f_b$. Thus the map is surjective.

The commutativity of the diagram is really the definition of the reduced norm. □

The last Lemma remains true on transition to relative restricted products. We then get an isomorphism

$$\mathfrak{J}(\mathrm{cent}(K\Gamma)) \cong \mathrm{Hom}_{\Omega_K}(R_\Gamma, \mathfrak{J}(\mathbb{Q}^c)),$$

restricting to

$$\mathrm{cent}(K\Gamma)^* \cong \mathrm{Hom}_{\Omega_K}(R_\Gamma, \mathbb{Q}^{c*}),$$

and

$$\mathrm{nrd}(\mathfrak{U}\mathfrak{A}) \cong \mathrm{Det}(\mathfrak{U}\mathfrak{A}).$$

From (1.13) we now at last deduce the required isomorphism of Theorem 1 (ii). Moreover the description (1.14), together with Lemma 1.6, shows that this isomorphism is indeed given as was stated, i.e., (X) corresponds to the class of $\mathrm{Det}(a)$, where $\mathrm{Det}(a)_\mathfrak{p} = \mathrm{Det}(a_\mathfrak{p})$. □

Finally, for Proposition I.2.2, note that if $f \in \mathrm{Hom}_{\Omega_\mathbb{Q}}(R_\Gamma, \mathfrak{J}(\mathbb{Q}^c))$, then $f(\chi + \bar{\chi})_\mathfrak{p}$ will be real and positive at all infinite prime divisors. Thus $f \in \mathrm{Hom}^+_{\Omega_\mathbb{Q}}(R_\Gamma, \mathfrak{J}(\mathbb{Q}^c))$ if, and only if, $f(\chi)_{\mathfrak{p}} > 0$ for all irreducible sympletic χ and all infinite prime divisors $\mathfrak{p}$. These are precisely the χ which correspond to components of $\mathbb{R}\Gamma$ of the form $\mathrm{M}_n(\mathbb{H})$. Proposition 2.2 now follows from Lemmas 1.3, 1.5, and 1.6 (note [2]). □

In conclusion we state, for completeness sake the well known finiteness result on class groups.

1.7. Proposition. $\mathrm{Cl}(\mathfrak{A})$ *is finite.*

There are various proofs, using the various descriptions of $\mathrm{Cl}(\mathfrak{A})$. Here we shall give a proof by reduction to number fields, using (1.13). If $\mathfrak{M}$ is a maximal order containing $\mathfrak{A}$ then

$$\mathrm{Cl}(\mathfrak{M}) = \prod_i \mathfrak{J}(\mathrm{cent}\, A_i)/((\mathrm{cent}\, A_i)^* \,\mathrm{nrd}(\mathfrak{U}\mathfrak{M}_i)),$$

where $K\Gamma = \prod A_i$ (product of simple algebras), $\mathfrak{M}_i$ the image of $\mathfrak{M}$ in A_i. Each of the factors in (1.16) is a quotient of the group of ideals in the number field cent A_i modulo the totally positive principal ideals, hence is finite. So $\mathrm{Cl}(\mathfrak{M})$ is finite. But $\mathrm{nrd}\,\mathfrak{U}\mathfrak{M}/\mathrm{nrd}\,\mathfrak{U}\mathfrak{A}$ surjects onto $\mathrm{Ker}[\mathrm{Cl}(\mathfrak{A}) \to \mathrm{Cl}(\mathfrak{M})]$, and is finite, as $\mathfrak{U}\mathfrak{M}/\mathfrak{U}\mathfrak{A} = \prod_{\mathfrak{p}} \mathfrak{M}_{\mathfrak{p}}^*/\mathfrak{A}_{\mathfrak{p}}^*$ is finite – the product extending over the finitely many prime divisors with $\mathfrak{M}_{\mathfrak{p}} \neq \mathfrak{A}_{\mathfrak{p}}$. Thus finally $\mathrm{Cl}(\mathfrak{A})$ is finite. □

§2. Localization

(Meaning local completion!) This section is entirely technical and of no great interest in itself. It is however needed to introduce the required localization procedures and to ensure their proper understanding. Throughout this section p is a rational prime divisor, possibly the infinite one. $\mathbb{Q}_p^c$ is an algebraic closure of $\mathbb{Q}_p$ (so that $\mathbb{Q}_\infty^c = \mathbb{C}$), not to be confused with $(\mathbb{Q}^c)_p = \mathbb{Q}^c \otimes_{\mathbb{Q}} \mathbb{Q}_p$, or for that matter with $(\mathbb{Q}^c)_{\mathfrak{p}} = \mathbb{Q}^c \otimes_L L_{\mathfrak{p}}$, where L is a number field, $\mathfrak{p}$ a prime divisor in L. Similarly $\mathfrak{U}(\mathbb{Q}_p^c)$ is the group of units of the ring $\mathbb{Z}_p^c$ of integers (valuation ring) in $\mathbb{Q}_p^c$, not to be confused with $\mathfrak{U}_p(\mathbb{Q}^c)$, the group of units of the ring of integers in $\mathbb{Q}^c \otimes_{\mathbb{Q}} \mathbb{Q}_p$, or for that matter with $\mathfrak{U}_{\mathfrak{p}}(\mathbb{Q}^c)$, the group of units of the ring of integers in $(\mathbb{Q}^c)_{\mathfrak{p}}$ (L, $\mathfrak{p}$ as above).

We consider representations $T: \Gamma \to \mathrm{GL}_n(\mathbb{Q}_p^c)$ and their associated characters χ, where now $\chi: \Gamma \to \mathbb{Q}_p^c$. In the usual way we get the additive group $R_{\Gamma,p} = R_\Gamma(\mathbb{Q}_p^c)$ of virtual $\mathbb{Q}_p^c$-characters, or equivalently the Grothendieck group of $\mathbb{Q}_p^c\Gamma$, on which the absolute Galois group $\Omega_{\mathbb{Q}_p} = \mathrm{Gal}(\mathbb{Q}_p^c/\mathbb{Q}_p)$ acts. For $[F:\mathbb{Q}_p] < \infty$, $F \subset \mathbb{Q}_p^c$, we are then interested in the Hom groups $\mathrm{Hom}_{\Omega_F}(R_{\Gamma,p}, G)$, $G = (\mathbb{Q}_p^c)^*$ or $G = \mathfrak{U}(\mathbb{Q}_p^c)$ and for infinite p also the corresponding $\mathrm{Hom}_{\Omega_F}^+$ – where we demand $f(\chi) > 0$, when χ is symplectic.

We can now define determinants just as before. Extend a given representation T of Γ as above, to an algebra homomorphism T: $\mathrm{M}_m(F\Gamma) \to \mathrm{M}_{mn}(\mathbb{Q}_p^c)$; composing with $\mathrm{GL}_{mn}(\mathbb{Q}_p^c) \to (\mathbb{Q}_p^c)^*$ we get a map

$$a \mapsto \mathrm{Det}_\chi(a), \qquad a \in \mathrm{GL}_m(F\Gamma), \qquad \chi \text{ the character of } T.$$

The map $\chi \mapsto \mathrm{Det}_\chi(a)$ then extends to an Ω_F-homomorphism $\mathrm{Det}(a)\colon R_{\Gamma,p} \to (\mathbb{Q}_p^c)^*$, and we thus get subgroups $\mathrm{Det}(F\Gamma^*)$ and $\mathrm{Det}(\mathfrak{A}^*)$ ($\mathfrak{A}$ any order in $F\Gamma$ over $\mathfrak{o}_F$, p finite) of $\mathrm{Hom}_{\Omega_F}(R_{\Gamma,p}, (\mathbb{Q}_p^c)^*)$.

Now let L be a number field, $j\colon L \to \mathbb{Q}_p^c$ an embedding of L corresponding to a prime divisor $\mathfrak{p}$ of L, and F the closure in $\mathbb{Q}_p^c$ of $\operatorname{Im} j$, i.e., $F \cong L_\mathfrak{p}$. We extend j to a homomorphism $j\colon \mathbb{Q}^c \to \mathbb{Q}_p^c$ and this gives rise to a homomorphism $\mathbb{Q}^c \otimes_L L_\mathfrak{p} = (\mathbb{Q}^c)_\mathfrak{p} \to \mathbb{Q}_p^c$, again denoted by j. On the other hand j also gives rise to an isomorphism $\chi \mapsto \chi^j$, of R_Γ onto $R_{\Gamma,p}$. Here $\chi^j(\gamma) = \chi(\gamma)^j$. There result homomorphisms

$$(2.1) \qquad j^*\colon \begin{cases} \mathrm{Hom}_{\Omega_L}(R_\Gamma, (\mathbb{Q}^c)_\mathfrak{p}^*) \to \mathrm{Hom}_{\Omega_F}(R_{\Gamma,p}, (\mathbb{Q}_p^c)^*), \\ \mathrm{Hom}_{\Omega_L}(R_\Gamma, \mathfrak{U}_\mathfrak{p}(\mathbb{Q}^c)) \to \mathrm{Hom}_{\Omega_F}(R_{\Gamma,p}, \mathfrak{U}(\mathbb{Q}_p^c)) \ (p \text{ finite}), \end{cases}$$

given by

$$(j^*f)(\chi) = f(\chi^{j^{-1}})^j.$$

Let now $T\colon \Gamma \to \mathrm{GL}_n(\mathbb{Q}^c)$ be a "global" representation, with character χ. We then get firstly an algebra homomorphism

$$T\colon \mathrm{M}_m(L_\mathfrak{p}\Gamma) \to \mathrm{M}_{mn}((\mathbb{Q}^c)_\mathfrak{p}),$$

and on composition with determinants a map

$$\mathrm{Det}_\chi\colon \mathrm{GL}_m(L_\mathfrak{p}\Gamma) \to (\mathbb{Q}^c)_\mathfrak{p}^*.$$

Secondly we get on applying j, a representation $T^j\colon \Gamma \to \mathrm{GL}_n(\mathbb{Q}_p^c)$, hence a homomorphism

$$T^j\colon \mathrm{M}_m(L_\mathfrak{p}\Gamma) \cong \mathrm{M}_m(F\Gamma) \to \mathrm{M}_{mn}(\mathbb{Q}_p^c),$$

yielding a determinant

$$\mathrm{Det}_{\chi^j}\colon \mathrm{GL}_m(L_\mathfrak{p}\Gamma) \to (\mathbb{Q}_p^c)^*.$$

Thus an element $a \in \mathrm{GL}_m(L_\mathfrak{p}\Gamma)$ will define two determinants, both denoted by $\mathrm{Det}(a)$, namely $R_\Gamma \to (\mathbb{Q}^c)_\mathfrak{p}^*$, and $R_\Gamma \to (\mathbb{Q}_p^c)^*$. Accordingly we get for any $\mathfrak{o}_L$-order $\mathfrak{A}$ in $L\Gamma$ homomorphisms

$$(2.2) \qquad \mathrm{Det}\colon \begin{cases} \mathfrak{A}_\mathfrak{p}^* \to \mathrm{Hom}_{\Omega_L}(R_\Gamma, (\mathbb{Q}^c)_\mathfrak{p}^*), \\ \mathfrak{A}_\mathfrak{p}^* \to \mathrm{Hom}_{\Omega_F}(R_{\Gamma,p}, (\mathbb{Q}_p^c)^*). \end{cases}$$

2.1. Lemma. *The maps* j^* *of* (2.1) *are isomorphisms and the diagram*

$$\begin{array}{ccc} & \mathfrak{A}_\mathfrak{p}^* & \\ {}^{\mathrm{Det}}\swarrow & & \searrow^{\mathrm{Det}} \\ \mathrm{Hom}_{\Omega_L}(R_\Gamma, (\mathbb{Q}^c)_\mathfrak{p}^*) & \xrightarrow[j^*]{} & \mathrm{Hom}_{\Omega_F}(R_{\Gamma,p}, (\mathbb{Q}_p^c)^*) \end{array}$$

commutes. □

Remark 1. The Lemma implies that j^* maps the images of the two maps Det of (2.2) isomorphically. Having established this we shall use the symbol $\mathrm{Det}(\mathfrak{A}_\mathfrak{p}^*)$ for the image in the appropriate Hom group, the correct group always being clear from the context.

Remark 2. Having established the required result we shall in future use the symbol R_Γ, rather than $R_{\Gamma,p}$, for the $\mathbb{Q}_p^c$ – characters – unless there is a danger of confusion.

The proof of the Lemma follows easily from standard change of ring techniques. We can throughout replace $\mathbb{Q}^c$ by a big enough number field E, Galois over $\mathbb{Q}$, and Ω_L by $\mathrm{Gal}(E/L)$ and analogously on the local side.

In conclusion we come back to the group $\mathfrak{K}_0T(\mathfrak{A})$, now in the local context. Thus F is again a local field, now an extension of finite degree of $\mathbb{Q}_p$, with p finite. $\mathfrak{A}$ is here an order in $F\Gamma$, and $\mathfrak{K}_0T(\mathfrak{A})$ is the Grothendieck group of $\mathfrak{A}$-modules X/Y, where X and Y are free $\mathfrak{A}$-modules, spanning the same $F\Gamma$-module $XF = YF$ (of finite rank of course). The same methods as those used in the preceding §1 then lead to an isomorphism

$$\mathfrak{K}_0T(\mathfrak{A}) \cong \mathrm{Hom}_{\Omega_F}(R_\Gamma, (\mathbb{Q}_p^c)^*)/\mathrm{Det}(\mathfrak{A}^*). \tag{2.3}$$

§3. Change in Basefield and Change in Group

We are now considering the class group $\mathrm{Cl}(\mathfrak{o}\Gamma)$ ($\mathfrak{o} = \mathfrak{o}_K$) and the associated Hom-groups in the context of a change in K or in Γ. The techniques for applications, which become available, are often also connected with the actions of character rings on class groups, and this will be dealt with as well. The superiority of the Hom description over the earlier (ideal theoretic or idele theoretic) descriptions of the class group, which had been given in terms of the simple components of the group ring, is essentially due to the following facts. Neither restriction to subgroups, nor induction are reflected directly in terms of maps on the respective sets of irreducible characters (or the simple components) – irreducibility is lost. But they are indeed naturally reflected in terms of homomorphisms on the R_Γ and thus also on the Hom groups. Similarly, on extension or contraction of basefield, the simple components – or equivalently the Galois orbits of irreducible characters – fail to be preserved. But again these operations have their natural expression in the language of the R_Γ and the Hom groups. Moreover the known action of K-characters on $\mathrm{Cl}(\mathfrak{o}\Gamma)$ (see below) has its evident counterpart in terms of Hom groups, but not in the older theory.

Proofs will be omitted in this section, but complete references will be given. For the general background of representation theory see [Se2], for Frobenius-functors and -modules see [Lm], [SE], [Ba]. We shall have to deal both with number fields, referring if necessary to the "global case", and with local fields.

Let $\Delta \to \Gamma$ be a homomorphism of groups (all groups are finite). Composing this with a representation of Γ we get a representation of Δ. In this way we obtain a homomorphism $R_\Gamma \to R_\Delta$ of rings. Specifically if Δ is a subgroup of Γ (with embedding as the map) we shall write this as

$$\mathrm{res}_\Delta^\Gamma = \mathrm{res}\colon R_\Gamma \to R_\Delta \qquad \text{(restriction).}$$

In this last case we also have induced representations, giving rise to a homomorphism of groups

$$\mathrm{ind}_\Delta^\Gamma = \mathrm{ind}\colon R_\Delta \to R_\Gamma \qquad \text{(induction).}$$

Going back to the previous situation let now $\Gamma \to \Sigma$ be the map onto a quotient group of Γ. This yields a homomorphism

$$\mathrm{inf}_\Sigma^\Gamma = \mathrm{inf}\colon R_\Sigma \to R_\Gamma \qquad \text{(inflation).}$$

Turning now to $\mathrm{Hom}_\Omega(R_\Gamma, G)$ for variable Γ, with Ω the appropriate "absolute" Galois group, we get contravariant homomorphisms

$$(3.1) \qquad \begin{cases} \mathrm{ind}_\Delta^\Gamma = \mathrm{ind}\colon \mathrm{Hom}_\Omega(R_\Delta, G) \to \mathrm{Hom}_\Omega(R_\Gamma, G), \\ \mathrm{res}_\Delta^\Gamma = \mathrm{res}\colon \mathrm{Hom}_\Omega(R_\Gamma, G) \to \mathrm{Hom}_\Omega(R_\Delta, G), \\ \mathrm{coinf}_\Sigma^\Gamma = \mathrm{coinf}\colon \mathrm{Hom}_\Omega(R_\Gamma, G) \to \mathrm{Hom}_\Omega(R_\Sigma, G), \end{cases}$$

induced, respectively, by res, ind, inf. (In keeping with the accepted terminology in the theory of Frobenius functors we have written ind, res in place of cores, coind respectively). Explicitly we have the formulae

$$(\mathrm{ind}\, f)(\chi) = f(\mathrm{res}\,\chi), \qquad \chi \in R_\Gamma, \qquad f \in \mathrm{Hom}_\Omega(R_\Delta, G),$$

$$(\mathrm{res}\, f)(\chi) = f(\mathrm{ind}\,\chi), \qquad \chi \in R_\Delta, \qquad f \in \mathrm{Hom}_\Omega(R_\Gamma, G).$$

We also note that the maps (3.1) commute with those coming from homomorphisms in the second variable G, a fact which we shall use without any further mention.

Now we turn to class groups, i.e., we are looking at the global case. ($\mathfrak{o} = \mathfrak{o}_K$, K a number field). If $\Delta \to \Gamma$ is, as before, a homomorphism of groups, and X is a locally free (resp. free) $\mathfrak{o}\Delta$-module, then $X \otimes_{\mathfrak{o}\Delta} \mathfrak{o}\Gamma$ is a locally free (resp. free) $\mathfrak{o}\Gamma$-module. The map $(X) \mapsto (X \otimes_{\mathfrak{o}\Delta} \mathfrak{o}\Gamma)$ defines a homomorphism $\mathrm{Cl}(\mathfrak{o}\Delta) \to \mathrm{Cl}(\mathfrak{o}\Gamma)$. If in particular $\Delta \to \Gamma$ is a subgroup embedding, we shall write this as

$$\mathrm{ind} = \mathrm{ind}_\Delta^\Gamma\colon \mathrm{Cl}(\mathfrak{o}\Delta) \to \mathrm{Cl}(\mathfrak{o}\Gamma).$$

As before, we use the symbol Σ for a quotient group of Γ and denote the associated map by

$$\mathrm{coinf} = \mathrm{coinf}_\Sigma^\Gamma\colon \mathrm{Cl}(\mathfrak{o}\Gamma) \to \mathrm{Cl}(\mathfrak{o}\Sigma).$$

Note that if $\Sigma = \Gamma/\Delta$ then in fact $X \otimes_{\mathfrak{o}\Gamma} \mathfrak{o}\Sigma \cong X^\Delta$ (fixed points). Finally if Δ is a subgroup of Γ then every locally free (resp. free) $\mathfrak{o}\Gamma$-module X retains this property when viewed as an $\mathfrak{o}\Delta$-module, and we thus get a homomorphism

$$\mathrm{res} = \mathrm{res}_\Delta^\Gamma \colon \mathrm{Cl}(\mathfrak{o}\Gamma) \to \mathrm{Cl}(\mathfrak{o}\Delta).$$

In the theorems of this section we shall formulate the results which refer to the basic isomorphism of Theorems 1 in terms of the associated surjection $\mathrm{Hom}_{\Omega_K}(R_\Gamma, \mathfrak{J}(\mathbb{Q}^c)) \to \mathrm{Cl}(\mathfrak{o}\Gamma)$.

Theorem 12 (note [3]). (i) *Let K be a number field, $\mathfrak{o} = \mathfrak{o}_K$. With the notation as above, the following diagrams commute:*

$$\begin{array}{ccc} \mathrm{Hom}_{\Omega_K}(R_\Delta, \mathfrak{J}(\mathbb{Q}^c)) & \to & \mathrm{Cl}(\mathfrak{o}\Delta) \\ {\scriptstyle\mathrm{ind}}\downarrow & & {\scriptstyle\mathrm{ind}}\downarrow \\ \mathrm{Hom}_{\Omega_K}(R_\Gamma, \mathfrak{J}(\mathbb{Q}^c)) & \to & \mathrm{Cl}(\mathfrak{o}\Gamma), \end{array}$$

$$\begin{array}{ccc} \mathrm{Hom}_{\Omega_K}(R_\Gamma, \mathfrak{J}(\mathbb{Q}^c)) & \to & \mathrm{Cl}(\mathfrak{o}\Gamma) \\ {\scriptstyle\mathrm{res}}\downarrow & & {\scriptstyle\mathrm{res}}\downarrow \\ \mathrm{Hom}_{\Omega_K}(R_\Delta, \mathfrak{J}(\mathbb{Q}^c)) & \to & \mathrm{Cl}(\mathfrak{o}\Delta), \end{array}$$

$$\begin{array}{ccc} \mathrm{Hom}_{\Omega_K}(R_\Gamma, \mathfrak{J}(\mathbb{Q}^c)) & \to & \mathrm{Cl}(\mathfrak{o}\Gamma) \\ {\scriptstyle\mathrm{coinf}}\downarrow & & {\scriptstyle\mathrm{coinf}}\downarrow \\ \mathrm{Hom}_{\Omega_K}(R_\Sigma, \mathfrak{J}(\mathbb{Q}^c)) & \to & \mathrm{Cl}(\mathfrak{o}\Sigma). \end{array}$$

Also for any prime divisor $\mathfrak{p}$ of K, the maps defined above restrict to maps

$$\begin{aligned} \mathrm{ind}\colon\ & \mathrm{Det}(\mathfrak{o}_\mathfrak{p}\Delta^*) \to \mathrm{Det}(\mathfrak{o}_\mathfrak{p}\Gamma^*), \\ \mathrm{res}\colon\ & \mathrm{Det}(\mathfrak{o}_\mathfrak{p}\Gamma^*) \to \mathrm{Det}(\mathfrak{o}_\mathfrak{p}\Delta^*), \\ \mathrm{coinf}\colon & \mathrm{Det}(\mathfrak{o}_\mathfrak{p}\Gamma^*) \to \mathrm{Det}(\mathfrak{o}_\mathfrak{p}\Sigma^*), \end{aligned}$$

and analogously with $\mathfrak{o}_\mathfrak{p}\Gamma^*$ replaced by $\mathfrak{U}(\mathfrak{o}\Gamma)$ etc.

(ii) *Let F be a local field. The maps defined above restrict to maps*

$$\begin{aligned} \mathrm{ind}\colon\ & \mathrm{Det}(F\Delta^*) \to \mathrm{Det}(F\Gamma^*), \\ \mathrm{res}\colon\ & \mathrm{Det}(F\Gamma^*) \to \mathrm{Det}(F\Delta^*), \\ \mathrm{coinf}\colon & \mathrm{Det}(F\Gamma^*) \to \mathrm{Det}(F\Sigma^*), \end{aligned}$$

and analogously with $F\Gamma$ replaced by $\mathfrak{o}_F\Gamma$. □

For the proof see [F17], [F18] where results of this nature appeared for the first time, and [F24] (Chaps. IV and V).

Next we turn to change of base field. We first have to give a general definition of the norm $\mathcal{N}$ which first occurred in I (2.15), (2.16). Let then for the moment N be an algebraic extension of finite degree of a field M. Let G, H be Ω_M-modules, with the group operation in G being written multiplicatively. Then on the one hand we have the inclusion map

$$(3.2) \qquad i_{M/N}\colon \mathrm{Hom}_{\Omega_M}(H,G) \to \mathrm{Hom}_{\Omega_N}(H,G),$$

and on the other hand we have the trace map in the opposite direction, which, in view of the multiplicative notation, we shall write a norm map

$$(3.3) \qquad \mathcal{N}_{N/M}\colon \mathrm{Hom}_{\Omega_N}(H,G) \to \mathrm{Hom}_{\Omega_M}(H,G),$$

where

$$\mathcal{N}_{N/M} f(h) = \prod_{\sigma} f(h^{\sigma^{-1}})^{\sigma}$$

for $f \in \mathrm{Hom}_{\Omega_N}(H,G)$ and $h \in H$. Here σ runs over a right transversal of Ω_N in Ω_M.

Now consider a subfield L of a number field K. If X is a locally free (resp. free) $\mathfrak{o}_L\Gamma$-module, then $X \otimes_{\mathfrak{o}_L} \mathfrak{o}_K$ is a locally free (resp. free) $\mathfrak{o}_K\Gamma$-module, and we thus get a homomorphism

$$(3.4) \qquad e_{L/K}\colon \mathrm{Cl}(\mathfrak{o}_L\Gamma) \to \mathrm{Cl}(\mathfrak{o}_K\Gamma).$$

Theorem 13 (note [3]). (i) *With K and L as above, the diagram*

$$\begin{array}{ccc} \mathrm{Hom}_{\Omega_L}(R_\Gamma, \mathfrak{J}(\mathbb{Q}^c)) & \to & \mathrm{Cl}(\mathfrak{o}_L\Gamma) \\ {\scriptstyle i_{L/K}}\downarrow & & {\scriptstyle e_{L/K}}\downarrow \\ \mathrm{Hom}_{\Omega_K}(R_\Gamma, \mathfrak{J}(\mathbb{Q}^c)) & \to & \mathrm{Cl}(\mathfrak{o}_K\Gamma) \end{array}$$

commutes, and $i_{L/K}$ maps $\mathrm{Det}(\mathfrak{o}_{L,\mathfrak{p}}\Gamma^*) \to \mathrm{Det}(\mathfrak{o}_{K,\mathfrak{p}}\Gamma^*)$ *for any prime divisor $\mathfrak{p}$ of L.*

Also $\mathcal{N}_{K/L}$ maps $\mathrm{Det}(\mathfrak{o}_{K,\mathfrak{p}}\Gamma^*) \to \mathrm{Det}(\mathfrak{o}_{L,\mathfrak{p}}\Gamma^*)$, *for any prime divisor $\mathfrak{p}$ of L, and gives rise to a homomorphism of classgroups, again denoted by $\mathcal{N}_{K/L}$, so that the diagram*

$$\begin{array}{ccc} \mathrm{Hom}_{\Omega_K}(R_\Gamma, \mathfrak{J}(\mathbb{Q}^c)) & \to & \mathrm{Cl}(\mathfrak{o}_K\Gamma) \\ {\scriptstyle \mathcal{N}_{K/L}}\downarrow & & {\scriptstyle \mathcal{N}_{K/L}}\downarrow \\ \mathrm{Hom}_{\Omega_L}(R_\Gamma, \mathfrak{J}(\mathbb{Q}^c)) & \to & \mathrm{Cl}(\mathfrak{o}_L\Gamma) \end{array}$$

commutes. If X is a locally free $\mathfrak{o}_K\Gamma$-module then in $\mathrm{Cl}(\mathfrak{o}_L\Gamma)$

$$(3.5) \qquad (X)_{\mathfrak{o}_L\Gamma} = (\mathcal{N}_{K/L}(X)_{\mathfrak{o}_K\Gamma}) \cdot (\mathfrak{o}_K\Gamma)^{r(X)}_{\mathfrak{o}_L\Gamma},$$

$r(X)$ the rank of X over $\mathfrak{o}_K\Gamma$.

(ii) *Let F be a local field, $F \supset E \supset \mathbb{Q}_p$. Then $i_{E/F}$ maps* $\mathrm{Det}(E\Gamma^*) \to \mathrm{Det}(F\Gamma^*)$ *and $\mathcal{N}_{F/E}$ maps* $\mathrm{Det}(F\Gamma^*) \to \mathrm{Det}(E\Gamma^*)$, *and analogously with F, E replaced by $\mathfrak{o}_F$, $\mathfrak{o}_E$.* □

For the proofs see [F17] or [F24] (Chaps. IV and V).

Theorem 2 follows immediately from the restriction formula (3.5). For, $\mathfrak{o}_K\Gamma$ is free over $\mathbb{Z}\Gamma$, hence the second factor on the right of (3.5) is now $= 1$.

We finally come to character action. In [Sw1] it was first proved that $\mathrm{Cl}(\mathfrak{o}\Gamma)$ ($\mathfrak{o} = \mathfrak{o}_K$) is a module over the subring $R_\Gamma(K)$ of R_Γ, of characters which correspond to representations of Γ defined over K. Explicitly, let X be a locally free $\mathfrak{o}\Gamma$-module and let χ be a character of Γ, corresponding to a $K\Gamma$-module of the form YK, where Y is an $\mathfrak{o}\Gamma$-module, free over $\mathfrak{o}$. Equivalently we can say that χ corresponds to a representation $\Gamma \to \mathrm{GL}_n(\mathfrak{o}) \to \mathrm{GL}_n(\mathbb{Q}^c)$. These χ will generate $R_\Gamma(K)$. With X and Y as above, the $\mathfrak{o}\Gamma$-module $X \otimes_{\mathfrak{o}} Y$, with diagonal action of Γ, is again locally free over $\mathfrak{o}\Gamma$ and $(X \otimes_{\mathfrak{o}} Y) \in \mathrm{Cl}(\mathfrak{o}\Gamma)$ only depends on (X) and on χ. This then gives $\mathrm{Cl}(\mathfrak{o}\Gamma)$ the structure of a $R_\Gamma(K)$-module. On the other hand (cf. [U2]) the groups $\mathrm{Hom}_{\Omega_K}(R_\Gamma, G)$ have a natural structure of $R_\Gamma(K)$-module, by setting

$$(3.6) \qquad \phi f(\chi) = f(\bar{\phi}\chi) \qquad (f \in \mathrm{Hom}_{\Omega_K}(R_\Gamma, G),\ \phi \in R_\Gamma(K),\ \chi \in R_\Gamma),$$

and the same is true also for local fields in place of number fields.

Theorem 14. (i) *With K, $\mathfrak{o}$ as above, the map*

$$\mathrm{Hom}_{\Omega_K}(R_\Gamma, \mathfrak{J}(\mathbb{Q}^c)) \to \mathrm{Cl}(\mathfrak{o}\Gamma)$$

is one of $R_\Gamma(K)$-modules. Also for any prime divisor $\mathfrak{p}$ of K, $\mathrm{Det}(\mathfrak{o}_\mathfrak{p}\Gamma^*)$ *is a $R_\Gamma(K)$-submodule of* $\mathrm{Hom}_{\Omega_K}(R_\Gamma, \mathfrak{J}(\mathbb{Q}^c))$.

(ii) *If F is a local field,* $\mathrm{Det}(F\Gamma^*)$ *and* $\mathrm{Det}(\mathfrak{o}_F\Gamma^*)$ *are $R_\Gamma(F)$-submodules of* $\mathrm{Hom}_{\Omega_F}(R_\Gamma, (\mathbb{Q}_p^c)^*)$.

For a proof see [U2] where these results appeared for the first time, or [F24] (Chap. V).

We note, as can easily be verified (see e.g., [F24]) that if Δ is a subgroup of Γ, then (see note [4])

$$(3.7) \qquad \begin{cases} \mathrm{res}\,\phi \cdot \mathrm{res} f = \mathrm{res}(\phi f), & f \in \mathrm{Hom}_{\Omega_F}(R_\Gamma, G), \quad \phi \in R_\Gamma(F), \\ \mathrm{ind}(\phi \cdot \mathrm{res} f) = (\mathrm{ind}\,\phi) f, & f \in \mathrm{Hom}_{\Omega_F}(R_\Gamma, G), \quad \phi \in R_\Delta(F), \\ \mathrm{ind}(\mathrm{res}\,\phi \cdot f) = \phi(\mathrm{ind} f), & f \in \mathrm{Hom}_{\Omega_F}(R_\Delta, G), \quad \phi \in R_\Gamma(F), \end{cases}$$

(where of course F may be local or global). One may reexpress this by saying that $\mathrm{Hom}_{\Omega_F}(R_\Gamma, G)$ is a Frobenius module over the Frobenius functor $R_\Gamma(K)$, and the same applies for example to $\mathrm{Det}(\mathfrak{o}_F\Gamma^*)$, or in the global situation to $D(\mathbb{Z}\Gamma)$ (cf. [U2]).

(For the definitions of Frobenius functors and Frobenius modules see the quoted literature ([Lm], [Ba], [SE]) – but in fact Eqs. (3.7) give the definitions, and nothing more is needed.)

§4. Reduction mod l and Some Computations

The purpose of this section is threefold. We shall establish a Theorem which will be needed subsequently. In order not to make the discussion too cumbersome we shall state and prove it in three separate parts. We shall also come back to the "problem on the map t", which was raised in I §7 and throw new light on it. Finally we shall – as illustration – give some explicit computations, both for the structure of the class group, and for the behaviour of t.

Throughout this section Γ is a given finite group, and l is a prime number. We shall write $R_\Gamma(l)$ for the Grothendieck group of $\mathbb{F}_l^c\Gamma$ (do not confuse with $R_{\Gamma,l}$ as introduced in §2 – the corresponding object for $\mathbb{Q}_l^c$). $R_\Gamma(l)$ can be viewed as the additive Abelian group, freely generated by symbols $[T]$, one for each equivalence class of irreducible representations T of Γ over $\mathbb{F}_l^c$. For an arbitrary representation $T\colon \Gamma \to \mathrm{GL}_n(\mathbb{F}_l^c)$ the element $[T]$ of $R_\Gamma(l)$ is defined as $\sum_i n_i[T_i]$, where the T_i run through the irreducible representations modulo equivalence, and n_i is the multiplicity with which T_i occurs in a triangularization of T. Note for subsequent use that $\operatorname{Det} T$ only depends on $[T]$.

For $E = \mathbb{Q}^c$ (global case) or $E = \mathbb{Q}_l^c$ (local case) let s_l be some homomorphism (hence surjection) $\mathfrak{o}_E \to \mathbb{F}_l^c$. Let $T\colon \Gamma \to \mathrm{GL}_n(E)$ be a representation, and suppose (as we may) that $T(\Gamma) \subset \mathrm{GL}_n(\mathfrak{o}_E)$. Then $s_l T$ is a representation $\Gamma \to \mathrm{GL}_n(\mathbb{F}_l^c)$, and the class $[s_l T] = d_l\chi$ only depends on s_l and on the character χ of T. This map extends to a homomorphism (the Brauer decomposition map) (cf. [Se2] §15, 16)

$$d_l\colon R_\Gamma \to R_\Gamma(l), \tag{4.1}$$

where R_Γ is the character group over $\mathbb{Q}^c$ or over $\mathbb{Q}_l^c$, respectively. d_l will still depend on the choice of s_l, but the subgroup $\operatorname{Ker} d_l$ is entirely independent of choices. In fact (cf. [Se2] Corollary 2 to Theorem 42)

$$\operatorname{Ker} d_l = \operatorname{Ker} d_{l,\Gamma} = [\chi \in R_\Gamma \mid \chi(\gamma) = 0 \text{ if } (\text{order}(\gamma), l) = 1]. \tag{4.2}$$

Moreover $\operatorname{Ker} d_l$ is an $\Omega_{\mathbb{Q}}$- (respectively an $\Omega_{\mathbb{Q}_l}$-) module.

Let $\mathfrak{L}$ be the radical in $\mathfrak{o}_E$ of the ideal (l). A congruence $x \equiv y \pmod{\mathfrak{L}}$, for x, y l-integers, means that in any extension F of finite degree of $\mathbb{Q}$, respectively of $\mathbb{Q}_l$, which contains x and y, we have $x \equiv y \pmod{\mathfrak{L}_F}$, where $\mathfrak{L}_F$ is the product of the maximal ideals of $\mathfrak{o}_F$ above l.

Theorem 15A. *Let $\chi \in \operatorname{Ker} d_l$, and let M be an extension of finite degree of $\mathbb{Q}$ (respectively of $\mathbb{Q}_l$). Let*

$$a \in \begin{cases} \mathfrak{o}_{M,l}\Gamma^* & (\textit{local case}),\\ \mathfrak{U}(\mathfrak{o}_M\Gamma) & (\textit{global case}). \end{cases}$$

Then

$$\operatorname{Det}_\chi(a) \equiv 1 \pmod{\mathfrak{L}}.$$

Remark. This is the first example of the method of congruences mentioned in I §1, as applied here to group determinants.

We shall first prove this part of the theorem before stating its second part. We can write $\chi = \psi - \phi$, where ψ and ϕ are characters of representations T_ψ, T_ϕ. Let $\mathfrak{l}$ be a maximal ideal above l in the field of values of $\operatorname{Det}_\psi(a)$ and $\operatorname{Det}_\phi(a)$.

We have to show that

$$\mathrm{Det}_{\psi}(a) \equiv \mathrm{Det}_{\phi}(a) \pmod{\mathfrak{l}}. \tag{4.3}$$

As $\operatorname{Ker} d_l$ is independent of the particular choice of s_l, this congruence may be interpreted as

$$s_l \,\mathrm{Det}_{\psi}(a) = s_l \,\mathrm{Det}_{\phi}(a),$$

or extending s_l to group-rings, as

$$\mathrm{Det}_{d_l\psi}(s_l a) = \mathrm{Det}_{d_l\phi}(s_l a).$$

But this is obvious, as $d_l\psi = d_l\phi$. □

Now we turn to the global case. For any number field L, let $\mathfrak{B}_l(L) = (\mathfrak{o}_L/\mathfrak{L}_L)^*$, where $\mathfrak{L}_L = \mathfrak{L} \cap \mathfrak{o}_L$ is the product of the maximal ideals in $\mathfrak{o}_L$ above l. Let $\mathfrak{B}_l(\mathbb{Q}^c)$ be the direct limit (union) of the $\mathfrak{B}_l(L)$. Its crucial property here is that, for all L,

$$\mathfrak{B}_l(\mathbb{Q}^c)^{\Omega_L} = \mathfrak{B}_l(L).$$

If $f\colon R_\Gamma \to \mathfrak{U}(\mathbb{Q}^c)$ is an $\Omega_{\mathbb{Q}}$-homomorphism, we shall denote by $r_l f$ the composite map

$$\operatorname{Ker} d_l \to R_\Gamma \xrightarrow{f} \mathfrak{U}(\mathbb{Q}^c) \xrightarrow{\text{residue class}} \mathfrak{B}_l(\mathbb{Q}^c).$$

We thus get a homomorphism

$$r_l\colon \mathrm{Hom}^+_{\Omega_{\mathbb{Q}}}(R_\Gamma, \mathfrak{U}(\mathbb{Q}^c)) \to \mathrm{Hom}_{\Omega_{\mathbb{Q}}}(\operatorname{Ker} d_l, \mathfrak{B}_l(\mathbb{Q}^c)). \tag{4.4}$$

This is only of interest for divisors l of order (Γ), for otherwise, by (4.2), $\operatorname{Ker} d_l = 0$. We shall write

$$\mathfrak{E}_l(\Gamma) = \mathrm{Hom}_{\Omega_{\mathbb{Q}}}(\operatorname{Ker} d_l, \mathfrak{B}_l(\mathbb{Q}^c))/r_l(\mathrm{Hom}^+_{\Omega_{\mathbb{Q}}}(R_\Gamma, \mathfrak{Y}(\mathbb{Q}^c))), \tag{4.5}$$

and more generally, for any finite set S of primes,

$$\mathfrak{E}_S(\Gamma) = \left(\prod_{l\in S} \mathrm{Hom}_{\Omega_{\mathbb{Q}}}(\operatorname{Ker} d_l, \mathfrak{B}_l(\mathbb{Q}^c))\right)\Big/ r_S(\mathrm{Hom}^+_{\Omega_{\mathbb{Q}}}(R_\Gamma, \mathfrak{Y}(\mathbb{Q}^c))), \tag{4.6}$$

where $r_S = \prod_{l\in S} r_l$. Note that $\mathrm{Hom}_{\Omega_{\mathbb{Q}}}(\operatorname{Ker} d_l, \mathfrak{B}_l(\mathbb{Q}^c))$ has order prime to l, hence so has $\mathfrak{E}_l(\Gamma)$.

Theorem 15B (See note [5]). *The maps r_S give rise to surjective homomorphisms*

$$h_S\colon D(\mathbb{Z}\Gamma) \to \mathfrak{E}_S(\Gamma).$$

(If S consists of just one prime l we write $h_S = h_l$.)

Proof. The map r_l factorizes through the projection

$$\mathrm{Hom}^+_{\Omega_\mathbb{Q}}(R_\Gamma, \mathfrak{U}(\mathbb{Q}^c)) \to \mathrm{Hom}_{\Omega_\mathbb{Q}}(R_\Gamma, \mathfrak{U}_l(\mathbb{Q}^c)).$$

It thus suffices to show that

$$\mathrm{Hom}_{\Omega_\mathbb{Q}}(R_\Gamma, \mathfrak{U}_l(\mathbb{Q}^c)) \to \mathrm{Hom}_{\Omega_\mathbb{Q}}(\mathrm{Ker}\, d_l, \mathfrak{V}_l(\mathbb{Q}^c)) \tag{4.7}$$

is surjective, and that $\mathrm{Det}(\mathbb{Z}_l\Gamma^*)$ lies in its kernel. But the latter follows from the first part of the theorem. Moreover $\mathfrak{V}_l(\mathbb{Q}^c)$ is an $\Omega_\mathbb{Q}$-direct summand of $\mathfrak{U}_l(\mathbb{Q}^c)$ – note that if F is a number field then $\mathfrak{U}_l(F) = \mathfrak{V}_l(F) \times \mathfrak{U}_l^{(1)}(F)$, the second factor being the "one-units" (see note [6]). Thus

$$\mathrm{Hom}_{\Omega_\mathbb{Q}}(R_\Gamma, \mathfrak{U}_l(\mathbb{Q}^c)) \to \mathrm{Hom}_{\Omega_\mathbb{Q}}(R_\Gamma, \mathfrak{V}_l(\mathbb{Q}^c)) \tag{4.8}$$

is surjective. Now (cf. [Se2] §15, 16) R_Γ contains the direct sum $\mathfrak{D}$ of $\mathrm{Ker}\, d_l$ and an $\Omega_\mathbb{Q}$-submodule $\mathrm{Im}\, e_l$, and the index of $\mathfrak{D}$ in R_Γ is a power of l, while on the other hand l acts by automorphisms on $\mathfrak{V}_l(\mathbb{Q}^c)$. Therefore indeed

$$\mathrm{Hom}_{\Omega_\mathbb{Q}}(R_\Gamma, \mathfrak{V}_l(\mathbb{Q}^c)) = \mathrm{Hom}_{\Omega_\mathbb{Q}}(\mathrm{Ker}\, d_l, \mathfrak{V}_l(\mathbb{Q}^c)) \times \mathrm{Hom}_{\Omega_\mathbb{Q}}(\mathrm{Im}\, e_l, \mathfrak{V}_l(\mathbb{Q}^c)),$$

and this yields, in conjunction with the surjectivity of (4.8), that of (4.7). □

Now we come to the third part of Theorem 15. Let $\mathfrak{M}$ be a maximal order in $\mathbb{Q}\Gamma$, containing $\mathbb{Z}\Gamma$. Denote the Jacobson radical of $\mathfrak{M}_l$ by $\mathfrak{r}_l$ and let $\mathfrak{A}_l = \mathbb{Z}_l\Gamma + \mathfrak{r}_l$. This is an order.

Given a finite set S of primes, we denote by $\mathfrak{A}(S)$ the order in $\mathbb{Q}\Gamma$ with local components

$$\mathfrak{A}(S)_l = \mathfrak{A}_l, \qquad \text{if } l \in S,$$

$$\mathfrak{A}(S)_p = \mathfrak{M}_p, \qquad p \text{ otherwise}.$$

$\mathfrak{A}(S)$ depends on certain choices but, as we shall presently see, its classgroup only depends on S.

If F is a number field, l a finite prime, we shall, as above, denote by $\mathfrak{U}_l^{(1)}(F)$ the group of ideles x, with component $x_\mathfrak{p} = 1$ if $\mathfrak{p}$ does not lie above l, and $x_\mathfrak{p} \equiv 1 \pmod{\mathfrak{p}}$ if $\mathfrak{p}$ lies above l. Finally we shall write $\mathfrak{W}_S^{(1)}(L)$ for the unit ideles of L, whose l-components, for each $l \in S$, are restricted to $\mathfrak{U}_l^{(1)}(L)$ and we shall denote by $\mathfrak{W}_S^{(1)}(\mathbb{Q}^c)$ the union or direct limit of the $\mathfrak{W}_S^{(1)}(L)$.

Theorem 15C.

$$\mathrm{Ker}\, r_S = \Big(\prod_S \mathrm{Det}(\mathbb{Z}_l\Gamma^*)\Big)\mathrm{Hom}^+_{\Omega_\mathbb{Q}}(R_\Gamma, \mathfrak{W}_S^{(1)}(\mathbb{Q}^c)) = \mathrm{Det}(\mathfrak{U}(\mathfrak{A}(S))),$$

and hence

$$\mathfrak{E}_S(\Gamma) = \mathrm{Ker}[\mathrm{Cl}(\mathfrak{A}(S)) \to \mathrm{Cl}(\mathfrak{M})].$$

This result is in [CN2] (see also [CN4]), modulo a formally different definition, in which the Jacobson radical is replaced by the radical of the central conductor of $\mathfrak{M}_l$ in $\mathbb{Z}_l\Gamma$. The resulting order is however the same as our $\mathfrak{A}_l$.

We shall outline the proof, following that in the quoted paper, assuming for simplicity that S consists of one prime number l. We shall, as we may, go over to the local fields above $\mathbb{Q}_l$, and introduce for this the appropriate notations. $\mathfrak{U}(\mathbb{Q}_l^c)$ is the group of units of $\mathbb{Z}_l^c$, $\mathfrak{V}(\mathbb{Q}_l^c)$ the multiplicative group of its residue class field, and $\mathfrak{U}^{(1)}(\mathbb{Q}_l^c)$ the kernel of $\mathfrak{U}(\mathbb{Q}_l^c) \to \mathfrak{V}(\mathbb{Q}_l^c)$. Also

$$\tilde{r}_l\colon \operatorname{Hom}_{\Omega_{\mathbb{Q}_l}}(R_\Gamma, \mathfrak{U}(\mathbb{Q}_l^c)) \to \operatorname{Hom}_{\Omega_{\mathbb{Q}_l}}(\operatorname{Ker} d_l, \mathfrak{V}(\mathbb{Q}_l^c))$$

is the local analogue of r_l, i.e., $\tilde{r}_l f$ is the composite map

$$\operatorname{Ker} d_l \to R_\Gamma \xrightarrow{f} \mathfrak{U}(\mathbb{Q}_l^c) \to \mathfrak{V}(\mathbb{Q}_l^c).$$

Finally $\mathfrak{A}_l$ is as defined earlier. The symbols $\operatorname{Det}(\mathfrak{A}_l^*)$, $\operatorname{Det}(\mathbb{Z}_l\Gamma^*)$ now stand for subgroups of $\operatorname{Det}(\mathfrak{M}_l^*) = \operatorname{Hom}_{\Omega_{\mathbb{Q}_l}}(R_\Gamma, \mathfrak{U}(\mathbb{Q}_l^c))$.

By Theorem 15A, and by the definition of $\mathfrak{U}^{(1)}(\mathbb{Q}_l^c)$,

$$\operatorname{Det}(\mathbb{Z}_l\Gamma^*) \operatorname{Hom}_{\Omega_{\mathbb{Q}_l}}(R_\Gamma, \mathfrak{U}^{(1)}(\mathbb{Q}_l^c)) \subset \operatorname{Ker} \tilde{r}_l. \tag{4.9}$$

Denote again by s_l any residue class map coming from $\mathbb{Z}_l^c \to \mathbb{F}_l^c$. Let $\{T_i\}$ run through the irreducible representations of Γ over $\mathbb{F}_l^c$, to within equivalence, and choose $\psi_i \in R_\Gamma$, with $d_l\psi_i = [T_i] \in R_\Gamma(l)$. The reduced norm $(\mathbb{F}_l\Gamma/\mathfrak{r}(\mathbb{F}_l\Gamma))^* \to C^*$ is surjective, where $C = \operatorname{cent}(\mathbb{F}_l\Gamma/\mathfrak{r}(\mathbb{F}_l\Gamma))$ and $\mathfrak{r}$ denotes the radical. On the other hand the map

$$\mathbb{Z}_l\Gamma^* \to (\mathbb{Z}_l\Gamma/\mathfrak{r}(\mathbb{Z}_l\Gamma))^* = (\mathbb{F}_l\Gamma/\mathfrak{r}(\mathbb{F}_l\Gamma))^*$$

is also surjective, hence so is the composition of the two maps. In terms of determinants, this implies that given $a \in \mathfrak{M}_l^*$, $\exists b \in \mathbb{Z}_l\Gamma^*$ with $\operatorname{Det}_{d_l\psi_i}(s_l b) = s_l \operatorname{Det}_{\psi_i}(a)$, for all i. But $\operatorname{Det}_{d_l\psi_i}(s_l b) = s_l \operatorname{Det}_{\psi_i}(b)$. Moreover if now $\operatorname{Det}(a) \in \operatorname{Ker} \tilde{r}_l$, then $s_l \operatorname{Det}_\chi(a) = 1 = s_l \operatorname{Det}_\chi(b)$ for all $\chi \in \operatorname{Ker} d_l$. As $\operatorname{Ker} d_l$ and the ψ_i generate R_Γ, we get $s_l \operatorname{Det}_\chi(ab^{-1}) = 1$ for all $\chi \in R_\Gamma$. Thus indeed $\operatorname{Det}(ab^{-1}) \in \operatorname{Hom}_{\Omega_{\mathbb{Q}_l}}(R_\Gamma, \mathfrak{U}^{(1)}(\mathbb{Q}_l^c))$. We have thus established the opposite inclusion of (4.9), and so

$$\operatorname{Ker} \tilde{r}_l = \operatorname{Det}(\mathbb{Z}_l\Gamma^*) \operatorname{Hom}_{\Omega_{\mathbb{Q}_l}}(R_\Gamma, \mathfrak{U}^{(1)}(\mathbb{Q}_l^c)). \tag{4.10}$$

Next, every element of $\mathfrak{A}_l$ is of form $x + y$, $x \in \mathbb{Z}_l\Gamma$, $y \in \mathfrak{r}_l$. If $x + y \in \mathfrak{A}_l^*$, then $x \in \mathfrak{M}_l^*$, hence $x \in \mathbb{Z}_l\Gamma^*$, and so $x + y = xu$, $u \in 1 + \mathfrak{r}_l$. $\operatorname{Det}(u)$ lies in the second factor of the right hand side of (4.10), and $\operatorname{Det}(x)$ in the first factor. What we have shown is that $\operatorname{Det}(\mathfrak{A}_l^*)$ is contained in the right hand side of (4.10).

For the converse we have to show that

$$\operatorname{Hom}_{\Omega_{\mathbb{Q}_l}}(R_\Gamma, \mathfrak{U}^{(1)}(\mathbb{Q}_l^c)) \subset \operatorname{Det}(\mathfrak{A}_l^*).$$

What this amounts to is, that given a local field F and a finite dimensional central simple F-algebra B with maximal order $\mathfrak{B}$, the reduced norm maps $1 + \mathfrak{p}_F\mathfrak{B}$ surjectively onto $1 + \mathfrak{p}_F$. This however is well known to be true. Therefore finally

$$\mathrm{Det}(\mathbb{Z}_l\Gamma^*)\,\mathrm{Hom}_{\Omega_{\mathbb{Q}_l}}(R_\Gamma, \mathfrak{U}^{(1)}(\mathbb{Q}_l^c)) = \mathrm{Det}(\mathfrak{A}_l^*). \tag{4.11}$$ □

Let now S be any finite set of primes, containing all prime divisors of order (Γ). As non-divisors of order (Γ) make no contribution, we get the same groups and maps for any such S. Accordingly we write $h_S = h_\Gamma$. By the last theorem, $\mathrm{Ker}\, h_\Gamma$ is a quotient of products of pro l-groups, l running through the prime divisors of order (Γ). Using Theorem 15C, and induction techniques (as described in §3) one can get fairly precise results on $\mathrm{Ker}\, h_\Gamma$ (cf. [CN2], [CN4], note [7]). We are in particular interested in the quaternion groups, whose definition we recall (see also I.1.14)

$$H_{4m} = gp[\sigma, \omega \mid \sigma^m = \omega^2,\ \omega^4 = 1,\ \omega^{-1}\sigma\omega = \sigma^{-1}]. \tag{4.12}$$

It should be kept in mind that we always use the term quaternion group in this "generalized" sense. By Theorem 4.1 in [CN4] we know that for $\Gamma = H_{4m}$, with m odd, $\mathrm{Ker}\, h_\Gamma$ is of odd order. It will be convenient to look instead at h_S with S the set of all odd prime divisors of m. Then of course $\mathrm{Ker}\, h_S \supset \mathrm{Ker}\, h_\Gamma$ and $\mathrm{Ker}\, h_S/\mathrm{Ker}\, h_\Gamma$ is easily seen to be a quotient of the odd order group $\mathrm{Hom}_{\Omega_\mathbb{Q}}(\mathrm{Ker}\, d_2, \mathfrak{B}_2(\mathbb{Q}^c))$. We thus get a useful Lemma, implicitly contained in [CN2].

4.1. Lemma. *If m is odd, $\Gamma = H_{4m}$, S the set of prime divisors of m, then h_S induces an isomorphism*

$$D(\mathbb{Z}\Gamma)_2 \cong \mathfrak{E}_S(\Gamma)_2 \qquad \text{(2-primary components).}$$ □

We shall outline below a technique for evaluating the group $\mathrm{Im}\, t_\Gamma$ (cf. I (6.3)) via the h-maps, specifically for Γ quaternion. As pointed out in I §7 this evaluation is of some significance. The actual scope of the technique of course depends on how h acts on $\mathrm{Im}\, t$. From 4.1 we get in particular

Corollary. *If $\Gamma = H_{4m}$, m odd then* $\mathrm{Im}\, t_\Gamma \cong \mathrm{Im}\, h_S t_\Gamma$. □

For other quaternion groups such a strong result is false. But in view of I, Proposition 7.2, it could suffice to show that if $\Gamma = H_{4m}$, $m = 2^r n$, n odd, $r \geqslant 1$, then the map

$$D(\mathbb{Z}\Gamma) \to \mathfrak{E}_S(\Gamma) \times D(\mathbb{Z}H_{2^s}) \qquad (s = r + 2)$$

maps $\mathrm{Im}\, t$ injectively.

Take for the moment again Γ as any finite group and let S be the set of all odd prime divisors of order (Γ). We then get a commutative diagram with exact row,

$$\begin{array}{ccccc} & & \mathrm{Hom}_{\Omega_\mathbb{Q}}(R_\Gamma^s/\mathrm{Tr}(R_\Gamma), \pm 1) & & \\ & & \downarrow r_S t'_\Gamma & \searrow h_S t_\Gamma & \\ \mathrm{Hom}^+_{\Omega_\mathbb{Q}}(R_\Gamma, \mathfrak{Y}(\mathbb{Q}^c)) & \to & \prod_{l \in S} \mathrm{Hom}_{\Omega_\mathbb{Q}}(\mathrm{Ker}\, d_l, \mathfrak{B}_l(\mathbb{Q}^c)) & \to & \mathfrak{E}_S(\Gamma) \to 1, \end{array} \tag{4.13}$$

with t'_Γ as defined in I (6.1), (6.2). As t'_Γ is given quite explicitly, and so are the r_l, the computation of $h_S t_\Gamma$ via the above commutative diagram is usually more feasible than that of t_Γ itself.

The second ingredient of the method is the use of evaluation maps. For $\chi \in R_\Gamma$, we define the function $\delta_l\chi$ on Γ by

(4.14) $\delta_l\chi(\gamma) = \chi(\gamma')$, if $\gamma = \gamma'\gamma''$, γ' of order prime to l, $\gamma''^{l^N} = 1$ some N.

Then $\delta_l\chi \in R_\Gamma$, and if $\chi \in R_\Gamma^s$ then $\delta_l\chi \in R_\Gamma^s$ (for l odd) (cf. [CN3], Propositions 3.1 and 3.5). Moreover $\chi - \delta_l\chi \in \operatorname{Ker} d_l$ and if χ runs through a generating set of R_Γ then $\chi - \delta_l\chi$ runs through one of $\operatorname{Ker} d_l$. If e.g., χ is irreducible symplectic the aim is then to replace the three Hom groups in (4.13) by their respective images under the evaluation maps $f \mapsto f(\chi)$, or $f \mapsto f(\chi - \delta_l\chi)$. We shall illustrate the methods when $\Gamma = H_{4m}$ (cf. (4.12)) is some quaternion group. No restrictions are imposed here on m, but we replace $\mathfrak{C}_S(\Gamma)$ by $\mathfrak{C}_l(\Gamma)$, where l is an odd prime divisor of m. This gives less sharp results, but is easier to handle.

We shall need a parametric description of the $\Omega_{\mathbb{Q}}$-orbits of irreducible non-Abelian characters of H_{4m}. This is given in terms of a parameter d, where $d > 2$ and $d \mid 2m$, and these d are in biunique correspondence with the $\Omega_{\mathbb{Q}}$-orbits. We call $d = d(\psi)$ the *parameter* of any character ψ in the corresponding orbit. Explicitly $d = d(\psi)$, if ψ is lifted from a character of a faithful irreducible representation of a quotient of H_{4m}, of order $2d$. This quotient is either dihedral or (generalized) quaternion and accordingly we call d *dihedral* or *quaternion*. (A harmless abuse of terminology: this property depends not only on d but also on m.) To give actual representations, a character ψ with parameter d corresponds to a representation T, where

$$(4.15)\quad \begin{cases} T(\sigma) = \begin{pmatrix} y & 0 \\ 0 & y^{-1} \end{pmatrix}, & y \text{ a primitive } d\text{-th root of } 1, \\ T(\omega) = \begin{pmatrix} 0 & 1 \\ \pm 1 & 0 \end{pmatrix}, & \text{with } \begin{cases} +1, & \text{if } d \text{ dihedral, i.e., } d \mid m \\ -1, & \text{if } d \text{ quaternion, i.e. } d \nmid m. \end{cases} \end{cases}$$

These are really the same representations as those of (I.1.13), and (I.1.14) respectively. Write $y = y'y''$, y' of order prime to l, y'' an l-power root of 1. Replacing, in (4.15), y by y' we get a representation corresponding to $\delta_l\psi$. This need no longer be irreducible, but it is symplectic precisely when ψ is. Moreover $\psi - \delta_l\psi$ is non-zero precisely when $l \mid d$.

We shall introduce some groups, defined in terms of congruences, which we shall then use to describe $\mathfrak{C}_l(H_{4m})$ and $\operatorname{Im} h_l t$. These are associated with the parameters d. Let $\mathbb{Q}^+(d)$ be the maximal real subfield of the field of d-th roots of unity over $\mathbb{Q}$. Write

$$\mathfrak{Y}'(\mathbb{Q}^+(d)) = \begin{cases} \text{group of global units in } \mathbb{Q}^+(d) = \mathfrak{Y}(\mathbb{Q}^+(d)), \\ \qquad d \text{ dihedral}, \\ \text{group of totally positive global units in } \mathbb{Q}^+(d) = \mathfrak{Y}^+(\mathbb{Q}^+(d)), \\ \qquad d \text{ quaternion}. \end{cases}$$

(Warning: The definition also involves m, but this is not indicated in our notation.)

Let

$$(4.16)\qquad \begin{cases} r'_d\colon \mathfrak{Y}'(\mathbb{Q}^+(d)) \to \mathfrak{B}_l(\mathbb{Q}^+(d)) & \text{(residue class map)},\\ \mathfrak{E}(d) = \operatorname{Cok} r'_d, & \end{cases}$$

and

$$(4.17)\qquad r''_d\colon \pm 1 \to \mathfrak{B}_l(\mathbb{Q}^+(d)) \text{ the residue class map, } t''_d \text{ its compositum with } \mathfrak{B}_l(\mathbb{Q}^+(d)) \to \mathfrak{E}(d).$$

We shall see that $\mathfrak{E}_l(H_{4m})$ is the product of the $\mathfrak{E}(d)$ for all values of the parameter d (i.e., $2 < d \mid 2m$), and $\operatorname{Im} h_l t$ that of the $\operatorname{Im} t''_d$ for all quaternion d. To compute $\mathfrak{E}(d)$ we have to consider the group of prime residue classes of integers in $\mathbb{Q}^+(d)$ modulo the product $\mathfrak{L}_d$ of prime ideals above l, and then factor out those classes occupied by units (d dihedral), or just totally positive units (d quaternion). We then get a simple criterion for $\operatorname{Im} t''_d$ (d quaternion), which we already know to be of order 1 or 2.

4.2. Proposition. *t''_d is null or injective, according to whether -1 is, or is not, congruent to a totally positive unit modulo $\mathfrak{L}_d$.* □

The following diagram (4.18) will be seen to effect the transition from Hom-groups (making up its "inner" part) to groups of numbers and residue classgroups (making up its "outer" part). The maps connecting the two parts are given by evaluation.

$$(4.18)\qquad \begin{array}{ccccc} & & \operatorname{Hom}_{\Omega_\mathbb{Q}}(R_\Gamma^s/\operatorname{Tr}(R_\Gamma), \pm 1) & \xrightarrow{e''} & \prod_d{}^* (\pm 1) \\ & & \downarrow{\scriptstyle r_l t'_\Gamma} & & \\ \operatorname{Hom}^+_{\Omega_\mathbb{Q}}(R_\Gamma, \mathfrak{Y}(\mathbb{Q}^c)) & \xrightarrow{r_l} & \operatorname{Hom}_{\Omega_\mathbb{Q}}(\operatorname{Ker} d_l, \mathfrak{B}_l(\mathbb{Q}^c)) & & \Big\downarrow{\scriptstyle \prod^* r''_d} \\ \downarrow{\scriptstyle e'} & & & \searrow{\scriptstyle e} & \\ \prod_d \mathfrak{Y}'(\mathbb{Q}^+(d)) & & \xrightarrow{\quad \prod r'_d \quad} & & \prod_d \mathfrak{B}_l(\mathbb{Q}^+(d)). \end{array}$$

$$(\Gamma = H_{4m})$$

Explanation: The symbol r_l stands for any map on Hom groups restricted from the original r_l (cf. (4.4)). The maps r'_d, r''_d are those of (4.16), (4.17). The unadorned product runs over all parameters d, with $l \mid d$, and $\prod^*$ is the subproduct restricted to quaternion parameters d. Moreover in the map $\prod^* r''_d$ it is to be understood, that the component going into $\mathfrak{B}_l(\mathbb{Q}^c(d))$, for d dihedral, is null. To define e, e', e'' we choose for each d, with $l \mid d$, an irreducible character ψ_d with $d(\psi_d) = d$. For $f \in \operatorname{Hom}^+_{\Omega_\mathbb{Q}}(R_\Gamma, \mathfrak{Y}(\mathbb{Q}^c))$, and $f \in \operatorname{Hom}_{\Omega_\mathbb{Q}}(\operatorname{Ker} d_l, \mathfrak{B}_l(\mathbb{Q}^c))$ the d-component of $e'(f)$, respectively of $e(f)$, is just the value $f(\psi_d - \delta_l \psi_d)$. This does lie in $\mathfrak{Y}'(\mathbb{Q}^+(d))$, and $\mathfrak{B}_l(\mathbb{Q}^+(d))$ respectively, as we shall show below. For $f \in \operatorname{Hom}_{\Omega_\mathbb{Q}}(R_\Gamma^s/\operatorname{Tr}(R_\Gamma), \pm 1)$ and d quaternion, $l \mid d$, the d-component of $e''(f)$ is $f(\psi_d - \delta_l \psi_d \operatorname{mod} \operatorname{Tr}(R_\Gamma))$.

4.3. Proposition. (i) *Diagram* (4.18) *commutes.*
(ii) *e', e'' are surjective, e is an isomorphism.*
(iii) *e gives rise to an isomorphism*

$$\mathfrak{E}_l(H_{4m}) \cong \prod_{l|d} \mathfrak{E}(d)$$

(iv) *The diagram* (with $\Gamma = H_{4m}$)

$$\begin{array}{ccc} \mathrm{Hom}_{\Omega_\mathbb{Q}}(R_\Gamma^s/\mathrm{Tr}(R_\Gamma), \pm 1) & \xrightarrow{e''} & \prod^*(\pm 1) \\ \downarrow h_l t_\Gamma & & \downarrow \prod^* t''_d \\ \mathfrak{E}_l(H_{4m}) & \longrightarrow & \prod^* \mathfrak{E}(d) \end{array}$$

commutes.

Proof. (i) is immediate from the definitions, and (iii), (iv) follow from (i) and (ii). It thus remains to prove (ii). Also the proof that e' and e are properly defined is still outstanding. For e' this is immediate. Indeed, $\mathbb{Q}^+(d)$ is the field of values of $\psi_d - \delta_l\psi_d$, and writing $\Omega = \mathrm{Gal}(\mathbb{Q}^c/\mathbb{Q}^+(d))$, we see that for all d, $f(\psi_d - \delta_l\psi_d) \in \mathfrak{Y}(\mathbb{Q}^c)^\Omega = \mathfrak{Y}(\mathbb{Q}^+(d))$, and analogously with $\mathfrak{Y}^+$ replacing $\mathfrak{Y}$, when d is quaternion. Although this is not quite so immediate, one verifies that similarly $\mathfrak{B}_l(\mathbb{Q}^c)^\Omega = \mathfrak{B}_l(\mathbb{Q}^{c\Omega}) = \mathfrak{B}_l(\mathbb{Q}^+(d))$. Thus indeed e and e' are properly defined. For e'' this is in any case obvious.

For the surjectivity of e', fix a value of d and an element $y \in \mathfrak{Y}'(\mathbb{Q}^+(d))$. Then there is a unique $f \in \mathrm{Hom}^+_{\Omega_\mathbb{Q}}(R_\Gamma, \mathfrak{Y}(\mathbb{Q}^c))$, so that $f(\psi_d) = y$, and $f(\psi) = 1$ if ψ is irreducible Abelian, or irreducible non-Abelian of parameter $\neq d$. Clearly $e(f)$ has d-component y, and all other components 1. For, $\delta_l\psi_d$ involves no irreducible characters in the $\Omega_\mathbb{Q}$-orbit of parameter d. This proves the surjectivity of e'. That of e'' is established similarly. For the bijectivity of e one uses the same procedure plus the fact that the $\psi_d - \delta_l\psi_d$, for all d with $l|d$, together with their distinct $\Omega_\mathbb{Q}$-conjugates generate $\mathrm{Ker}\, d_l$ freely over $\mathbb{Z}$. □

As an application we prove

4.4. Proposition. *For $\Gamma = H_{4l^r}$ $(r \geqslant 1)$, $\mathfrak{E}_l(\Gamma) = \mathfrak{E}_S(\Gamma)$ is an $\mathbb{F}_2$-space of dimension r. If $l \equiv 1 \pmod 4$ then t_Γ is null. If $l \equiv -1 \pmod 4$ then t_Γ is injective* (see note [8]).

Proof. l is totally ramified in $\mathbb{Q}^+(l^j) = \mathbb{Q}^+(2l^j)$. Denote the prime divisor above l by $\mathfrak{L}$. We know classically that all the prime residue classes mod $\mathfrak{L}$ are occupied by the global units $(w^r - w^{-r})/(w - w^{-1})$ $(r = 1, \ldots, l-1)$, where w is a primitive l-th root of unity. Thus all squares of classes are occupied by totally positive units. Conversely, if y is a totally positive unit in $\mathbb{Q}^+(l^j)$, and if the field of l^j-th roots of unity is $\mathbb{Q}^+(l^j)(\sqrt{a})$, then, by the product formula for the Hilbert symbol,

$$1 = \prod_{\text{all } \mathfrak{p}} \left(\frac{a, y}{\mathfrak{p}}\right) = \left(\frac{a, y}{\mathfrak{L}}\right) \prod_{\text{inf } \mathfrak{p}} \left(\frac{a, y}{\mathfrak{p}}\right) = \left(\frac{a, y}{\mathfrak{L}}\right) = \left(\frac{y}{\mathfrak{L}}\right)$$

(quadratic symbol). Thus the classes of totally positive units are precisely the squares. Hence for $d = l^j$, i.e., d dihedral, we have $\mathfrak{E}(d) = 1$, and for $d = 2l^j$, i.e., d quaternion, $\mathfrak{E}(d) = \mathbb{F}_l^*/\mathbb{F}_l^{*2}$. The result on $\mathfrak{E}_l(\Gamma)$ follows now from Proposition 4.3. Moreover by Proposition 4.2 and by the above, t''_d $(d = 2l^j)$ will be injective if, and only if, $(-1/l) = -1$. If now $l \equiv -1 \pmod 4$, then – as $\mathfrak{E}_l(\Gamma) = \prod \mathfrak{E}(d)$ for quaternion d only – we conclude from Proposition 4.3 again that $h_l t_\Gamma$ is surjective, hence (e.g., by comparing dimensions over $\mathbb{F}_2$) bijective. Hence t_Γ is injective. On the other hand, if $l \equiv 1 \pmod 4$ then the t''_d are null and so by Proposition 4.3, $h_l t_\Gamma$ is null. By Lemma 4.1, t_Γ is null. □

The examples we have obtained, when t_Γ is null, show that in the basic equation $t_\Gamma W_{N/K} = U_{N/K}$, the class $U_{N/K}$ will not necessarily determine the values of the symplectic root numbers. To get a full interpretation of these values one therefore would need additional algebraic structure on N/K. This observation, first made in [F11], was the original motivation for the development of a Hermitian theory (cf. [F20], [F24], [F25]), which yields invariants for the object (called a Hermitian module) consisting of the module $\mathfrak{o}_N$ and the Γ-invariant trace form on N. In the particular case where the problem first arose, i.e., for $\Gamma = H_{4l^r}$, $l \equiv 1 \pmod 4$ a complete characterization of $W_{N/K}$ by properties of the Hermitian module associated with N/K was obtained at the time. Recently Ph. Cassou-Noguès and M. J. Taylor have established such a characterization in full generality both locally and globally, thus proving a conjecture of the author (cf. [CN-T1], [CN-T2]). This topic lies however outside the scope of this book. (See also note [3] to Chap. III, and the Appendix.)

§5. The Logarithm for Group Rings

In this section the integral logarithm for certain local group rings will be introduced and its main properties will be derived. All this comes from [Ty 6], with some later improvements, mostly suggested by M. Taylor. Throughout l is a fixed prime, Γ a finite l-group, M a non-ramified extension of $\mathbb{Q}_l$ of finite degree and $\mathfrak{o}$ its ring of integers.

Both here and in §6 we shall solely have to look at virtual characters in $\mathbb{Q}_l^c$. Exceptionally the symbols R_Γ etc. will be used in this sense in both sections, i.e. R_Γ is to be read as $R_{\Gamma,l} = R_\Gamma(\mathbb{Q}_l^c)$. The results of this section are crucial in the proofs of both Theorem 10 and Theorem 11. The underlying "philosophy" in these applications is that of a two pronged attack on certain problems, one prong being transition to Abelian groups, the other being provided by the logarithm. To illustrate this we consider the commutative diagram, whose rows and columns are exact (as we shall prove).

$$\begin{array}{ccccccccc}
1 \to & 1+\mathfrak{a} & \to & \mathfrak{o}\Gamma^* & \to & (\mathfrak{o}\Gamma^{ab})^* & \to 1 \\
& \downarrow & & \downarrow & & \downarrow & \\
1 \to & \mathrm{Det}(1+\mathfrak{a}) & \to & \mathrm{Det}(\mathfrak{o}\Gamma^*) & \to & \mathrm{Det}((\mathfrak{o}\Gamma^{ab})^*) & \to 1. \\
& \downarrow & & \downarrow & & \downarrow & \\
& 1 & & 1 & & 1 &
\end{array} \tag{5.1}$$

Here Γ^{ab} is the commutator quotient of Γ, and $\mathfrak{a} = \mathfrak{a}(\mathfrak{o}\Gamma) = \mathfrak{a}_M$ is the commutator ideal of $\mathfrak{o}\Gamma$, generated by the additive commutators $ab - ba$, $a, b \in \mathfrak{o}\Gamma$, i.e., $\mathfrak{a}$ is the kernel of the $\mathfrak{o}$-algebra homomorphism $\mathfrak{o}\Gamma \to \mathfrak{o}\Gamma^{ab}$, coming from the quotient map $\Gamma \to \Gamma^{ab}$. The top row of (5.1) is given by the corresponding homomorphisms on groups of units. $\mathfrak{a}$ is contained in the radical $\mathfrak{r}$ of $\mathfrak{o}\Gamma$ and the $\mathfrak{r}$-adic topology on $\mathfrak{o}\Gamma$ is the same as the l-adic one, whence $\mathfrak{o}\Gamma$ is complete under it. It follows that $(\mathfrak{o}\Gamma)^* \to (\mathfrak{o}\Gamma^{ab})^*$ is surjective, i.e., the top row is exact. Each column takes its source in the top row into its Det-image in the appropriate Hom group and is thus surjective. It is clear that the diagram commutes. The right hand column is bijective (see below). The exactness of the columns and the exactness of the top row now imply exactness of the bottom row.

The role of the integral logarithm is illustrated in the analysis of the group $\mathrm{Det}((\mathfrak{o}\Gamma)^*)$. Here one uses the bottom exact sequence of (5.1). The structure of $\mathrm{Det}((\mathfrak{o}\Gamma^{ab})^*)$ is fairly well known, via its isomorphism with $(\mathfrak{o}\Gamma^{ab})^*$. At the other end it will be seen that the logarithm establishes an isomorphism of $\mathrm{Det}(1 + \mathfrak{a})$ with an $\mathfrak{o}$-lattice of transparent structure.

The Det isomorphism for Abelian groups, used above, is quite general and easy to establish. Indeed let F be any field (say of characteristic zero). We claim that if Δ is a finite Abelian group then

$$F\Delta^* \to \mathrm{Det}(F\Delta^*) \tag{5.2}$$

is injective (hence an isomorphism). Let ϕ run through the Abelian characters of Δ, i.e., the homomorphisms $\Delta \to F^{c*}$. Each ϕ extends to an algebra homomorphism $f_\phi\colon F\Delta \to F^c$ over F and $\prod f_\phi$ is an injection $F\Delta \to \prod_\phi F^c$ (product of copies of F^c indexed by ϕ). We thus get an injection $\prod f_\phi^*\colon F\Delta^* \to \prod_\phi F^{c*}$, which factorizes through the map (5.2) – whence (5.2) is injective. The required homomorphism $\mathrm{Det}(F\Delta^*) \to \prod_\phi F^{c*}$ is given by evaluation: $\mathrm{Det}(a) \mapsto \{\mathrm{Det}_\phi(a)\}_\phi$. Indeed trivially $\mathrm{Det}_\phi(a) = f_\phi(a)$.

Now we come to the definitions. We let

$$\Psi = \Psi_M\colon M\Gamma \to M\Gamma \tag{5.3}$$

be the semilinear map of M-modules which takes each $\gamma \in \Gamma$ into γ^l, and on M is the Frobenius endomorphism (over $\mathbb{Q}_l$). Ψ commutes with $\mathrm{Gal}(M/\mathbb{Q}_l)$ acting on $M\Gamma$ via the coefficients. Next let $\mathfrak{C} = \mathfrak{C}_\Gamma$ be the set of conjugacy classes of Γ, and for any ring R denote by $R\mathfrak{C}$ the free R-module on $\mathfrak{C}$. Write

$$c\colon M\Gamma \to M\mathfrak{C} \tag{5.4}$$

for the M-linear map which takes each group element γ into its conjugacy class $c(\gamma)$. Again c commutes with $\mathrm{Gal}(M/\mathbb{Q}_l)$. Denote by

$$\mathfrak{r} = \mathfrak{r}(\mathfrak{o}\Gamma)$$

the (Jacobson) radical of $\mathfrak{o}\Gamma$. The series

$$\mathscr{L}_0(1 - r) = -l \sum_{n=1}^{\infty} (r^n/n) + \sum_{n=1}^{\infty} (\Psi(r^n)/n), \qquad r \in \mathfrak{r} \tag{5.5}$$

converge in the l-adic topology, defining a set map

$$\mathcal{L}_0\colon 1+\mathfrak{r}\to M\Gamma,$$

commuting with the action of $\mathrm{Gal}(M/\mathbb{Q}_l)$. The logarithm is then defined to be the composite map

$$\mathcal{L} = c\circ\mathcal{L}_0 \tag{5.6}$$

Thus $\mathcal{L}=\mathcal{L}_M$ is a map $1+\mathfrak{r}\to M\mathfrak{C}$, commuting with $\mathrm{Gal}(M/\mathbb{Q}_l)$. Its basic properties are formulated in the following two Theorems. Note that $1+\mathfrak{r}$ is a subgroup of $\mathfrak{o}\Gamma^*$.

Theorem 16 (cf. [Ty6]). *$\mathcal{L}\colon 1+\mathfrak{r}\to M\mathfrak{C}$ is a homomorphism of groups, and*

$$\operatorname{Im}\mathcal{L}\subset l\mathfrak{o}\mathfrak{C}.$$ □

Remark. The fact that the image lies in the integral lattice above justifies calling $\mathcal{L}$ an integral logarithm.

Theorem 17 (cf. [Ty6]). (i) *There is a unique homomorphism*

$$\Lambda\colon \mathrm{Hom}_{\Omega_M}(R_\Gamma,\mathfrak{U}(\mathbb{Q}_l^c))\to M\mathfrak{C}$$

so that the diagram

$$\begin{array}{ccc} & 1+\mathfrak{r} & \\ {\scriptstyle\mathrm{Det}}\swarrow & & \searrow{\scriptstyle\mathcal{L}} \\ \mathrm{Det}(1+\mathfrak{r}) & \xrightarrow[\Lambda']{} & M\mathfrak{C}\end{array}$$

commutes, the bottom row being the restriction of Λ. Λ commutes with the actions of $\mathrm{Gal}(M/\mathbb{Q}_l)$.

(ii) $\operatorname{Ker}\Lambda=\mathrm{Hom}_{\Omega_M}(R_\Gamma,\mu(\mathbb{Q}_l^c))$ *(the torsion subgroup of* $\mathrm{Hom}_{\Omega_M}(R_\Gamma,\mathfrak{U}(\mathbb{Q}_l^c))$). *Here $\mu(\mathbb{Q}_l^c)$ is the group of roots of unity in $\mathbb{Q}_l^c$.*

(iii) *Λ yields an isomorphism*

$$\mathrm{Det}(1+\mathfrak{a})\cong lc(\mathfrak{a}).$$ □

The maps in (5.1) again commute with the actions of $\mathrm{Gal}(M/\mathbb{Q}_l)$. Thus we have the

Corollary 1. *From* (5.1) *and the isomorphism* (iii) *above, we get an exact sequence*

$$0\to lc(\mathfrak{a}_M)\to\mathrm{Det}(\mathfrak{o}\Gamma^*)\to\mathrm{Det}((\mathfrak{o}\Gamma^{ab})^*)\to 1 \tag{5.7}$$

of $\mathrm{Gal}(M/\mathbb{Q}_l)$*-modules. Here* $\mathrm{Det}((\mathfrak{o}\Gamma^{ab})^*)$ *may be replaced by* $(\mathfrak{o}\Gamma^{ab})^*$. □

Corollary 2. *Suppose $f\in\mathrm{Hom}_{\Omega_M}(R_\Gamma,\mathfrak{U}(\mathbb{Q}_l^c))$ has the property that firstly its image under the homomorphism into $\mathrm{Hom}_{\Omega_M}(R_{\Gamma^{ab}},\mathfrak{U}(\mathbb{Q}_l^c))$ is of form $\mathrm{Det}(a)$, $a\in(\mathfrak{o}\Gamma^{ab})^*$, and* *secondly for some* $a' \in (\mathfrak{o}\Gamma)^*$ *with image* a *under* $\mathfrak{o}\Gamma \to \mathfrak{o}\Gamma^{ab}$, $\Lambda(f \cdot \mathrm{Det}(a')^{-1}) \in lc(\mathfrak{a})$. *Then* $f = g \cdot \mathrm{Det}(b)$, $b \in \mathfrak{o}\Gamma^*$, $g \in \mathrm{Hom}_{\Omega_M}(R_\Gamma, \mu(\mathbb{Q}_l^c))$.

Proof. By Theorem 17 (iii), $\exists a'' \in 1 + \mathfrak{a}$ with $\Lambda(f \cdot \mathrm{Det}(a'a'')^{-1}) = 0$. By Theorem 17 (ii) $f \cdot \mathrm{Det}(a'a'')^{-1} = g$, with g as above. □

Changing fields, we clearly have in the obvious notation,

$$c(\mathfrak{a}_M) = c(\mathfrak{a}_{\mathbb{Q}_l}) \otimes_{\mathbb{Z}_l} \mathfrak{o}_M. \tag{5.8}$$

Corollary 3. $\mathrm{Det}(1 + \mathfrak{a})$ *is torsion free.*

This was actually known to C. T. C. Wall (cf. [Wa3]).

Corollary 4. $\Lambda(\mathrm{Det}\, \mathfrak{o}_M\Gamma^*) = \Lambda(\mathrm{Det}(1 + \mathfrak{r}))$.

For, the map $\sum a_\gamma \gamma \to \sum a_\gamma \bmod \mathfrak{p}_M$ $(a_\gamma \in \mathfrak{o}_M)$ gives rise to a surjection of $\mathfrak{o}_M\Gamma^*$ onto the group of $\mathrm{N}_M\mathfrak{p}_M - 1$st roots of unity, which splits. Now apply Theorem 17 (ii). □

The remainder of this section is taken up with the proofs of the two Theorems. Our first step is a description of the image $\mathscr{L}(1 - r)$, $r \in \mathfrak{r}$, in terms of values of determinants at certain virtual characters. Here R_Γ will be the additive group of virtual characters $\Gamma \to M^c$. If $\chi \in R_\Gamma$ we define $\Psi_l\chi$ by $\Psi_l\chi(\gamma) = \chi(\gamma^l)$. Ψ_l defines an endomorphism $R_\Gamma \to R_\Gamma$ (the Adams operation). Note that the field $\mathbb{Q}_l(\chi)$ of values of χ on Γ is totally ramified, hence linearly disjoint from M. We thus can define a unique automorphism f on $M(\chi)$, by demanding it to be the Frobenius on M (over $\mathbb{Q}_l$) and the identity on $\mathbb{Q}_l(\chi)$. We extend f to $M\Gamma$ via M. Note here that $\mathbb{Q}_l(\Psi_l\chi) \subset \mathbb{Q}_l(\chi)$. Finally we view R_Γ also as a group of M-linear maps $M\mathfrak{C}_\Gamma \to \mathbb{Q}_l^c$ by

$$\chi(c(\gamma)) = \chi(\gamma).$$

In the sequel log is always the l-adic logarithm, defined for all x in the valuation ideal of $\mathbb{Q}_l^c$ by

$$\log(1 - x) = -\sum_{n=1}^{\infty} (x^n/n).$$

We shall then establish an important identity.

5.1. Proposition. For $\chi \in R_\Gamma$, $r \in \mathfrak{r}(\mathfrak{o}_M\Gamma)$,

$$\chi(\mathscr{L}(1 - r)) = \log[\mathrm{Det}_{l\chi}(1 - r) \cdot \mathrm{Det}_{-\Psi_l\chi}(1 - r^f)].$$

Proof. It suffices to prove this when χ is an actual character, corresponding to a representation $T\colon \Gamma \to \mathrm{GL}_n(\mathbb{Q}_l^c)$. The characteristic polynomial $\mathrm{Det}(XI_n - T(r))$ then factorizes in the form $\prod_j (X - a_j)$, where the a_j lie in the valuation ideal of

$\mathbb{Q}_l^c$. Hence

$$\log[\operatorname{Det}_\chi(1-r)] = \log\left(\prod_j (1-a_j)\right)$$
$$= -\sum_j \sum_{n=1}^{\infty} (a_j^n/n)$$
$$= -\sum_{n=1}^{\infty} \chi c(r^n)/n$$
$$= \chi c\left[-\sum_{n=1}^{\infty} (r^n/n)\right],$$

as

$$\chi c(r^n) = \operatorname{trace}(r^n) = \sum_j a_j^n.$$

Applying this to $l\chi$ in place of χ, and also to $\Psi_l\chi$, r^f in place of χ, r, we get

$$\log[\operatorname{Det}_{l\chi}(1-r)] = l\chi c\left(-\sum_{n=1}^{\infty} (r^n/n)\right) = \chi c\left(-l\sum_{n=1}^{\infty} (r^n/n)\right),$$

$$\log[\operatorname{Det}_{\Psi_l\chi}(1-r^f)^{-1}] = (\Psi_l\chi)c\left(\sum_{n=1}^{\infty} r^{nf}/n\right) = \chi c\left(\sum_{n=1}^{\infty} ((\Psi_M r^n)/n)\right).$$

This implies the proposition. □

There are two immediate consequences. Firstly it follows that, for all χ and all x, $y \in 1 + \mathfrak{r}$

$$\chi(\mathscr{L}(xy)) = \chi(\mathscr{L}(x)) + \chi(\mathscr{L}(y)).$$

But evaluation, even when restricted to all irreducible characters, is injective on $\mathbb{Q}_l^c\mathfrak{C}$. Hence we can conclude that

$$\mathscr{L}(xy) = \mathscr{L}(x) + \mathscr{L}(y).$$

Thus $\mathscr{L}$ is indeed a homomorphism, as asserted in Theorem 16.

Next observe that the kernel of the Det map on $1 + \mathfrak{r}$ is stable under the action of f. Thus, by the proposition, $\chi(\mathscr{L}(1 - r)) = 0$ for all χ, if $1 - r \in \operatorname{Ker}\operatorname{Det}$, hence $\operatorname{Ker}\mathscr{L} \supset \operatorname{Ker}\operatorname{Det}$. Therefore indeed there is a unique homomorphism Λ': $\operatorname{Det}(1 + \mathfrak{r}) \to M\mathfrak{C}$, which makes the diagram in Theorem 17 (i) commute. As $\mathscr{L}$ and Det commute with the action of $\operatorname{Gal}(M/\mathbb{Q}_l)$, so does Λ' by uniqueness. Now we use the fact that if $\mathfrak{M}$ is a maximal order in $M\Gamma$, containing $\mathfrak{o}\Gamma$, then $\mathfrak{M}^*/1 + \mathfrak{r}$ is finite. Therefore $\operatorname{Det}\mathfrak{M}^*/\operatorname{Det}(1 + \mathfrak{r})$ is finite, hence by Proposition I.2.2, so is $\operatorname{Hom}_{\Omega_M}(R_\Gamma, \mathfrak{U}(\mathbb{Q}_l^c))/\operatorname{Det}(1 + \mathfrak{r})$. On the other hand $M\mathfrak{C}$ is divisible and torsion free. Hence Λ' has a unique extension to a homomorphism Λ: $\operatorname{Hom}_{\Omega_M}(R_\Gamma, \mathfrak{U}(\mathbb{Q}_l^c)) \to M\mathfrak{C}$, which must again commute with the action of $\operatorname{Gal}(M/\mathbb{Q}_l)$.

Remark. The explicit description of $\mathscr{L}$ in Proposition 5.1, together with Theorem 16, yields a logarithmic congruence

$$\log \mathrm{Det}_{l\chi}(1-r) \equiv \log \mathrm{Det}_{\Psi_l\chi}(1-r^f) \pmod{l},$$

and there are in fact further such congruences (cf. [Ty1]). It is as yet unknown if this can be strengthened to a multiplicative congruence

$$\mathrm{Det}_\chi(1-r)^l \cdot \mathrm{Det}_{-\Psi_l\chi}(1-r^f) \equiv 1 \pmod{l},$$

although no counter-example is known either.

Next we shall prove Theorem 17 (ii). As the range of Λ is torsion free, the torsion subgroup $\mathrm{Hom}_{\Omega_M}(R_\Gamma, \mu(\mathbb{Q}_l^c))$ of $\mathrm{Hom}_{\Omega_M}(R_\Gamma, \mathfrak{U}(\mathbb{Q}_l^c))$ is contained in $\mathrm{Ker}\,\Lambda$. We have to prove the opposite inclusion. As moreover $\mathrm{Det}(1+\mathfrak{r})$ is of finite index in $\mathrm{Hom}_{\Omega_M}(R_\Gamma, \mathfrak{U}(\mathbb{Q}_l^c))$, it will suffice to show that

$$\mathrm{Ker}\,\Lambda' \subset \mathrm{Hom}_{\Omega_M}(R_\Gamma, \mu(\mathbb{Q}_l^c)). \tag{5.9}$$

The group Γ will now be indicated in our notation, as it will vary – Eq. (5.9) to be established by induction on order (Γ).

We shall use the equation in Proposition 5.1 to compute $\Lambda'_\Gamma \mathrm{Det}(1-r)$. First suppose that $\Gamma = 1$. If $\mathrm{Det}(1-r) \in \mathrm{Ker}\,\Lambda'_1$, then the element $x = \log \mathrm{Det}_\varepsilon(1-r)$ satisfies the equation $lx = x^f$, which implies $x = 0$. Thus $\mathrm{Det}_\varepsilon(1-r) \in \mu(\mathbb{Q}_l^c)$. (Here $\varepsilon(1) = 1$.)

Now let $\Gamma \neq 1$. For the induction step write $\bar{\Gamma} = \Gamma/\Delta$, where Δ is a central subgroup of Γ of order l. Consider an element $\mathrm{Det}(1-r) \in \mathrm{Ker}\,\Lambda'_\Gamma$. Then

$$\log \mathrm{Det}_{l\chi}(1-r) = \log \mathrm{Det}_{\Psi_l\chi}(1-r^f), \qquad \text{for all } \chi. \tag{5.10}$$

Write $\bar{r}$ for the image of r in $\mathfrak{o}\bar{\Gamma}$. Then $\mathrm{Det}(1-\bar{r}) \in \mathrm{Ker}\,\Lambda'_{\bar{\Gamma}}$. By induction hypothesis $\log \mathrm{Det}_\phi(1-\bar{r}) = 0$ for all $\phi \in R_{\bar{\Gamma}}$. But for all $\chi \in R_\Gamma$, the virtual character $\Psi_l\chi$ is inflated from $R_{\bar{\Gamma}}$. Therefore $\log \mathrm{Det}_{\Psi_l\chi}(1-r) = 0$. This is true for all r with $\mathrm{Det}(1-r) \in \mathrm{Ker}\,\Lambda'_\Gamma$, and so also with r replaced by r^f. Hence by (5.10), $\log \mathrm{Det}_{l\chi}(1-r) = 0$, i.e., $\mathrm{Det}_\chi(1-r)^l \in \mu(\mathbb{Q}_l^c)$, and thus $\mathrm{Det}_\chi(1-r) \in \mu(\mathbb{Q}_l^c)$. This then yields the required result and proves Theorem 17 (ii).

Our next step is the proof that the restriction of Λ to $\mathrm{Det}(1+\mathfrak{a})$ is injective, i.e., by Theorem 17 (ii), that

$$\mathrm{Det}(1+\mathfrak{a}) \text{ is torsion free.} \tag{5.11}$$

Here we use a result of C. T. C. Wall (cf. [Wa3]), which when translated into the Det language states that

$$\mathrm{Det}(\mathfrak{o}\Gamma^*) \cap \mathrm{Hom}(R_\Gamma, \mu(\mathbb{Q}_l^c)) = \mathrm{Det}(\Gamma)\,\mathrm{Det}(\mu(M)), \tag{5.12}$$

($\mu(M)$ the group of roots of unity in M). Thus suppose $\mathrm{Det}(1+a) \in \mathrm{Hom}(R_\Gamma, \mu(\mathbb{Q}_l^c))$, with $a \in \mathfrak{a}$. By (5.12), $\mathrm{Det}(1+a) = \mathrm{Det}(\gamma)\,\mathrm{Det}(y)$, $\gamma \in \Gamma$,

$y \in \mu(M)$. Evaluate this at the identity character ε, i.e., the augmentation. Then $\mathrm{Det}_\varepsilon(1+a) = \varepsilon(1+a) = 1$, $\mathrm{Det}_\varepsilon(\gamma) = 1$, $\mathrm{Det}_\varepsilon(y) = y$. Thus $y = 1$. So $\mathrm{Det}(1+a) = \mathrm{Det}(\gamma)$. Evaluating at any Abelian character ϕ, we get $1 = \mathrm{Det}_\phi(1+a) = \phi(\gamma)$. Therefore $\gamma \in (\Gamma, \Gamma)$ the commutator subgroup of Γ. But then $\mathrm{Det}(\gamma) = 1$, i.e., $\mathrm{Det}(1+a) = 1$. We have thus established (5.11).

We are now turning to the proof that $\mathrm{Im}\, \mathscr{L} \subset l\mathfrak{o}\mathfrak{C}$, and with this Theorem 16 will be established. This inclusion relation is a consequence of

(5.13) $$c(lr^n/n) \in l\mathfrak{o}\mathfrak{C}, \quad \text{if} \quad (n,l) = 1 \quad \text{and} \quad r \in \mathfrak{r},$$

and

(5.14) $$c((r^{nl}/n) - \Psi_M(r^n)/n) \in l\mathfrak{o}\mathfrak{C}, \quad \text{if} \quad r \in \mathfrak{r}.$$

(5.13) is obvious, (5.14) is a consequence of

5.2. Lemma. *Let $r \in \mathfrak{r}$ and let l^{e-1} be the highest power of l dividing n. Then*

$$c(r^{nl} - \Psi_M(r^n)) \in l^e \mathfrak{o}\mathfrak{C}.$$

This Lemma is the consequence of a non-commutative generalization of the binomial congruence

$$\left(\sum X_i\right)^l \equiv \sum X_i^l \pmod{l}$$

holding in commutative rings. One first of all works in the free associative $\mathfrak{o}$-algebra A on non-commuting variables X_γ $(\gamma \in \Gamma)$, and subsequently specializes. For $j \leqslant e$, let S_j be the set of monomials in the X_γ of degree l^j. The cyclic group Σ of order l^e on a given generator σ then acts as permutation group of S_j, with σ acting by cyclic permutation, thus:

$$(X_{\gamma(1)} \cdots X_{\gamma(l^j)})^\sigma = (X_{\gamma(l^j)} X_{\gamma(1)} \cdots).$$

The free $\mathfrak{o}$-module $\mathfrak{o}(S_j)$ on S_j is then naturally embedded in A. Define c^* to be the canonical map $S_j \to S_j/\Sigma$ (set of Σ-orbits of S_j), extended to a homomorphism of $\mathfrak{o}(S_j)$ onto the free $\mathfrak{o}$-module $\mathfrak{o}(S_j/\Sigma)$ on S_j/Σ. Observe that if $s \in S_{e-1}$ then $s^l \in S_e$. Let then Ψ^* be the semilinear map $\mathfrak{o}(S_{e-1}) \to \mathfrak{o}(S_e)$, which takes $s \in S_{e-1}$ into s^l, and acts on $\mathfrak{o}$ via the Frobenius over $\mathbb{Q}_l$. Then we get

5.3. Lemma. *Let $u \in \mathfrak{o}(S_0)$. Then*

$$c^*(u^{l^e} - \Psi^*(u^{l^{e-1}})) \in l^e \mathfrak{o}(S_e/\Sigma). \qquad \square$$

Before giving the proof of this Lemma, we shall show that it implies Lemma 5.2, and hence Theorem 16. Indeed, let here $t: A \to \mathfrak{o}\Gamma$ be the $\mathfrak{o}$-algebra homomorphism with $t(X_\gamma) = \gamma$. Then we get a commutative diagram of homomorphisms of $\mathfrak{o}$-modules

$$\begin{array}{ccccc}
\mathfrak{o}(S_{e-1}) & \xrightarrow{\Psi^*} & \mathfrak{o}(S_e) & \xrightarrow{c^*} & \mathfrak{o}(S_e/\Sigma) \\
\downarrow t & & \downarrow t & & \downarrow t' \\
\mathfrak{o}(\Gamma) & \xrightarrow{\Psi} & \mathfrak{o}(\Gamma) & \xrightarrow{c} & \mathfrak{o}(\mathfrak{C})
\end{array}$$

with unique t'. If in Lemma 5.2, $n = l^{e-1}m$, $(m, l) = 1$, and $r^m = \sum_{\gamma \in \Gamma} r_\gamma \gamma$ with $r_\gamma \in \mathfrak{o}$, then put $u = \sum r_\gamma X_\gamma$. As $t(u) = r^m$, the required relation in Lemma 5.2 now follows by Lemma 5.3 and the commutativity of our diagram. □

Proof of 5.3. Write

$$u^{l^{e-1}} = \sum_{s \in S_{e-1}} a_s s, \qquad a_s \in \mathfrak{o}. \tag{5.15}$$

A monomial $s \in S_{e-1}$ is fixed under the subgroup Σ_d of Σ of order l^d, precisely when it has period l^{e-d} in the X_γ. As s is of length l^{e-1}, this means that $s = s_1^{l^{d-1}}$, $s_1 \in S_{e-d}$, but then in (5.15) $a_s = b_s^{l^{d-1}}$. Now let, for each s, $\Sigma_{d(s)}$ be the actual stabilizer of s. Observe also that the orbit of s under Σ has precisely $l^{e-d(s)}$ elements. Thus we get

$$c^*\Psi^*(u^{l^{e-1}}) = \sum{}^* \Psi(b_s)^{l^{d(s)-1}} l^{e-d(s)} c^*(s^l). \tag{5.16}$$

Here $\sum^*$ is the sum over a set of representatives of S_{e-1}/Σ in S_{e-1}. On the other hand a monomial in S_e is fixed under the subgroup Σ_1 of Σ, precisely when it is of the form s^l, $s \in S_{e-1}$. We thus get from (5.15) that

$$u^{l^e} = \sum_{s \in S_{e-1}} a_s^l s^l + \sum{}' a'_{s'} s',$$

with $\sum'$ extending over monomials s' which have a Σ-orbit of l^e elements. As $a_s = b_s^{l^{d(s)-1}}$, we now get, by the same argument as before, that

$$c^*(u^{l^e}) \equiv \sum{}^* b_s^{l^{d(s)}} l^{e-d(s)} c^*(s^l) \pmod{l^e \mathfrak{o}(S_e/\Sigma)}.$$

But

$$b^{l^d} \equiv \Psi(b^{l^{d-1}}) \pmod{l^d \mathfrak{o}}$$

for $b \in \mathfrak{o}$. Comparing with (5.16) we get

$$c^*(u^{l^e}) \equiv c^*\Psi^*(u^{l^{e-1}}) \pmod{l^e \mathfrak{o}(S_e/\Sigma)},$$

as we had to show. □

We are now left with the proof that

$$\mathscr{L}(1 + \mathfrak{a}) = lc(\mathfrak{a}). \tag{5.17}$$

The proof is based on

5.4. Lemma. *Let z be a central element of order l in Γ, and a commutator. Then*

$$\mathscr{L}(1 + (1 - z)\mathfrak{o}\Gamma) = lc((1 - z)\mathfrak{o}\Gamma).$$ □

We shall first show that the lemma implies (5.17). In the sequel z is always an element as given in the lemma, and we write

$$\bar{\Gamma} = \Gamma/\langle z \rangle.$$

The cyclic group $\langle z\rangle$ acts on $R\Gamma$ and on $R\mathfrak{C}_\Gamma$, for $R = \mathfrak{o}, M$ and one verifies quickly that

(5.18) $$\operatorname{Ker}[R\mathfrak{C}_\Gamma \to R\mathfrak{C}_{\bar{\Gamma}}] = (1-z)R\mathfrak{C}_\Gamma.$$

As the map c preserves $\langle z\rangle$-action, it follows on replacing above R by $l\mathfrak{o}$, that

(5.19) $$l\mathfrak{o}\mathfrak{C}_\Gamma \cap (1-z)M\mathfrak{C}_\Gamma = lc((1-z)\mathfrak{o}\Gamma).$$

The proof of (5.17) now proceeds by induction on order (Γ), i.e., assuming it for $\bar{\Gamma}$. Throughout, maps induced from $\Gamma \to \bar{\Gamma}$ are denoted by $x \mapsto \bar{x}$.

First we shall show that $\mathscr{L}_\Gamma(1+\mathfrak{a}) \supset lc_\Gamma(\mathfrak{a})$. Let then $a \in \mathfrak{a}$. As the map $\mathfrak{a}(\mathfrak{o}\Gamma) \to \mathfrak{a}(\mathfrak{o}\bar{\Gamma})$ is surjective, $\exists x \in \mathfrak{a}(\mathfrak{o}\Gamma)$, with $\mathscr{L}_{\bar{\Gamma}}(1+\bar{x}) = lc_{\bar{\Gamma}}(\bar{a})$, or $\overline{\mathscr{L}_\Gamma(1+x)} = \overline{lc_\Gamma(a)}$. Write

$$y = lc_\Gamma(a) - \mathscr{L}_\Gamma(1+x).$$

By Theorem 16 (already established), and as $x \in \mathfrak{r}(\mathfrak{o}\Gamma)$,

$$y \in \operatorname{Ker}[M\mathfrak{C}_\Gamma \to M\mathfrak{C}_{\bar{\Gamma}}] \cap l\mathfrak{o}\mathfrak{C}_\Gamma,$$

whence by (5.18), (5.19), $y \in lc((1-z)\mathfrak{o}\Gamma)$. Now we use Lemma 5.4, to get $y = \mathscr{L}(1+u)$ with $u \in (1-z)\mathfrak{o}\Gamma \subset \mathfrak{a}(\mathfrak{o}\Gamma)$. Therefore finally

$$lc_\Gamma(a) = \mathscr{L}_\Gamma((1+x)(1+u)) \in \mathscr{L}_\Gamma(1+\mathfrak{a}).$$

Next we shall prove that $lc_\Gamma(\mathfrak{a}) \supset \mathscr{L}_\Gamma(1+\mathfrak{a})$.

Let then $a \in \mathfrak{a}(\mathfrak{o}\Gamma)$. By induction hypothesis

$$\mathscr{L}_\Gamma(1+a) - lc(b) \in \operatorname{Ker}[M\mathfrak{C}_\Gamma \to M\mathfrak{C}_{\bar{\Gamma}}]$$

for some $b \in \mathfrak{a}(\mathfrak{o}\Gamma)$, i.e., by Theorem 16 and (5.18)

$$\mathscr{L}_\Gamma(1+a) - lc(b) \in l\mathfrak{o}\mathfrak{C}_\Gamma \cap (1-z)M\mathfrak{C}_\Gamma,$$

and so by (5.19) finally $\mathscr{L}_\Gamma(1+a) \in lc(\mathfrak{a}(\mathfrak{o}\Gamma))$. □

Proof of Lemma 5.4. Let $x \in \mathfrak{o}\Gamma$. As $\Psi_M((1-z)x) = 0$, we have

$$\mathscr{L}_0(1-(1-z)x) = -l\sum_{n=1}^{\infty}(1-z)^n x^n/n$$
$$\equiv -l(1-z)x - (1-z)^l x^l \bmod l(1-z)^2x^2.$$

But $(1-z)^l \in l(1-z)\mathfrak{o}\Gamma$, and hence

$$\mathscr{L}_0(1-(1-z)x) \in l(1-z)\mathfrak{o}\Gamma.$$

Applying c, we get the inclusion

$$\mathscr{L}(1+(1-z)\mathfrak{o}\Gamma) \subset lc((1-z)\mathfrak{o}\Gamma).$$

For the opposite inclusion one first shows by a standard approximation argument that

$$lc((1-z)\mathfrak{r}) \subset \mathscr{L}(1+(1-z)\mathfrak{o}\Gamma),$$

where $\mathfrak{r}$ is again the radical of $\mathfrak{o}\Gamma$. It then remains to be shown that

$$c((1-z)\mathfrak{o}\Gamma) \subset c((1-z)\mathfrak{r}). \tag{5.20}$$

It is here that we use the property of z being a commutator. Write then

$$z = \alpha^{-1}\beta^{-1}\alpha\beta, \qquad \alpha, \beta \in \Gamma.$$

Let $v = \sum_\gamma v_\gamma \gamma$ $(v_\gamma \in \mathfrak{o})$. We have to show that

$$c((1-z)v) \in c((1-z)\mathfrak{r}).$$

Rewrite v as $\sum_\gamma v_\gamma(\gamma-\alpha) + \sum_\gamma v_\gamma \alpha$. Now $\alpha(1-z) = \alpha - \beta^{-1}\alpha\beta \in \operatorname{Ker} c$, and $(1-z)(\gamma-\alpha) \in (1-z)\mathfrak{r}$. Therefore indeed

$$c((1-z)v) = \sum_\gamma v_\gamma c((1-z)(\gamma-\alpha)) \in c((1-z)\mathfrak{r}).$$ □

Remark. We shall subsequently have to make use of some further technical Lemmas on the logarithm $\mathscr{L}$. These will be stated at the appropriate places. Proofs will not be given – they are rather straightforward and uninteresting. The technique is that exhibited in part of the present section.

We conclude this section with an evaluation rule for Λg.

5.5. Proposition. *If*

$$g \in \operatorname{Hom}_{\Omega_M}(R_\Gamma, \mathfrak{U}(\mathbb{Q}_l^c)),$$

then, for all $\chi \in R_\Gamma$,

$$\chi(\Lambda g) = \log[g(l\chi) \cdot g(-\Psi_l \chi)^f], \tag{5.21}$$

where f is the Frobenius on M, and the identity on $\mathbb{Q}_l(\chi)$.

Proof. By Proposition 5.1 and Theorem 17 (i), (5.21) holds, if g is of form $\operatorname{Det}(1-r)$, with $r \in \mathfrak{r}$. But, for some positive integer h, g^h is certainly of this form. Thus (5.21) holds if both sides are multiplied by h. But, trivially, we can divide by h. □

The inclusion map $\Delta \to \Gamma$ of a subgroup gives rise to a linear map P_Γ^Δ: $M\mathfrak{C}_\Delta \to M\mathfrak{C}_\Gamma$. With respect to the induction map of (3.1) we now have the

Corollary. *Let* $g \in \mathrm{Hom}_{\Omega_M}(R_\Delta, \mathfrak{U}(\mathbb{Q}_l^c))$. *Then*

$$\Lambda_\Gamma \operatorname{ind}_\Delta^\Gamma g = P_\Gamma^\Delta \Lambda_\Delta g.$$

Proof. Use the formula of the Proposition to evaluate at each $\chi \in R_\Gamma$. □

§6. Galois Properties of the Determinant

We shall now give an outline of all the main steps in the proof of the fixed point theorem for determinants, Theorem 10. This global theorem is seen to follow easily from a local analogue (Theorem 10A), which really lies at the core of the matter. It was clear early on, even for a much weaker Galois module structure result (cf. [Ty5]), that some such fixed point theorem would be needed. The version given first in [Ty6] in fact sufficed to prove the basic conjecture (i.e., Theorem 9) under somewhat more restrictive conditions, and one of the advances which led to the proof of Theorem 9 in full generality in [Ty7], was an improvement of the fixed point theorem, obtained there.

We shall essentially follow the strategy of the proof in [Ty6] and [Ty7], but after the initial reduction will keep within an entirely local framework, and also some of the individual steps are different. At one stage (proof of III) the underlying formal procedures will be discussed in more detail than is necessary for the immediate purpose, as they will also have to be used later in Chap. IV (§3 and §4). Both the theory of the logarithm in §5, as well as the theorems of §3 will play a crucial role. Throughout Γ is a finite group and l a fixed prime number. R_Γ is always the ring of virtual characters of Γ for representations over $\mathbb{Q}_l^c$ in the present section. Keep this difference from the usual meaning in mind!

Theorem 10A. (Cf. [Ty7].) *Let M be a tame Galois extension of $\mathbb{Q}_l$ of finite degree, with $H = \mathrm{Gal}(M/\mathbb{Q}_l)$. Then*

$$(\mathrm{Det}(\mathfrak{o}_M\Gamma^*))^H = \mathrm{Det}(\mathbb{Z}_l\Gamma^*). \tag{6.1}$$ □

(Recall here remarks on notation made in I §6, prior to the statement of Theorem 10.) Here we shall use the localization procedure described in §2. In particular recall Lemma 2.1!

We shall first show that Theorem 10A implies Theorem 10. In the notation of that theorem (see in particular I(6.6)) note that, with $\mathfrak{l}$ running over the prime divisors of F above l,

$$\mathrm{Det}(\mathfrak{O}_{F,l}\Gamma^*) = \prod_{\mathfrak{l}} \mathrm{Det}(\mathfrak{o}_{\mathfrak{l},1}\Gamma^*),$$

so that an element x of $\mathrm{Det}(\mathfrak{O}_{F,l}\Gamma^*)$ is a "vector" with components $x_\mathfrak{l} \in \mathrm{Det}(\mathfrak{o}_{\mathfrak{l},1}\Gamma^*)$.

Now $\mathrm{Det}(\mathbb{Z}_l\Gamma^*)$ is embedded in each component, but its embedding in $\mathrm{Det}(\mathfrak{O}_{F,l}\Gamma^*)$, with which we are concerned in Theorem 10, is the diagonal one. We thus have to show that if x is fixed under $G = \mathrm{Gal}(F/\mathbb{Q})$, then its components $x_{\mathfrak{l}}$ firstly all lie in $\mathrm{Det}(\mathbb{Z}_l\Gamma^*)$ and secondly all coincide. The first assertion follows from Theorem 10A; for if x is fixed under G, then its $\mathfrak{l}$-component is fixed under the decomposition group of $\mathfrak{l}$ in G, i.e., under the Galois group $\mathrm{Gal}(F_{\mathfrak{l}}/\mathbb{Q}_l)$ as embedded in G, and thus for each $\mathfrak{l}$, $x_{\mathfrak{l}} \in \mathrm{Det}(\mathbb{Z}_l\Gamma^*)$. But G acts on the subgroup $\prod_{\mathfrak{l}} \mathrm{Det}(\mathbb{Z}_l\Gamma^*)$ (one factor for each $\mathfrak{l}$) of $\mathrm{Det}(\mathfrak{O}_{F,l}\Gamma^*)$ by permuting the factors transitively, and this implies the second assertion above: all components of x coincide. Thus indeed $(\mathrm{Det}(\mathfrak{O}_{F,l}\Gamma^*))^G \subset \mathrm{Det}(\mathbb{Z}_l\Gamma^*)$. The opposite inclusion is obvious. □

Now we turn to the proof of Theorem 10A. Note here too that we only have to show that the left hand side of (6.1) is contained in the right hand side, the opposite inclusion being obvious.

We shall first show that it suffices to prove Theorem 10A for M non-ramified. This was the original version in [Ty6] before the final improvement in [Ty7]. Let then H_0 be the inertia group of H, so that $M_0 = M^{H_0}$ is the maximal non-ramified subfield of $M/\mathbb{Q}_l$. We shall prove that

$$(6.2) \qquad (\mathrm{Det}(\mathfrak{o}_M\Gamma^*))^{H_0} \subset \mathrm{Det}(\mathfrak{o}_{M_0}\Gamma^*)$$

(equality is then obvious). This implies that

$$(\mathrm{Det}(\mathfrak{o}_M\Gamma^*))^{H} \subset (\mathrm{Det}(\mathfrak{o}_{M_0}\Gamma^*))^{H/H_0} \subset \mathrm{Det}(\mathbb{Z}_l\Gamma^*),$$

the last inclusion coming from the non-ramified version of Theorem 10A.

Let $\mathfrak{l}_M$ be the maximal ideal of $\mathfrak{o}_M$. Every residue class mod $\mathfrak{l}_M$ is then represented by an element of $\mathfrak{o}_{M_0}$. It follows easily that

$$\mathfrak{o}_M\Gamma^* = (\mathfrak{o}_{M_0}\Gamma^*)(1 + \mathfrak{l}_M\mathfrak{o}_M\Gamma).$$

Going over to determinants and taking fixed points, we have

$$(6.3) \qquad (\mathrm{Det}(\mathfrak{o}_M\Gamma^*))^{H_0} = (\mathrm{Det}(\mathfrak{o}_{M_0}\Gamma^*))(\mathrm{Det}(1 + \mathfrak{l}_M\mathfrak{o}_M\Gamma))^{H_0}.$$

Apply the norm operator $\mathcal{N}_{M/M_0}$ to $\mathrm{Det}(1 + \mathfrak{l}_M\mathfrak{o}_M\Gamma)$. By Theorem 13, the image will lie in $\mathrm{Det}(\mathfrak{o}_{M_0}\Gamma^*)$. But restricting to $(\mathrm{Det}(1 + \mathfrak{l}_M\mathfrak{o}_M\Gamma))^{H_0}$, the norm operator is just $x \mapsto x^{[M:M_0]}$, and $(\mathrm{Det}(1 + \mathfrak{l}_M\mathfrak{o}_M\Gamma))^{H_0}$ being a pro-l-group, while $([M:M_0], l) = 1$, the map $x \mapsto x^{[M:M_0]}$ is an automorphism of $(\mathrm{Det}(1 + \mathfrak{l}_M\mathfrak{o}_M\Gamma))^{H_0}$. Thus this group is indeed a subgroup of $\mathrm{Det}(\mathfrak{o}_{M_0}\Gamma^*)$, and this, in conjunction with (6.3), now implies (6.2). □

From now on we shall assume the field M in Theorem 10A to be non-ramified. The proof will now proceed by stages, restricting Γ first to special types of groups.

I. (6.1) *holds if* Γ *is an* l*-group.*

Proof. It is here that the logarithmic methods of §5 come into play. From the exact sequences (5.7) for $\mathbb{Q}_l$, and for M (see also (5.8)) and taking fixed points in the latter, we get a commutative diagram with exact rows and columns

(6.4)
$$\begin{array}{ccccccc} & & 0 & & 1 & & 1 \\ & & \downarrow & & \downarrow & & \downarrow \\ 0 \to & & lc(\mathfrak{a}(\mathbb{Z}_l\Gamma)) & \to & \mathrm{Det}(\mathbb{Z}_l\Gamma^*) & \to & \mathrm{Det}((\mathbb{Z}_l\Gamma^{ab})^*) \to 1 \\ & & \downarrow & & \downarrow & & \downarrow \\ 0 \to & & (lc(\mathfrak{a}(\mathfrak{o}_M\Gamma)))^H & \to & (\mathrm{Det}(\mathfrak{o}_M\Gamma^*))^H & \to & (\mathrm{Det}((\mathfrak{o}_M\Gamma^{ab})^*))^H \\ & & \downarrow & & & & \downarrow \\ & & 0 & & & & 1 \end{array}$$

That the right hand column is exact follows from the fact that for any M and any Abelian group Δ, we have an isomorphism $\mathfrak{o}_M\Delta^* \cong \mathrm{Det}(\mathfrak{o}_M\Delta^*)$, compatible with all automorphisms of $\mathfrak{o}_M$. That the left hand column is exact follows from the fact that (see (5.8))

$$c(\mathfrak{a}(\mathfrak{o}_M\Gamma)) = c(\mathfrak{a}(\mathbb{Z}_l\Gamma)) \otimes_{\mathbb{Z}_l} \mathfrak{o}_M$$

with H acting on the right hand group via the tensor factor $\mathfrak{o}_M$. It now follows that the inclusion represented by the middle column of (6.4) is an isomorphism, i.e., we have (6.1). □

We shall need a slight strengthening of I, for l-groups Γ. We now take in place of $\mathfrak{o}_M$ a base ring $\mathfrak{O}$ which is the product of (finitely many) rings $\mathfrak{o}_{M_i}$, $M_i/\mathbb{Q}_l$ non-ramified, and we let G be a finite group of automorphisms of $\mathfrak{O}$, so that $\mathfrak{O}^G = \prod \mathfrak{o}_{F_j}$ is of the same type. Taking products and using (5.8), we get an exact sequence, like (5.7), and then a commutative diagram with exact rows and columns, analogous to (6.4):

$$\begin{array}{ccccccc} & & 0 & & 1 & & 1 \\ & & \downarrow & & \downarrow & & \downarrow \\ 0 \to & & lc(\mathfrak{a}(\mathbb{Z}_l\Gamma)) \otimes_{\mathbb{Z}_l} \mathfrak{O}^G & \to & \mathrm{Det}(\mathfrak{O}^G\Gamma^*) & \to & \mathrm{Det}((\mathfrak{O}^G\Gamma^{ab})^*) \to 1 \\ & & \downarrow & & \downarrow & & \downarrow \\ 0 \to & & (lc(\mathfrak{a}(\mathbb{Z}_l\Gamma)) \otimes_{\mathbb{Z}_l} \mathfrak{O})^G & \to & (\mathrm{Det}(\mathfrak{O}\Gamma^*))^G & \to & (\mathrm{Det}((\mathfrak{O}\Gamma^{ab})^*))^G \\ & & \downarrow & & & & \downarrow \\ & & 0 & & & & 1 \end{array}$$

As before we conclude:

Ia. *In the above situation* $\mathrm{Det}(\mathfrak{O}^G\Gamma^*) = \mathrm{Det}(\mathfrak{O}\Gamma^*)^G$. □

A $\mathbb{Q}$-p-elementary group Γ is a semidirect product

(6.5) $$\Gamma = \Sigma \rtimes \Pi$$

of a normal cyclic subgroup Σ of order prime to p and a p-group Π.

II. *If* Γ *is* $\mathbb{Q}$-*p*-*elementary with* $p \neq l$, *then* $(\mathrm{Det}(\mathfrak{o}_M\Gamma^*))^H/\mathrm{Det}(\mathbb{Z}_l\Gamma^*)$ *is a pro l-group.*

Proof. Let Σ_l be the l-Sylow group of Σ. It is also the normal l-Sylow group of Γ and we put $\Gamma' = \Gamma/\Sigma_l$. Let $\mathfrak{L}$ be the maximal ideal of the ring $\mathbb{Z}_l^c$ of integers in $\mathbb{Q}_l^c$, and denote by $\mathrm{Det}^1(\mathfrak{o}_M\Gamma^*)$ the subgroup of $\mathrm{Det}(\mathfrak{o}_M\Gamma^*)$ of maps f, with $f(\chi) \equiv 1 \pmod{\mathfrak{L}}$ for all $\chi \in R_\Gamma$ – and analogously for Γ'. The map coinf (cf. (3.1)) gives rise to a homomorphism

$$\mathrm{Det}(\mathfrak{o}_M\Gamma^*)/\mathrm{Det}^1(\mathfrak{o}_M\Gamma^*) \to \mathrm{Det}(\mathfrak{o}_M\Gamma'^*)/\mathrm{Det}^1(\mathfrak{o}_M\Gamma'^*). \tag{6.6}$$

We shall prove that this is an isomorphism.

The proof is based on the fact that

$$R_\Gamma = \mathrm{inf}_{\Gamma'}^{\Gamma}(R_{\Gamma'}) + \mathrm{Ker}\, d_{l,\Gamma}, \tag{6.7}$$

(for $\mathrm{Ker}\, d_{l,\Gamma}$ see (4.1), (4.2)). This will be established below. Suppose then that $f = \mathrm{Det}(a)$, $a \in \mathfrak{o}_M\Gamma^*$, gets mapped into $\mathrm{Det}^1(\mathfrak{o}_M\Gamma'^*)$. Following (6.7), write a given $\chi \in R_\Gamma$ as $\chi = \mathrm{inf}_{\Gamma'}^{\Gamma}(\chi') + \phi$. By hypothesis on f, $f(\mathrm{inf}_{\Gamma'}^{\Gamma}(\chi')) \equiv 1 \pmod{\mathfrak{L}}$, and by Theorem 15A, $f(\phi) \equiv 1 \pmod{\mathfrak{L}}$. Thus indeed always $f(\chi) \equiv 1 \pmod{\mathfrak{L}}$, i.e., $f \in \mathrm{Det}^1(\mathfrak{o}_M\Gamma^*)$. We have now shown that (6.6) is injective. But (6.6) is also surjective. For, $\mathfrak{o}_M\Gamma^* \to \mathfrak{o}_M\Gamma'^*$ is surjective, as $\mathrm{Ker}[\mathfrak{o}_M\Gamma \to \mathfrak{o}_M\Gamma']$ lies in the radical of $\mathfrak{o}_M\Gamma$.

To prove (6.7), it suffices to write any irreducible character in the form $\chi = \mathrm{inf}_{\Gamma'}^{\Gamma}(\chi') + \phi$, $\phi \in \mathrm{Ker}\, d_{l,\Gamma}$. For this we use a description of the irreducible characters of Γ, involving Abelian characters α of Σ (cf. [Se2], Proposition 25). Let Π_0 be the subgroup of Π which fixes α (under conjugation). Then mod $\mathrm{Ker}\,\alpha$, $\Sigma \cdot \Pi_0$ is a direct product, and so given an irreducible character ρ of Π_0 we get an irreducible character of $\Sigma \cdot \Pi_0$, which by abuse of notation we write as $\alpha \otimes \rho$. Now χ is the induced of such an $\alpha \otimes \rho$. Replace α by the character α' of Σ, with $\alpha'(\sigma) = \alpha(\sigma)$ if σ is of order prime to l, $\alpha'(\sigma) = 1$ if σ is of order a power of l. Let χ'' be induced by the character of $\Sigma \cdot \Pi_0$ which we may write as $\alpha' \otimes \rho$. Then $\chi - \chi'' \in \mathrm{Ker}\, d_{l,\Gamma}$, and $\chi'' = \mathrm{inf}_{\Gamma'}^{\Gamma}(\chi')$, as $\mathrm{Ker}\,\chi'' \supset \Sigma_l$.

We shall prove II by showing that

$$\mathrm{Det}(\mathfrak{o}_M\Gamma^*) \subset \mathrm{Det}(\mathbb{Z}_l\Gamma^*)\,\mathrm{Hom}_{\Omega_M}(R_\Gamma, U^1(\mathbb{Q}_l^c)), \tag{6.8}$$

where $U^1(\mathbb{Q}_l^c)$ is the subgroup of $U(\mathbb{Q}_l^c)$ of elements u with $u \equiv 1 \pmod{\mathfrak{L}}$. This is a pro-$l$-group and so the Hom group on the right of (6.8) is also a pro-l-group. This then implies II.

Let then $f \in \mathrm{Det}(\mathfrak{o}_M\Gamma^*)^H$. Thus $f \in \mathrm{Hom}_{\Omega_{\mathbb{Q}_l}}(R_\Gamma, U(\mathbb{Q}_l^c))$. Note now that coinf maps this latter group surjectively onto $\mathrm{Hom}_{\Omega_{\mathbb{Q}_l}}(R_{\Gamma'}, U(\mathbb{Q}_l^c)) = \mathrm{Det}(\mathbb{Z}_l\Gamma'^*)$ – the last equation holds because l does not divide the order of Γ'. But as $\mathbb{Z}_l\Gamma^* \to \mathbb{Z}_l\Gamma'^*$ is surjective, so is the map coinf on Det. In other words,

$$\mathrm{coinf}\, f = \mathrm{coinf}\,\mathrm{Det}(b), \qquad b \in \mathbb{Z}_l\Gamma^*. \tag{6.9}$$

Put $f = \mathrm{Det}(a)$, $a \in \mathfrak{o}_M \Gamma^*$. Then $\mathrm{Det}(ab^{-1}) \in \mathrm{Ker\,coinf}$. Hence, by the isomorphism (6.6),

$$f = \mathrm{Det}(b)h, \qquad h \in \mathrm{Hom}_{\Omega_M}(R_\Gamma, U^1(\mathbb{Q}_l^c)).$$

This yields (6.8).

Remark. The proof of II in [Ty6] is different from that given here.

III. *If Γ is a $\mathbb{Q}$-l-elementary group then* (6.1) *holds.*

We shall adjourn the proof of III to the end of this section, and, assuming III, prove that

IV. *If Γ is a $\mathbb{Q}$-p-elementary group then* (6.1) *holds.*

Proof. By III, we may assume that $p \neq l$.

Suppose now that there is an integer $s \neq 0$, a set $\mathcal{S}$ of subgroups Δ_i, with

$$(\mathrm{Det}(\mathfrak{o}_M \Delta_i^*))^H = \mathrm{Det}(\mathbb{Z}_l \Delta_i^*), \tag{6.10}$$

and virtual characters $\theta_i \in R_{\Delta_i}(\mathbb{Q})$, so that the identity character ε of Γ has a representation

$$s\varepsilon = \sum_i \mathrm{ind}_{\Delta_i}^\Gamma(\theta_i). \tag{6.11}$$

Under these hypotheses consider an element $b \in \mathfrak{o}_M \Gamma^*$, with $\mathrm{Det}(b)$ fixed under H. Then

$$\begin{aligned}\mathrm{Det}(b)^s &= (s\varepsilon)\,\mathrm{Det}(b) \qquad \text{(character action, see §3)},\\ &= \prod_i (\mathrm{ind}_{\Delta_i}^\Gamma(\theta_i) \cdot \mathrm{Det}(b)) \qquad \text{(by (6.11))},\\ &= \prod_i \mathrm{ind}_{\Delta_i}^\Gamma(\theta_i \,\mathrm{res}_{\Delta_i}^\Gamma \mathrm{Det}(b)) \qquad \text{(by (3.7))}.\end{aligned}$$

By Theorem 12,

$$\mathrm{res}_{\Delta_i}^\Gamma \mathrm{Det}(b) = \mathrm{Det}(b_i), \qquad b_i \in \mathfrak{o}_M \Delta_i^*.$$

As $\mathrm{res}_{\Delta_i}^\Gamma$ commutes with H-action, $\mathrm{Det}(b_i)$ is fixed under H. The same is then true for $\theta_i\,\mathrm{Det}(b_i)$, because $\theta_i \in R_{\Delta_i}(\mathbb{Q})$. Moreover, by Theorem 14,

$$\theta_i\,\mathrm{Det}(b_i) = \mathrm{Det}(b_i'), \qquad b_i' \in \mathfrak{o}_M \Delta_i^*.$$

As we have seen, $\mathrm{Det}(b_i')$ is fixed under H. By (6.10) we may assume that $b_i' \in \mathbb{Z}_l \Delta_i^*$. By Theorem 12, $\mathrm{ind}_{\Delta_i}^\Gamma \mathrm{Det}(b_i') \in \mathrm{Det}(\mathbb{Z}_l \Gamma^*)$. As we now have

$$\mathrm{Det}(b)^s = \prod_i \mathrm{ind}_{\Delta_i}^\Gamma \mathrm{Det}(b_i'), \qquad b_i' \in \mathbb{Z}_l \Delta_i^*, \qquad \text{all } i,$$

we conclude that $\mathrm{Det}(b)^s \in \mathrm{Det}(\mathbb{Z}_l\Gamma^*)$. We have thus shown, that under hypotheses (6.10), (6.11) we obtain

(6.12) $$((\mathrm{Det}(\mathfrak{o}_M\Gamma^*))^H)^s \subset \mathrm{Det}(\mathbb{Z}_l\Gamma^*).$$

Now take $\mathscr{S}$ to be the set of $\mathbb{Q}$-l-elementary subgroups of Γ. Then, by III, (6.10) will hold for $\Delta_i \in \mathscr{S}$, and also (6.11) holds for some s prime to l (cf. [Se2] Theorem 28). By (6.12) raising to a power s annihilates the group

$$(\mathrm{Det}(\mathfrak{o}_M\Gamma^*))^H/\mathrm{Det}(\mathbb{Z}_l\Gamma^*),$$

while by II it is an automorphism of that group. Thus this group collapses, i.e., we do get IV. □

Proof of (6.1) *for any* Γ. Let $\mathscr{S}$ now be the set of all $\mathbb{Q}$-elementary subgroups of Γ. Then (6.10) holds by IV, and (6.11) holds with $s = 1$ (cf. [Se2] Theorem 27). Thus we deduce (6.12) for $s = 1$, which gives us (6.1). □

We now turn to the preparations for the proof of III. The techniques involved will also be used in Chap. IV and will be developed now to the full extent that they will be needed. In the sequel Γ is $\mathbb{Q}$-l-elementary, i.e., in (6.5) Π is an l-group and Σ cyclic of order prime to l. The ultimate aim is to reduce everything to l-groups.

We shall need some definitions and notations. For any positive integer m, let $h_m(X) \in \mathbb{Z}[X]$ be the polynomial whose roots are the primitive m-th roots of unity, and view $h_m(X)$ also as a polynomial in $S[X]$, for any commutative ring S. Define

(6.13) $$S[m] = S[X]/(h_m(X)),$$

i.e., we have

(6.13a) $$S[m] = S \otimes_{\mathbb{Z}} \mathbb{Z}[m],$$

and $\mathbb{Z}[m]$ is the ring of integers in the m-th cyclotomic field $\mathbb{Q}[m] = \mathbb{Q}(m)$. *Warning*. In general $M[m] \neq M(m)$.

Now let $m \mid \mathrm{order}\,(\Sigma)$. If σ is a generator of Σ, which we fix once and for all, the map

(6.14) $$\sigma \mapsto \sigma_m = X \bmod h_m(X)$$

extends to a surjective homomorphism $S\Sigma \to S[m]$ of rings. The action of Π on Σ yields an action on $\mathbb{Z}[m]$, hence on $S[m]$, independent of the choice of σ, and via this action we define the twisted group ring $S[m] \circ \Pi$. Additively this is the free right $S[m]$-module on Π, and multiplication is given by

$$\pi_1 s_1 \cdot \pi s = (\pi_1\pi)s_1^\pi s, \quad \pi_1, \pi \in \Pi, \quad s_1, s \in S[m], \quad s_1^\pi \text{ the image of } s_1 \text{ under } \pi.$$

The group ring $\mathbb{Z}_l\Sigma$ splits into a direct product

(6.15) $$\mathbb{Z}_l\Sigma = \prod \mathbb{Z}_l[m] \qquad (\text{product over } m \mid \mathrm{order}\,(\Sigma)),$$

giving rise to products

$$\text{(6.15a)}\qquad \begin{cases} M\Gamma = \prod_m (M[m]\circ\Pi), \\ \mathfrak{o}_M\Gamma = \prod_m (\mathfrak{o}_M[m]\circ\Pi), \end{cases}$$

$$\text{(6.15b)}\qquad \begin{cases} \mathrm{Det}(M\Gamma^*) = \prod_m \mathrm{Det}((M[m]\circ\Pi)^*), \\ \mathrm{Det}(\mathfrak{o}_M\Gamma^*) = \prod_m \mathrm{Det}((\mathfrak{o}_M[m]\circ\Pi)^*). \end{cases}$$

We shall view the $\mathrm{Det}((\mathfrak{o}_M[m]\circ\Pi)^*)$ in the standard way as subgroups of $\mathrm{Det}(\mathfrak{o}_M\Gamma^*)$. For the same values of m, let $R_\Gamma^{(m)}$ be the subgroup of R_Γ, generated by the irreducible characters whose restrictions to Σ are sums of Abelian characters of exact order m. Clearly R_Γ is the direct sum of the $R_\Gamma^{(m)}$, and so

$$\text{(6.16)}\qquad \mathrm{Hom}_{\Omega_M}(R_\Gamma,\cdot) = \prod_m \mathrm{Hom}_{\Omega_M}(R_\Gamma^{(m)},\cdot).$$

The decomposition (6.16) is the "same" as that in (6.15)–(6.15b). More precisely we get, via the two product expansions, isomorphisms

$$\text{(6.17)}\qquad \begin{cases} \mathrm{Det}(M\Gamma^*)_{|R_\Gamma^{(m)}} \cong \mathrm{Det}((M[m]\circ\Pi)^*), \\ \mathrm{Det}(\mathfrak{o}_M\Gamma^*)_{|R_\Gamma^{(m)}} \cong \mathrm{Det}((\mathfrak{o}_M[m]\circ\Pi)^*), \end{cases}$$

where the subscript denotes restriction to $R_\Gamma^{(m)}$. All the product expansions and isomorphisms clearly respect change of basefield and the action of H.

We now fix our attention on one divisor d of order (Σ). We shall write Φ for the set of Abelian characters of Σ with values in $\mathbb{Q}_l^{c*}$ of precise order d. We let Π_1 be the elementwise stabilizer in Π of Φ or, which is the same, of $M[d]$. We put

$$\Xi = \Pi/\Pi_1, \qquad \Delta = \Sigma\cdot\Pi_1 \subset \Gamma \qquad (\text{so } \Delta = \Sigma \rtimes \Pi_1).$$

Then Ξ acts (via Π) faithfully as a group of automorphisms of the rings $M[d]$, $\mathfrak{o}_M[d]$, and on the set Φ. Again, via Π, it acts by automorphisms of R_{Π_1} and of $R_\Delta^{(d)}$, hence also on the corresponding Hom groups – assuming trivial action on $\mathbb{Q}_l^{c*}$ etc. These actions commute with those of Galois groups. For instance H and Ξ act on $\mathfrak{o}_M[d] = \mathfrak{o}_M \otimes_{\mathbb{Z}} \mathbb{Z}[d]$ via the first, and the second tensor factor respectively.

Our earlier discussion applies of course with Δ in place of Γ. Now, however, the twisted group ring $\mathfrak{o}_M[d]\circ\Pi_1$ is just the ordinary group ring $\mathfrak{o}_M[d]\Pi_1$. Its group of units $(\mathfrak{o}_M[d]\Pi_1)^*$ appears simultaneously also as a factor of the group $\mathfrak{o}_M\Delta^*$. Accordingly we get two distinct determinants and we have to keep them apart notationally. Let then $y\in\mathfrak{o}_M[d]\Pi_1^*$. Viewing it as contained in $\mathfrak{o}_M\Delta^*$, we get for every $\psi\in R_\Delta^{(d)}$ a determinant in $\mathbb{Q}_l^{c*}$, now to be denoted by $\Delta-\mathrm{Det}_\psi(y)$. For varying y we get, via (6.17), a subgroup

$$\text{(6.18a)}\qquad \Delta-\mathrm{Det}((\mathfrak{o}_M[d]\Pi_1)^*) \subset \mathrm{Hom}(R_\Delta^{(d)}, \mathbb{Q}_l^{c*}).$$

On the other hand, if $T\colon \Pi_1 \to \mathrm{GL}_n(\mathbb{Q}_l^c)$ is a representation with character χ, and we extend T to an algebra homomorphism of $M[d]\Pi_1$ into n by n matrices over $\mathbb{Q}_l^c[d]$, then we obtain a determinant $\Pi_1 - \mathrm{Det}_\chi(y) \in \mathbb{Q}_l^c[d]^*$, and the same also for virtual characters χ. Thus we have also a subgroup

$$\text{(6.18b)} \qquad \Pi_1 - \mathrm{Det}((\mathfrak{o}_M[d]\Pi_1)^*) \subset \mathrm{Hom}(R_{\Pi_1}, \mathbb{Q}_l^c[d]^*).$$

We shall show that the two pairs of groups in (6.18) are canonically isomorphic.

Every pair (χ, ϕ), $\chi \in R_{\Pi_1}$, $\phi \in \Phi$ defines a virtual character $\chi \times \phi \in R_\Delta^{(d)}$, where $(\chi \times \phi)(\pi\sigma) = \chi(\pi)\phi(\sigma)$, for $\pi \in \Pi_1$, $\sigma \in \Sigma$. The map $(\chi, \phi) \mapsto \chi \times \phi$ defines a bijection from pairs (χ, ϕ) of irreducible characters to the irreducible characters in $R_\Delta^{(d)}$. We thus obtain a natural isomorphism

$$\text{(6.19)} \qquad \mathrm{Hom}(R_\Delta^{(d)}, G) \cong \mathrm{Hom}(R_{\Pi_1}, \mathrm{Map}(\Phi, G))$$

for variable G, given by

$$f \mapsto f_1, \qquad f_1(\chi)(\phi) = f(\chi \times \phi).$$

(This is really the natural isomorphism defining Hom and $\otimes$ as adjoints: replace Map by Hom, Φ by $R_\Sigma^{(d)}$, so that in fact $R_\Delta^{(d)} = R_{\Pi_1} \otimes_{\mathbb{Z}} R_\Sigma^{(d)}$.) If G is an Ω-module ($\Omega = \Omega_{\mathbb{Q}_l}, \Omega_M, \Xi$) and Ω acts on Hom groups by conjugation, then (6.19) preserves Ω-action.

For the next Lemma note that the set $\mathrm{Map}(\Phi, \mathbb{Q}_l^c)$ has the structure of an algebra, by pointwise operation, e.g., $h_1 \cdot h_2(\phi) = h_1(\phi)h_2(\phi)$.

6.1. Lemma. (i) *For each $\phi \in \Phi$ there is a unique*

$$g_\phi \in \mathrm{Alg}(\mathbb{Q}_l^c[d], \mathbb{Q}_l^c) \qquad \textit{(set of algebra homomorphisms)},$$

so that ϕ is the composite map

$$\Sigma \to \mathbb{Z}_l\Sigma \to \mathbb{Z}_l[d] \subset \mathbb{Q}_l^c[d] \xrightarrow{g_\phi} \mathbb{Q}_l^c.$$

The map $\phi \mapsto g_\phi$ is a bijection

$$\Phi \to \mathrm{Alg}(\mathbb{Q}_l^c[d], \mathbb{Q}_l^c).$$

(ii) *Associating with $y \in \mathbb{Q}_l^c[d]$ the map $\phi \mapsto g_\phi(y)$ gives an isomorphism*

$$\mathbb{Q}_l^c[d] \cong \mathrm{Map}(\Phi, \mathbb{Q}_l^c)$$

of $\mathbb{Q}_l^c$-algebras, and of $\Omega_{\mathbb{Q}_l} \times \Xi$ modules.

Proof. There is a bijection $\phi \mapsto h_\phi$, where $h_\phi\colon \mathbb{Z}[d] \to \mathbb{Q}_l^c$ is a ring homomorphism. Now g_ϕ is just

$$\mathbb{Z}[d] \otimes_{\mathbb{Z}} \mathbb{Q}_l^c \xrightarrow{h_\phi \otimes 1} \mathbb{Q}_l^c.$$

Clearly $h_\phi \mapsto g_\phi$ is a bijection.

The map in (ii) clearly preserves all structures as stated. We shall show that it is injective, and comparison of vector space dimensions over $\mathbb{Q}_l^c$ implies then bijectivity. Let then σ_d be the image in $\mathbb{Z}[d]$ of a generator σ of Σ; then $\{\sigma_d^j\}$ for certain values of j is a $\mathbb{Z}$-basis, and so $\mathrm{Det}(g_\phi(\sigma_d^j))_{\phi,j} = \mathrm{Det}(\phi(\sigma^j))_{\phi,j} \neq 0$. This implies the required injectivity. □

Combining Lemma 6.1 (ii) with (6.19) we get an isomorphism

$$\mathrm{Hom}(R_\Delta^{(d)}, \mathbb{Q}_l^{c*}) \cong \mathrm{Hom}(R_{\Pi_1}, \mathbb{Q}_l^c[d]^*). \tag{6.20}$$

Explicitly $f \mapsto f'$ where

$$g_\phi(f'(\chi)) = f(\chi \times \phi).$$

Now recall the two inclusions (6.18a), (6.18b).

6.2. Lemma. (6.20) *gives rise to an isomorphism*

$$\Delta - \mathrm{Det}((\mathfrak{o}_M[d]\Pi_1)^*) \cong \Pi_1 - \mathrm{Det}((\mathfrak{o}_M[d]\Pi_1)^*).$$

Proof. Let $y \in \mathfrak{o}_M[d]\Pi_1^*$. We have to show that (6.20) maps $\Delta - \mathrm{Det}(y)$ into $\Pi_1 - \mathrm{Det}(y)$. Write, with the σ_d^j as in the proof of Lemma 6.1,

$$y = \sum_{j,\pi} a_{j,\pi}\sigma_d^j \pi, \qquad a_{j,\pi} \in \mathfrak{o}_M \qquad (\pi \text{ running over } \Pi_1).$$

Suppose χ is the character of a representation T of Π_1. Then $\chi \times \phi$ is that of a representation $T \times \phi$. We then get

$$\begin{aligned}
\Delta - \mathrm{Det}_{\chi\times\phi}(y) &= \mathrm{Det}\left(\sum_{j,\pi} a_{j,\pi}\, \phi(\sigma^j) T(\pi)\right) \\
&= \mathrm{Det}\left(\sum_\pi \left[g_\phi\left(\sum_j a_{j,\pi}\sigma_d^j\right)\right] T(\pi)\right) \\
&= g_\phi\left(\mathrm{Det} \sum_\pi \left(\sum_j a_{j,\pi}\, \sigma_d^j\right) T(\pi)\right) \\
&= g_\phi(\Pi_1 - \mathrm{Det}_\chi(y)).
\end{aligned}$$

This is the required result. □

Remark. It is clear that the isomorphisms of (6.20) and of the last Lemma preserve the actions of H and of Ξ.

For subsequent use we shall need yet another result like Lemma 6.2. On the way we have to establish a refinement of Lemma 6.1 (ii). Let $L = M(\Phi)$ be the field of values of Φ, over M, i.e., the field $M(d)$ of primitive d-th roots of 1 over M.

6.3. Lemma. *The isomorphism of Lemma* 6.1 (ii) *gives rise to an isomorphism*

$$\mathfrak{o}_M[d] \cong \mathrm{Map}_{\Omega_M}(\Phi, \mathfrak{o}_L),$$

of $\mathfrak{o}_M$*-algebras and of H- and* Ξ*-modules.*

Proof. Preservation of structure is again obvious.

The isomorphism of Lemma 6.1 (ii) preserves integrality and thus restricts to an isomorphism

$$\mathbb{Z}_l^c[d] \cong \mathrm{Map}(\Phi, \mathbb{Z}_l^c)$$

of maximal $\mathbb{Z}_l^c$-orders, and preserves Ω_M-structure. Now we take fixed points under Ω_M and observe that, as Φ is elementwise fixed under Ω_L, we have

$$\mathrm{Map}_{\Omega_M}(\Phi, \mathbb{Z}_l^c) = \mathrm{Map}_{\Omega_M}(\Phi, \mathfrak{o}_L). \qquad \square$$

From the last Lemma we derive an isomorphism of group rings of Π_1:

(6.21) $$\mathfrak{o}_M[d](\Pi_1) \cong \mathrm{Map}_{\Omega_M}(\Phi, \mathfrak{o}_L)(\Pi_1) = \mathrm{Map}_{\Omega_M}(\Phi, \mathfrak{o}_L\Pi_1).$$

Next we note a variant of (6.19), interchanging the two groups, namely an isomorphism

(6.22) $$\begin{cases} \mathrm{Hom}(R_\Delta^{(d)}, G) \cong \mathrm{Map}(\Phi, \mathrm{Hom}(R_{\Pi_1}, G)), \\ \text{where} \quad f \mapsto f_2, \quad f_2(\phi)(\chi) = f(\chi \times \phi). \end{cases}$$

6.4. Lemma. (6.22) *gives by restriction rise to an isomorphism*

$$\Delta - \mathrm{Det}((\mathfrak{o}_M[d]\Pi_1)^*) \cong \mathrm{Map}_{\Omega_M}(\Phi, \Pi_1 - \mathrm{Det}(\mathfrak{o}_L\Pi_1^*)).$$

Proof. Let $f = \Delta - \mathrm{Det}(y)$, $y \in (\mathfrak{o}_M[d]\Pi_1)^*$. The computation in the proof of Lemma 6.2 also shows that

$$f_2(\phi)(\chi) = \Pi_1 - \mathrm{Det}_\chi(g_\phi y).$$

As g_ϕ maps $(\mathfrak{o}_M[d]\Pi_1)^*$ into $\mathfrak{o}_L\Pi_1^*$, we conclude that (6.22) maps

$$\Delta - \mathrm{Det}((\mathfrak{o}_M[d]\Pi_1)^*) \to \mathrm{Map}_{\Omega_M}(\Phi, \Pi_1 - \mathrm{Det}(\mathfrak{o}_L\Pi_1^*)).$$

Being the restriction of an isomorphism, this is injective.

To establish surjectivity, let

$$f_2 \in \mathrm{Map}_{\Omega_M}(\Phi, \mathrm{Hom}(R_{\Pi_1}, \mathbb{Q}_l^{c*})),$$

$$f_2(\phi)(\chi) = \Pi_1 - \mathrm{Det}_\chi(y_\phi), \qquad y_\phi \in \mathfrak{o}_L\Pi_1^*.$$

Now, for all $\phi \in \Phi$, all $\omega \in \Omega_M$ we have

$$\Pi_1 - \mathrm{Det}_\chi(y_\phi^{(\omega)}) = (\Pi_1 - \mathrm{Det}_{\chi^{\omega^{-1}}}(y_\phi))^\omega = f_2(\phi^\omega)(\chi) = \Pi_1 - \mathrm{Det}_\chi(y_{\phi^\omega}).$$

Note also that if ω fixes ϕ, then it must fix y_ϕ.

We may thus choose the map $\phi \mapsto y_\phi$ to be an Ω_M-map. By (6.21), $y_\phi = g_\phi z$ for some $z \in \mathfrak{o}_M[d]\Pi_1^*$, as we had to show. □

Now we return to our original group Γ, keeping d fixed, as above. We consider the sequence

$$0 \to R_\Delta^{(d)} I \to R_\Delta^{(d)} \xrightarrow{\mathrm{ind}_\Delta^\Gamma} R_\Gamma^{(d)} \to 0,$$

where I is the augmentation ideal of $\mathbb{Z}\Xi$. By Mackey's irreducibility criterion (cf. [Se2], Corollary to Proposition 23), $\mathrm{ind}_\Delta^\Gamma$ above is surjective. The sequence is now easily seen to be exact. Thus $\mathrm{res}_\Delta^\Gamma$ yields an isomorphism

$$\text{(6.23)} \qquad \mathrm{Hom}_{\Omega_M}(R_\Gamma^{(d)}, \mathbb{Q}_l^{c*}) \cong \mathrm{Hom}_{\Omega_M}(R_\Delta^{(d)}, \mathbb{Q}_l^{c*})^\Xi,$$

which in turn gives rise to an injective homomorphism

$$\text{(6.24)} \qquad (\mathrm{Det}(\mathfrak{o}_M\Gamma^*))_{|R_\Gamma^{(d)}} \to ((\mathrm{Det}(\mathfrak{o}_M\Delta^*))_{|R_\Delta^{(d)}})^\Xi,$$

of H-modules. Now compose the isomorphism

$$(\mathrm{Det}(\mathfrak{o}_M\Delta^*))_{|R_\Delta^{(d)}} \cong \Delta - \mathrm{Det}((\mathfrak{o}_M[d]\Pi_1)^*)$$

(cf. (6.17)) with the isomorphism of Lemma 6.2, and take Ξ - fixed point. There results an isomorphism

$$\text{(6.25)} \qquad ((\mathrm{Det}(\mathfrak{o}_M\Delta^*))_{|R_\Delta^{(d)}})^\Xi \cong (\Pi_1 - \mathrm{Det}((\mathfrak{o}_M[d]\Pi_1)^*))^\Xi$$

of H-modules. The composite e of (6.24), (6.25) is thus injective. We aim to prove

6.5. Lemma. *e is an isomorphism.*

Proof. We define in the usual manner the norm operator $\mathcal{N}_\Xi$ on the Ξ-module $\mathrm{Det}(\mathfrak{o}_M\Delta^*)$ by $\mathcal{N}_\Xi f = \prod_\zeta f^\zeta$ (product over Ξ). Explicitly let $\{\rho\}$ be a right transversal of Π_1 in Π, i.e., of Δ in Γ. If $f = \mathrm{Det}(z)$, $z \in \mathfrak{o}_M\Delta^*$, then for $\psi \in R_\Delta^{(d)}$,

$$(\mathcal{N}_\Xi f)(\psi) = \prod_\rho \mathrm{Det}_\psi(\rho^{-1}z\rho).$$

If in particular ψ is the character of a representation T of Δ, and we define the induced representation T' of Γ via the given transversal $\{\rho\}$, then

$$T'(z) = \mathrm{diag}_\rho\, T(\rho^{-1}z\rho).$$

Therefore $(\mathcal{N}_\Xi f)(\psi) = \mathrm{Det}_{\mathrm{ind}\,\psi}(z)$, and so we see that

$$\mathcal{N}_\Xi((\mathrm{Det}(\mathfrak{o}_M\Delta^*))_{|R_\Delta^{(d)}}) \subset \mathrm{Im}[(\mathrm{Det}(\mathfrak{o}_M\Gamma^*))_{|R_\Gamma^{(d)}} \to (\mathrm{Det}(\mathfrak{o}_M\Delta^*))_{|R_\Delta^{(d)}}].$$

Hence also

$$\mathcal{N}_\Xi(\mathrm{Det}((\mathfrak{o}_M[d]\Pi_1)^*)) \subset \mathrm{Im}\, e, \tag{6.26}$$

where we shall from now on again omit the perfix Π_1. To prove the Lemma it will thus suffice to show that

$$\mathcal{N}_\Xi(\mathrm{Det}((\mathfrak{o}_M[d]\Pi_1)^*)) = (\mathrm{Det}((\mathfrak{o}_M[d]\Pi_1)^*))^\Xi. \tag{6.27}$$

Let $\mathfrak{a}$ be the commutator ideal of $\mathfrak{o}_M\Pi_1$, and observe that $\mathfrak{o}_M[d]$ is a product of copies of $\mathfrak{o}_L$. Tensoring the sequence (5.7) with $\mathfrak{o}_L$, and taking products we get the exact sequence of Ξ-modules, which is the top row in the commutative diagram

$$\begin{array}{ccccccc} 0 \to & lc(\mathfrak{a}) \otimes_{\mathfrak{o}_M} \mathfrak{o}_M[d] & \to & \mathrm{Det}((\mathfrak{o}_M[d]\Pi_1)^*) & \to & \mathrm{Det}((\mathfrak{o}_M[d]\Pi_1^{ab})^*) & \to 1 \\ & \downarrow{\scriptstyle t_\Xi} & & \downarrow{\scriptstyle \mathcal{N}_\Xi} & & \downarrow{\scriptstyle \mathcal{N}_\Xi} & \\ 0 \to & l(c(\mathfrak{a}) \otimes_{\mathfrak{o}_M} \mathfrak{o}_M[d])^\Xi & \to & (\mathrm{Det}((\mathfrak{o}_M[d]\Pi_1)^*))^\Xi & \to & (\mathrm{Det}((\mathfrak{o}_M[d]\Pi^{ab})^*))^\Xi. & \end{array} \tag{6.28}$$

The bottom row is that of the groups of fixed points, hence exact. The left hand column is written as the Ξ-trace rather than norm, because of the additive nature of the groups. We shall show

6.6. Lemma. *The extreme columns of* (6.28) *are surjective.*

This clearly implies (6.27), hence Lemma 6.5. □

Proof of 6.6. $\mathbb{Z}_l[d]$ is a free $\mathbb{Z}_l\Xi$-module. The same is then true for $c(\mathfrak{a}) \otimes_{\mathfrak{o}_M} \mathfrak{o}_M[d] = c(\mathfrak{a}) \otimes_{\mathbb{Z}_l} \mathbb{Z}_l[d]$. Therefore t_Ξ is surjective. Similarly the Ξ-trace $\mathfrak{r}^k \to (\mathfrak{r}^k)^\Xi$ is surjective, $\mathfrak{r}$ the radical of $\mathfrak{o}_M[d]\Pi_1^{ab}$. By looking at the subquotients

$$\mathfrak{r}^k/\mathfrak{r}^{k+1}\ (\text{additive}) \cong 1 + \mathfrak{r}^k/1 + \mathfrak{r}^{k+1}\ (\text{multiplicative})$$

we conclude that $\mathcal{N}_\Xi\colon 1 + \mathfrak{r} \to 1 + \mathfrak{r}^\Xi$ is surjective. On the other hand $(\mathfrak{o}_M[d]\Pi_1^{ab}/\mathfrak{r})^*$ is of order prime to l, while Ξ is an l-group. Thus $\mathcal{N}_\Xi$ acts again surjectively, and thus the right hand column of (6.28) is surjective. □

Proof of III. By (6.15b) and (6.17), we shall have to prove that, for all m,

$$(\mathrm{Det}((\mathfrak{o}_M[m] \circ \Pi)^*))^H = \mathrm{Det}((\mathbb{Z}_l[m] \circ \Pi)^*).$$

Fixing $m = d$, as above, and using Lemma 6.5 this reduces to showing that

$$((\mathrm{Det}((\mathfrak{o}_M[d]\Pi_1)^*))^\Xi)^H = (\mathrm{Det}((\mathbb{Z}_l[d]\Pi_1)^*))^\Xi. \tag{6.29}$$

By Ia,

$$(\mathrm{Det}((\mathfrak{o}_M[d]\Pi_1)^*))^H = \mathrm{Det}((\mathbb{Z}_l[d]\Pi_1)^*),$$

and, on taking fixed Ξ points, this yields (6.29). □

Notes to Chapter II

[1] For surveys of the theory of classgroups of integral group rings up to about 1974 see [U1] and [Re], with a fairly complete bibliography in the former.

The initial development of the theory is due to Jacobinski (cf. [J1], [J2]). This approach was ideal theoretic, i.e., via groups of fractional ideals in the central components of the group ring. The development of $\mathfrak{K}$-theory brought a new point of view into the subject, introduced mainly by Swan (see e.g., [SE]). In this context the group $\mathrm{Cl}(\mathfrak{o}\Gamma)$ may be defined either essentially as the kernel of the rank map (see II (1.1)), or as the cokernel of locally free modules, modulo the free ones, as done here in I (2.1). Considerable progress in actual computations of classgroup was made via new techniques of fibre products and associated Mayer-Vietoris sequences, principally by Reiner and Ullom (see in particular [RU1], [RU2]). The idelic formulation, as given here in II (1.13), was first developed by the author in [F10], where however a slightly different method of proof was used. For similar points of view see also [Wa1], [Wa2] and [Wi1], and for a discussion of the connection between the $\mathfrak{K}$-theoretic and the idelic approach see [Sw3]. One of the strong points of the idelic formulation was that it included a quantitative criterion for the validity of the "cancellation law" (see Chap. I, Note 2, for the definition) and that it helped to free computational methods from the old restriction to cases where the cancellation law would be assumed.

The Hom-description was first motivated by, and indeed became indispensable in the applications to Galois module structure – see in particular Theorem 4. It was first properly developed in the appendix to [F17], although some of the ideas can already be detected in earlier papers by the author, e.g., [F15]. For more recent surveys see [Ri1] and [F24].

In II §3 it will become clear that the Hom language is well suited to the study of functorial properties of the classgroup. Moreover it provides powerful new tools for computations, and in a sense generalizes and stream-lines the old fibre product techniques. An example is the theory of $\mathfrak{E}$-groups, discussed in II §4 (see [F17] (Appendix), as well as [CN2], [CN4]). For another successful application, this time to groups Γ of prime power order (where the $\mathfrak{E}$-groups become trivial) see [Ty1]. For such groups Γ the quoted paper determines exactly the subgroup of $\mathrm{Cl}(\mathbb{Z}\Gamma)$ generated by the Swan-modules (see I §2, Example 2) – an old problem, which had resisted earlier attacks.

In many aspects the theory extends from group rings to arbitrary orders in semi-simple algebras, and in fact many of the quoted papers deal with these. There are actually two distinct possible generalizations, either in terms of projectives, the variant preferred by $\mathfrak{K}$-theorists and algebraists in general, or in terms of locally free modules, and it is the latter which carry the real interest from an arithmetic point of view. (For $\mathfrak{o}\Gamma$, $\mathfrak{o}$ a ring of algebraic integers, Swan had proved that projectives are locally free).

[2] We can now give an outline of S. Chase's proof of the stable self-duality of $\mathfrak{o}_N$ as a $\mathbb{Z}\Gamma$-module (cf. Corollary 3 to Theorem 9 in I §6). Recall that N/K is a tame Galois extension of number fields, with Galois group Γ. What one has to prove is that, in $\mathrm{Cl}(\mathbb{Z}\Gamma)$, we have $(\mathfrak{D}_N^{-1})(\mathfrak{o}_N)^{-1} = 1$, where $\mathfrak{D}_N$ is the absolute different. One sees easily, for instance by looking at representatives in $\mathrm{Hom}_{\Omega_\mathbb{Q}}(R_\Gamma, \mathfrak{J}(\mathbb{Q}^c))$, that $(\mathfrak{D}_N^{-1}) = (\mathfrak{D}_{N/K}^{-1})$, where $\mathfrak{D}_{N/K}$ is now the relative different of N/K. One thus has to show that the element $[\mathfrak{D}_{N/K}^{-1}/\mathfrak{o}_N]$ of $\mathfrak{K}_0T(\mathbb{Z}\Gamma)$ lies in the kernel of the map $\mathfrak{K}_0T(\mathbb{Z}\Gamma) \to \mathrm{Cl}(\mathbb{Z}\Gamma)$. Now $\mathfrak{D}_{N/K}^{-1}/\mathfrak{o}_N$ is indeed a genuine $\mathfrak{o}_K\Gamma$-module, and $\mathfrak{o}_K$-torsion module, the product of its local components $\mathfrak{D}_{N/K,\mathfrak{p}}^{-1}/\mathfrak{o}_{N,\mathfrak{p}}$. What we shall show is that the element $[\mathfrak{D}_{N/K,\mathfrak{p}}^{-1}/\mathfrak{o}_{N,\mathfrak{p}}]$ already lies in $\mathrm{Ker}[\mathfrak{K}_0T(\mathbb{Z}\Gamma) \to \mathrm{Cl}(\mathbb{Z}\Gamma)]$ – recall here that $\mathfrak{K}_0T(\mathbb{Z}\Gamma) = \coprod_p \mathfrak{K}_0T(\mathbb{Z}_p\Gamma)$. Let $\mathfrak{P}$ be a prime divisor of N above $\mathfrak{p}$, and let Δ be the decomposition group of $\mathfrak{P}$ in Γ. Write, for simplicity's sake, $E = N_\mathfrak{P}$, $F = K_\mathfrak{p}$. Then $\mathfrak{D}_{N/K,\mathfrak{p}}^{-1}/\mathfrak{o}_{N,\mathfrak{p}}$ is the Γ-module induced by the Δ-module $\mathfrak{D}_{E/F}^{-1}/\mathfrak{o}_E$, $\mathfrak{D}_{E/F}^{-1}$ again the relative different. It will thus suffice to show that the element $[\mathfrak{D}_{E/F}^{-1}/\mathfrak{o}_E]$ of the group $\mathfrak{K}_0T(\mathbb{Z}_p\Delta)$, viewed as a subgroup of $\mathfrak{K}_0T(\mathbb{Z}\Delta)$, lies in the kernel of the map into $\mathrm{Cl}(\mathbb{Z}\Delta)$. Now $\mathfrak{D}_{E/F}^{-1} = \mathfrak{p}_E^{1-e}$, $e = e(E/F)$ the ramification index. Multiplying up, we get $[\mathfrak{D}_{E/F}^{-1}/\mathfrak{o}_E] = [\mathfrak{p}_E/\mathfrak{p}_F] = [\mathfrak{o}_E/\mathfrak{p}_F] - [\mathfrak{o}_E/\mathfrak{p}_E]$, where we view the ideals of $\mathfrak{o}_F$ also as ideals of $\mathfrak{o}_E$. Now $\mathfrak{o}_E/\mathfrak{p}_F$ as a $\mathbb{Z}\Delta$-module is the sum of $f(F/\mathbb{Q}_p)$ copies of $\mathbb{F}_p\Delta$, and $[\mathfrak{o}_E/\mathfrak{p}_E]$ the sum of $f(F/\mathbb{Q}_p)$ copies of $[\mathbb{F}_p\Delta/(\mathbb{F}_p\Delta)I]$, where I is the augmentation ideal of $\mathbb{F}_p\Delta_0$ and Δ_0 is the inertia group and where $f(F/\mathbb{Q}_p)$ is the residue class degree. In their turn, $\mathbb{F}_p\Delta$ and $\mathbb{F}_p\Delta/(\mathbb{F}_p\Delta)I$ are the Δ-modules induced from the Δ_0-modules $\mathbb{F}_p\Delta_0$, and $\mathbb{F}_p\Delta_0/I \cong \mathbb{F}_p$, respectively. It will thus suffice if we show that

$$[\mathbb{F}_p\Delta_0] - [\mathbb{F}_p] \in \mathrm{Ker}[\mathfrak{K}_0T(\mathbb{Z}\Delta_0) \to \mathrm{Cl}(\mathbb{Z}\Delta_0)].$$

Now, as $p \nmid \mathrm{order}\ (\Delta_0)$, the Swan-module $S(p)$, i.e., the $\mathbb{Z}\Delta_0$ ideal generated by p and $\sum_{\delta\in\Delta_0}\delta$ is locally free (see I §2 Example 2). One then easily verifies that

$$[\mathbb{Z}\Delta_0/S(p)] = [\mathbb{F}_p\Delta_0] - [\mathbb{F}_p].$$

Thus the image of $[\mathbb{F}_p\Delta_0] - [\mathbb{F}_p]$ in $\mathrm{Cl}(\mathbb{Z}\Delta_0)$ is $(\mathbb{Z}(\Delta_0))^{-1}(S(p)) = (S(p))$. But as Δ_0 is cyclic, $S(p)$ is known to be free (cf. [SW2]), i.e., $S(p) = 1$. This then gives the result. As Chase has observed, a more careful analysis of the various steps will actually yield an isomorphism.

[3] Although no proofs are given for the theorems in II §3, we shall indicate briefly what is the most natural approach to these. The starting point is the local isomorphism (2.3), or better the associated map

(a) $$\mathrm{Hom}_{\Omega_F}(R_\Gamma, (\mathbb{Q}_p^c)^*) \to \mathfrak{K}_0T(\mathfrak{o}_F\Gamma).$$

One then defines restriction, induction, base ring extension etc. for torsion modules

X/Y, with X and Y free of finite rank over $\mathfrak{o}_F\Gamma$. These, together with the corresponding operations defined on the Hom groups are then seen to give rise to homomorphisms of maps (a). This implies already that the Det groups also get mapped into each other, and the rest of the proof is then straightforward, by going over to the global $\mathfrak{K}_0 T$, viewed as a product of local ones. Analogously one also proceeds for character action, considered later in §3.

[4] There are analogues to (3.7) with respect to change of basefield, using the ring embeddings $R_\Gamma(L) \to R_\Gamma(K)$ $(L \subset K)$ and the trace maps $t_{K/L}\colon R_\Gamma(K) \to R_\Gamma(L)$ etc.

[5] Write $\mathfrak{E}_S(\Gamma) = \mathfrak{E}(\Gamma)$, if S is the set of all prime divisors of order (Γ). This is the "biggest" of the $\mathfrak{E}$-groups, in that for any other S the map $D(\mathbb{Z}\Gamma) \to \mathfrak{E}_S(\Gamma)$ factorizes through a surjection $\mathfrak{E}(\Gamma) \to \mathfrak{E}_S(\Gamma)$. In fact the map $\mathfrak{E}(\Gamma) \to \prod_{l \in S} \mathfrak{E}_l(\Gamma)$ is surjective, but as shown in [CN1], not in general injective.

[6] In this note let Γ be an l-group. The groups $\mathfrak{U}_l^{(1)}(F)$ are pro-l-groups, hence so is $\mathrm{Hom}_{\Omega_\mathbb{Q}}(R_\Gamma, \mathfrak{U}_l^{(1)}(\mathbb{Q}^c))$. $\mathrm{Ker}\, h_l$ is a quotient of this, hence an l-group. On the other hand, $\mathfrak{E}_l(\Gamma)$ is a quotient of $\mathrm{Hom}_{\Omega_\mathbb{Q}}(\mathrm{Ker}\, d_l, \mathfrak{V}_l(\mathbb{Q}^c))$. As the group $\mathfrak{V}_l(\mathbb{Q}^c)$ is clearly of order prime to l, so is $\mathfrak{E}_l(\Gamma)$ – this is also true for arbitrary Γ. One knows (cf. [RU1]) that, for Γ an l-group, $D(\mathbb{Z}\Gamma)$ is of order a power of l and so one sees that $\mathfrak{E}_l(\Gamma) = 1$. One can in fact show directly, as we shall do here, that $\mathfrak{E}_l(\Gamma) = 1$ in this case, whence one deduces in turn that $D(\mathbb{Z}\Gamma) = \mathrm{Ker}\, h_l$ is an l-group.

$\mathrm{Ker}\, d_l$ is now freely generated by the virtual characters $\psi^* = \psi - \deg(\psi)\,\varepsilon$, where ψ runs through the irreducible characters of the l-group Γ, other than the identity character ε. Choosing one ψ out of each $\Omega_\mathbb{Q}$-orbit, and evaluating $f \mapsto f(\psi^*)$, we see that

$$\mathrm{Hom}_{\Omega_\mathbb{Q}}(\mathrm{Ker}\, d_l, \mathfrak{V}_l(\mathbb{Q}^c)) \cong \prod \mathfrak{V}_l(\mathbb{Q}(\psi)),$$

(product over their representative ψ) and $\mathfrak{E}_l(\Gamma)$ is a quotient of this group. Here the field $\mathbb{Q}(\psi)$ of values of ψ is a subfield of a cyclotomic field $\mathbb{Q}(l^m)$, in which l ramifies completely. Thus $\mathfrak{V}_l(\mathbb{Q}(\psi)) = (\mathfrak{o}_{\mathbb{Q}(\psi)}/\mathfrak{L}_{\mathbb{Q}(\psi)})^*$, $\mathfrak{L}_{\mathbb{Q}(\psi)}$ the unique prime ideal above l. For $l = 2$ this group is $= 1$, and so $\mathfrak{E}_l(\Gamma) = 1$. For l odd, $\mathfrak{E}_l(\Gamma)$ is the product of groups

$$\mathrm{Cok}[\mathfrak{Y}(\mathbb{Q}(\psi)) \to (\mathfrak{o}_{\mathbb{Q}(\psi)}/\mathfrak{L}_{\mathbb{Q}(\psi)})^*].$$

By a result of Roquette and Witt, the fields $\mathbb{Q}(\psi)$ are actual cyclotomic fields, and for these the above Cokernel is always trivial. Thus indeed $\mathfrak{E}_l(\Gamma) = 1$. (Alternatively one can reduce to the case when Γ is Abelian and then get the same conclusion quite trivially).

The fact that $\mathfrak{E}_l(\Gamma) = 1$ shows that e.g., in the context of quaternion 2-groups, as considered in I §7, the $\mathfrak{E}$-groups can be of no help.

[7] The $\mathfrak{E}$-groups were first introduced in [F17] as a tool for proving general results connecting Galois module structure and root numbers. Effectively the group $\mathfrak{E}_l(\Gamma)$ is implicit already in the proofs for H_{4l^r} in [F8], and in [F17] an equation was established for all Γ, which we may now write as $h_l t W_{N/K} \mathrel{\bar{=}} h_l U_{N/K}$. The

approach used there is the first example of the method of congruences. For the basic formalism, congruences on determinants were used, as also here in Theorem 15, and the result connecting $W_{N/K}$ and $U_{N/K}$ depended on congruences for resolvents and for Galois Gauss sums. The latter congruences still appear as an essential step in the proof of the main conjecture. To be more precise, the principal arithmetic result proved in this connection in [F17] was that, for virtual characters in Ker d_l, there is a good approximation at l to Galois Gauss sums, by a Galois homomorphism y_l into global roots of unity. The crucial step in strengthening these results was based on the proof that the y_l for varying l could be replaced by one Galois homomorphism y, serving for all l simultaneously (cf. [CN3]). The existence proof for such a y, given the y_l as in [F17], is based on a sophisticated analysis in [CN3] of idempotent operators on R_Γ, defined via Adams operations. Subsequently, in [FT] a canonical y was defined quite explicitly, and this will make its inevitable reappearance in Chapter IV.

[8] We shall discuss some further special results on t_Γ for quaternion groups. The parameter $d(\psi)$, introduced in §4, is independent of the particular group H_{4m}, i.e., if ψ is lifted from a character ψ' of a quotient H_{4n} of H_{4m}, then $d(\psi) = d(\psi')$. The maps t''_d of §4 will now be written as $t''_{d,l}$, as the prime l which entered the definition, and which was fixed once and for all in §4, may now be allowed to vary. Again the $t''_{d,l}$ are preserved on transition from a quotient group. For $d \equiv 0 \pmod 4$ we can also introduce formally maps

$$t''_{d,2}\colon \mathrm{Hom}_{\Omega_\mathbb{Q}}(R^s_\Gamma/\mathrm{Tr}(R_\Gamma), \pm 1) \to D(\mathbb{Z}\Gamma_2),$$

where $\Gamma = H_{4m}$, $m = 2^r m'$, m' odd, $r \geqslant 1$, $\Gamma_2 = H_{2^{r+2}}$ and $d = 2^{r+1}$. These are the composites of the two maps on either side of the commutative diagram

$$\begin{array}{ccc} & D(\mathbb{Z}\Gamma) & \\ {}^{t_\Gamma}\nearrow & & \searrow \\ \mathrm{Hom}_{\Omega_\mathbb{Q}}(R^s_\Gamma/\mathrm{Tr}(R_\Gamma), \pm 1) & & D(\mathbb{Z}\Gamma_2) \\ \searrow & & \nearrow {}^{t_{\Gamma_2}} \\ & \mathrm{Hom}_{\Omega_\mathbb{Q}}(R^s_{\Gamma_2}/\mathrm{Tr}(R_{\Gamma_2}), \pm 1). & \end{array}$$

It is clear that $t''_{d,2}$ is again invariant under transition from quotient groups. For a sharper analysis of the map t_Γ in the case when $\Gamma = H_{4m}$ has a quotient H_{2^n}, the map $t''_{d,2}$ will have to be considered as well. Here we just recall that, by the results of I §7 (cf. Proposition 7.2), $t''_{4,2}$ is injective, and $t''_{2^n,2}$ is null for $n \geqslant 3$.

A quaternion character ψ of parameter d has associated with it a quaternion algebra $\mathbb{H}_d$, central over $\mathbb{Q}(\psi) = \mathbb{Q}^+(d)$, which only depends on d. One can now easily show:

I. *The localization of* $\mathbb{H}_d$ *at any finite prime divisor* $\mathfrak{p}$ *of* $\mathbb{Q}^+(d)$ *splits, except precisely in the following cases*: (i) $d = 4$, $\mathfrak{p} = (2)$, (ii) $d = 2l^j, j \geqslant 1, l \equiv -1 \pmod 4$, $\mathfrak{p}$ *the prime divisor above* l.

From I Proposition 7.2, and II Proposition 4.4 we conclude that

II. *Whenever for all* $d|2m$ *the localization of* $\mathbb{H}_d$ *at some finite prime divisor does not split, then* t_Γ *is injective for* $\Gamma = H_{4m}$.

The converse is also true if we restrict ourselves to groups H_{2^n} and H_{4l^r}, the only groups for which we have complete results. As we shall see below, this converse is not true in general. A more precise question to ask will refer to the injectivity or otherwise of $t''_{d,l}$. Note that (cf. Proposition 4.4) if $l \equiv 1 \pmod 4$, then -1 is congruent to a totally positive unit modulo the prime ideal above l in $\mathbb{Q}^+(2l^j)$. This congruence can be lifted to bigger fields, and therefore, by Proposition 4.2, $t''_{d,l}$ will always be null. Analogously for $t''_{d,2}$, for $8|d$.

For the construction of examples we have thus necessarily to consider primes $\equiv -1 \pmod 4$, and allow m to involve two distinct primes. We shall give two examples.

First let l_1, l_2 be distinct primes $\equiv -1 \pmod 4$. We assume that

(a) $$\left(\frac{l_1}{l_2}\right) = 1 = -\left(\frac{l_2}{l_1}\right),$$

and take $m = l_1^{s_1} l_2^{s_2}$, $s_i > 0$. We get

III. *For any* d *of form* $2l_1^{r_1} l_2^{r_2}$, *with* $r_1 > 0$, $r_2 > 0$, t''_{d,l_1} *is null,* t''_{d,l_2} *is injective.*

Proof. Both l_1 and l_2 are ramified in the quadratic subfield $\mathbb{Q}(\sqrt{l_1 l_2})$ of $\mathbb{Q}^+(d)$. Denote by $\mathfrak{l}_i$ and $\mathfrak{L}_i$ the products of prime ideals above l_i in these two fields respectively. Then -1 will be congruent to a totally positive unit in $\mathbb{Q}^+(d) \bmod \mathfrak{L}_i$ if and only if it is so in $\mathbb{Q}(\sqrt{l_1 l_2})$ modulo $\mathfrak{l}_i$. This follows from the fact that $[\mathbb{Q}^+(d) \colon \mathbb{Q}(\sqrt{l_1 l_2}]$ is odd.

Let then ε be the fundamental unit > 1 of $\mathbb{Q}(\sqrt{l_1 l_2})$. ε is totally positive. If $(-\varepsilon/\mathfrak{l}_i) = 1$ (quadratic residue symbol), then $(\varepsilon/\mathfrak{l}_i) = -1$, i.e., $\varepsilon^{(l_i - 1)/2} \equiv -1 \pmod{\mathfrak{l}_i}$. Conversely if $-1 \equiv \varepsilon^r \pmod{\mathfrak{l}_i}$ then r must be odd, and hence $1 = ((-\varepsilon)^r/\mathfrak{l}_i) = (-\varepsilon/\mathfrak{l}_i)$. Thus, by Proposition 4.2, we have to show that

(b) $$\left(\frac{-\varepsilon}{\mathfrak{l}_1}\right) = 1 = \left(\frac{\varepsilon}{\mathfrak{l}_2}\right).$$

The field $F = \mathbb{Q}(\sqrt{l_1 l_2}, \sqrt{-l_1})$ is non-ramified over $\mathbb{Q}(\sqrt{l_1 l_2})$ except at infinity, and the class number of $\mathbb{Q}(\sqrt{l_1 l_2})$ is odd. Thus $\mathfrak{l}_1 = (\lambda)$, $\lambda \in \mathbb{Q}(\sqrt{l_1 l_2})$, and $-l_1 = \mu\lambda^2$, μ a unit in $\mathbb{Q}(\sqrt{l_1 l_2})$. As F is imaginary, $\mu = -\varepsilon^q$, and as 2 is non-ramified in F, i.e., $F \neq \mathbb{Q}(\sqrt{l_1 l_2}, \sqrt{-1})$, the exponent q must be odd, i.e., may be taken as 1. In other words $F = \mathbb{Q}(\sqrt{l_1 l_2}, \sqrt{-\varepsilon})$. Now $(-\varepsilon/\mathfrak{l}_i) = 1$ precisely if $\mathfrak{l}_i$ splits in F, i.e., l_i splits in $\mathbb{Q}(\sqrt{-l_j})$ $(1 \leqslant i, j \leqslant 2, i \neq j)$, i.e., $(l_i/l_j) = 1$. Thus, by (a), we have (b) and hence III.

Next let $l \equiv -1 \pmod 4$ and consider $d = 4l^j$ $(j \geqslant 1)$. We have

IV. $t''_{d,l}$ *is injective or null according to whether* $(2/l) = 1$ or -1.

Proof. We now consider the field $\mathbb{Q}(\sqrt{l})$. Let ε be its fundamental unit > 1. Again $[\mathbb{Q}^+(d): \mathbb{Q}(\sqrt{l})]$ is odd, and the same argument as for III reduces the proof to showing that

$$\text{(c)} \qquad \left(\frac{2}{l}\right) = \left(\frac{\varepsilon}{\mathfrak{l}}\right),$$

where $\mathfrak{l}$ is the prime ideal above l in $\mathbb{Q}(\sqrt{l})$. For, we clearly have $(-\varepsilon/\mathfrak{l}) = -(\varepsilon/\mathfrak{l})$.

To establish (c) we put $F = \mathbb{Q}(\sqrt{l}, \sqrt{2})$ and show that $F = \mathbb{Q}(\sqrt{l}, \sqrt{\varepsilon})$, which yields (c) by the same type of argument as that used before.

From IV we can deduce yet another case when t_Γ is injective.

V. *Let* $\Gamma = H_{8l^r}$, $r \geqslant 1$, $l \equiv -1 \pmod 4$, $(2/l) = 1$. *Then* t_Γ *is injective.*

III. Resolvents, Galois Gauss Sums, Root Numbers, Conductors

§1. Preliminaries

The main purpose of this chapter is the proof of Theorems 5 and 7 and of Proposition I.6.2. We shall proceed essentially by reduction to local fields. On the way we shall then have to derive results, on resolvents and Galois Gauss sums etc., which are of independent interest.

The present section serves to set up the formal apparatus which will be used subsequently. Hitherto we have managed with a rather straightforward approach to group characters associated with Galois extensions, but now it has become imperative to put up with the tedium involved in a more formalized point of view.

Let in the sequel F be a number field or a local field. We shall denote by F^t its maximal tame extension in F^c; for F Archimedean this is to mean $F^t = F^c$. We fix once and for all a Galois extension $\tilde{F}$ of F, which for applications will be either F^c or F^t, and we shall be concerned with the set $\mathscr{F}$ of finite Galois extensions E of F, with $E \subset \tilde{F}$. For the moment the symbol Ω will stand for $\mathrm{Gal}(\tilde{F}/F)$. We also fix an algebraically closed field A of characteristic zero, the field over which group representations are to be considered. If F is a number field then the only relevant choice for A is $A = \mathbb{Q}^c = F^c$. If, however, F is local, then we have to consider both $A = F^c$, and $A = \mathbb{Q}^c$. For $E \in \mathscr{F}$, $\Gamma = \mathrm{Gal}(E/F)$, we shall now write $R_{E/F} = R_\Gamma = R_\Gamma(A)$; A will not be indicated in our notation unless there is danger of confusion. We shall also consider representations $T: \Omega \to \mathrm{GL}_m(A)$ with open kernel, i.e., factorizing through a quotient map $\Omega \to \mathrm{Gal}(E/F)$ $(E \in \mathscr{F})$ of Galois groups, and their associated characters, and denote the additive group of corresponding virtual characters by $R_{(F)}$. (Ring properties of the groups of virtual characters and their maps will be neglected here.) Inflation of characters from $R_{E/F}$ to $R_{(F)}$ is an injective homomorphism. Alternatively $R_{(F)}$ is the direct limit of the direct system of groups $R_{E/F}$ $(E \in \mathscr{F})$ and inflation maps $R_{E/F} \to R_{E'/F}$, coming from the natural quotient maps $\mathrm{Gal}(E'/F) \to \mathrm{Gal}(E/F)$ $(E' \supset E)$ of Galois groups. From now on we shall view the $R_{E/F}$ as embedded in $R_{(F)}$. The elements of $R_{(F)}$ will be called (virtual) *Galois characters* of F, if $\tilde{F} = F^c$, and *tame* (virtual) *Galois characters* of F, if $\tilde{F} = F^t$. Analogous definitions can be made, and the subsequent discussion applies – mutatis mutandis – if we restrict ourselves to symplectic characters; we then use the notation $R^s_{E/F}$ and $R^s_{(F)}$.

An arithmetic character invariant of the type of interest to us will usually be given in the first place as a family $g(E/F, \cdot)$ $(E \in \mathscr{F})$ of homomorphisms of the $R_{E/F}$

into a fixed Abelian group G; we shall write the images as $g(E/F, \chi)$. If this family is compatible with the given direct system $\{R_{E/F}\}_E$, i.e., if $g(E/F, \chi) = g(E'/F, \chi')$, whenever $E' \supset E$, χ' the image of χ under the inflation map $R_{E/F} \to R_{E'/F}$, then there is a unique homomorphism $g: R_{(F)} \to G$, so that $g(E/F, \cdot)$ is its restriction to $R_{E/F}$, for all $E \in \mathscr{F}$. We then call the family $g(E/F, \cdot)$ *inflation invariant* and say that it *defines the homomorphism* g *on Galois characters* (when $\tilde{F} = F^c$), or *on tame Galois characters* (when $\tilde{F} = F^t$), respectively.

We shall now connect up our present discussion with that in I §5. (Recall in particular (I.5.15)). The field $\tilde{F}$ remains unchanged if we replace F by any field $L \in \mathscr{F}$. We shall write $\mathscr{F}_L$ for the set of fields E, $E \subset \tilde{F}$ with E a finite Galois extension of L. We now consider firstly a family $h(L, \cdot)$ $(L \in \mathscr{F})$ of homomorphisms from $R_{(L)}$ into a fixed Abelian group G, writing again $h(L, \chi)$ for the images. Secondly we also consider a double parameter family $g(E/L, \cdot)$ of homomorphisms $R_{E/L} \to G$, where $L \in \mathscr{F}$, $E \in \mathscr{F}_L$. We are interested in the validity of the equations

(1.1.a) $$h(L, \chi) = h(M, \operatorname{ind} \chi)$$

where $M, L \in \mathscr{F}, M \subset L$, and $\operatorname{ind}: R_{(L)} \to R_{(M)}$ is induction.

In the formal discussion of induction in II §3 we worked inside a finite group. It is however sufficient to assume that we induce from a subgroup of finite index, in our case from $\operatorname{Gal}(\tilde{F}/L)$ to $\operatorname{Gal}(\tilde{F}/M)$. Similarly we are interested in the validity of the equations

(1.1.b) $$g(E/L, \chi) = g(E/M, \operatorname{ind} \chi)$$

where $M, L \in \mathscr{F}, M \subset L, E \in \mathscr{F}_M \cap \mathscr{F}_L$ and $\operatorname{ind}: R_{E/L} \to R_{E/M}$ is again induction. If (1.1.a) holds for all L, M, χ (if (1.1.b) holds for all E, L, M, χ, respectively) we call the family $h(L, \cdot)$ (the family $g(E/L, \cdot)$, respectively) *fully inductive*. If these equations are at least valid under the further restriction that $\deg(\chi) = 0$, then we call the above families of homomorphisms *inductive in degree zero*. (For the $g(E/L, \cdot)$ this restates the definition already given in I §5.)

1.1. Lemma. *Suppose that* (i) *the family* $g(E/L, \cdot)$ *is inductive in degree zero and* (ii) *for all* L, E, E' *with* $L \in \mathscr{F}, E, E' \in \mathscr{F}_L, E' \supset E$, *and all Abelian characters* $\eta \in R_{E/L}$, *and* $\eta' = \inf \eta \in R_{E'/L}$ *the image of* η *under inflation, we have* $g(E/L, \eta) = g(E'/L, \eta')$ (*i.e.*, $g(E/L, \cdot)$ *is inflation invariant for Abelian characters*). *Then for each* $L \in \mathscr{F}$ *the family* $g(E/L, \cdot)$ $(E \in \mathscr{F}_L)$ *is inflation invariant.*

Proof. This follows from a strengthened version of Brauer's induction theorem (cf. [Se2] (Ex 2, p. 96)). This asserts that if χ is a (virtual) character of a finite group Γ, then

$$\chi - \deg(\chi)\varepsilon = \sum_i n_i \operatorname{ind}_{\Delta_i}^{\Gamma}(\eta_i - \varepsilon_i)$$

where the Δ_i are subgroups of Γ, η_i is an Abelian character of Δ_i, ε_i its identity character and ε the identity character of Γ. If now Γ is a quotient of a finite group Γ', with Δ_i' the inverse image of Δ_i and if α' is always the image of α under inflation, then

$$\chi' - \deg(\chi')\varepsilon' = \sum_i n_i \operatorname{ind}_{\Delta_i'}^{\Gamma'}(\eta_i' - \varepsilon_i').$$

Suppose then that for each finite group Σ, h_Σ is a homomorphism $R_\Sigma \to G$, and that h_Σ is inductive in degree zero and inflation invariant for Abelian characters. Then one verifies quickly that indeed $h_{\Gamma'}(\chi') = h_\Gamma(\chi)$. □

The proof of the last lemma is based on a principle, which we shall keep on using – without any further reference:

Let $h(\Delta, \cdot), h'(\Delta, \cdot)$ be families of homomorphisms $R_\Delta \to G$, defined for all subgroups Δ of a finite group Γ, which coincide on Abelian characters and are inductive on degree zero. Then they coincide. □

1.2. Lemma. *Suppose that* (i) *for each $L \in \mathscr{F}$ the family $g(E/L, \cdot)$ ($E \in \mathscr{F}_L$) is inflation-invariant* (*and thus has as direct limit a homomorphism g_L: $R_{(L)} \to G$*), *and* (ii) *the family $g(E/L, \cdot)$ is fully inductive* (*resp. inductive in degree zero*). *Then g_L is fully inductive* (*resp. inductive in degree zero*).

Proof. Obvious. □

We shall now give some examples. These are all well known. In each case the field A, over which representations are defined, is $\mathbb{Q}^c$ – irrespective of whether F is global or local. In all these examples we work with arbitrary Galois characters over the base field – no restriction of tame ramification is imposed (i.e., we choose $\tilde{F} = F^c$ etc.).

1) F is a local field, $\tilde{L}(s, E/F, \chi)$ the local L-function (cf. I (5.2), (5.4), (5.16), (5.18)). In the Archimedean case both inflation invariance and full inductivity can be read off from the definitions. So we now look at non-Archimedean fields. The map $\chi \mapsto \mathrm{n}\chi$ (non-ramified part, as defined in I §5, in particular (I.5.17)) is defined in a natural manner on the whole of $R_{(F)}$, irrespective of any intermediate field E. This implies then directly the inflation invariance of the L-function. Full inductivity is a consequence of Proposition 1.3, which follows later in this section. See note [1].

For an alternative approach to this problem, we look at the class field theoretic map $\chi \mapsto \theta_\chi$ from Abelian characters of Galois groups over F to multiplicative characters of F (see I §5). By the definition of θ_χ and the properties of the Artin symbol, $\theta_\chi = \theta_{\inf \chi}$. In other words we get a map $\chi \mapsto \theta_\chi$ from $R_{(F)}$ to the multiplicative characters of F. (In fact this is an isomorphism of multiplicative groups, and sums up an essential part of local class field theory.) As $\tilde{L}(s, E/F, \chi) = \tilde{L}(s, \theta_\chi)$ for χ Abelian, it now follows that the L-function is inflation invariant. By inductivity (see note [1] to Proposition 1.3) and by Lemma 1.2, we now get local L-functions $\tilde{L}(s, \chi)$ ($\chi \in R_{(F)}$), without reference to intermediate fields E. By Lemma 1.2, these are fully inductive.

2) Putting the local bits together we get the same results for global L-functions $\tilde{L}(s, \chi)$.

3) From the functional equation it follows that the Artin root numbers $W(\chi)$ may be viewed, for any given number field K, as a function of $R_{(K)}$. It is then fully inductive.

4) For the local root number $W(E/F, \chi)$ for Archimedean fields (cf. (I.5.16)) the inflation invariance can be read off from the definition. Thus also the root number

$W_\infty(N/K, \chi)$ at infinity for number fields (cf. (I.5.20)) has the same property. Both are seen to be inductive in degree zero.

5) Conductors (and their absolute norms) are inflation invariant. The global case follows from the local case, when the property can again be read off from the definition. We have actually stated the defining formula only for tame extensions (cf. (I.5.26)) – the case of interest to us – and in this case the verification is trivial.

6) By the preceding examples and the definition (I.5.22), the global Galois Gauss sum is inflation invariant, although our proof only covers the tame case.

For an alternative proof of the inflation invariance of the global Galois Gauss sum, we use both equations in (I.5.22). From the inflation-invariance of the global L-function and the functional equation we get that of both $A(N/K, \chi)$ and $W(N/K, \chi)$, hence that of $\mathbf{N}_K\mathfrak{f}(N/K, \chi)$ (via the first equation), and in conjunction with the same property of $W_\infty(N/K, \chi)$ this will suffice.

We shall from now on, and without further mention, write the character invariants, discussed above, as functions of Galois characters of the base field.

We now come to the induction properties of the non-ramified part $\mathrm{n}\chi$ of a Galois character χ (see (I.5.17)), no restriction of tameness being required here. We consider non-Archimedean local fields E, F with $E \supset F$. The derivation of the inductivity of local non-Archimedean L-functions is given in note [1].

Proposition 1.3. *Let* $\chi \in R_{(E)}$.

(i) $$\deg(\mathrm{nind}\,\chi) = f(E/F)\deg(\mathrm{n}\chi),$$

where ind *is the induction map* $R_{(E)} \to R_{(F)}$, *and* $f(E/F)$ *is the residue class degree.*

(ii) *Suppose* χ *is irreducible,* $\mathrm{n}\chi = 0$. *Then* $\mathrm{nind}\,\chi = 0$.

(iii) *Suppose* χ *is irreducible, non-ramified, i.e., Abelian and* $\chi = \mathrm{n}\chi$. *If* E/F *is non-ramified then* $\mathrm{nind}\,\chi = \mathrm{ind}\,\chi$. *If* E/F *is totally ramified then* $\mathrm{nind}\,\chi = \eta$ *where* (a) η *is Abelian,* (b) η *is non-ramified,* (c) $\mathrm{res}\,\eta = \chi$, res *the restriction map* $R_{(F)} \to R_{(E)}$, (d) η *is an irreducible component of* $\mathrm{ind}\,\chi$. *Properties* (b) *and* (d), *or* (b) *and* (c) *determine* η *uniquely.*

Proof. It will be convenient to work in the context of a Galois extension E'/F of finite degree, with $E' \supset E$, and write $\Gamma = \mathrm{Gal}(E'/F)$, $\Delta = \mathrm{Gal}(E'/E)$, with Γ_0, Δ_0 the inertia groups.

By checking on irreducible characters, one verifies that, for $\chi \in R_\Delta$,

$$\deg(\mathrm{n}\chi) = \langle \mathrm{res}^{\Delta}_{\Delta_0}\chi, \varepsilon_{\Delta_0}\rangle_{\Delta_0}, \tag{1.1}$$

where ε_{Δ_0} is the identity character on Δ_0 (i.e., $\varepsilon_{\Delta_0}(\delta) = 1$ for all $\delta \in \Delta_0$), and where $\langle\ ,\ \rangle_{\Delta_0}$ is the standard scalar product on R_{Δ_0}, so that for any virtual character ψ and irreducible character ϕ, $\langle\psi, \phi\rangle_{\Delta_0}$ is the multiplicity of ϕ in ψ. Applying the above formula to Γ, we get

$$\begin{aligned}\deg(\mathrm{nind}^{\Gamma}_{\Delta}\chi) &= \langle \mathrm{res}^{\Gamma}_{\Gamma_0}\mathrm{ind}^{\Gamma}_{\Delta}\chi, \varepsilon_{\Gamma_0}\rangle_{\Gamma_0}\\ &= \langle \chi, \mathrm{res}^{\Gamma}_{\Delta}\mathrm{ind}^{\Gamma}_{\Gamma_0}\varepsilon_{\Gamma_0}\rangle_{\Delta},\end{aligned}$$

by applying Frobenius reciprocity twice. But, by direct verification, or by Mackey's formula,

$$\begin{aligned}\operatorname{res}_\Delta^\Gamma \operatorname{ind}_{\Gamma_0}^\Gamma \varepsilon_{\Gamma_0} &= ([\Gamma:\Delta]/[\Gamma_0:\Delta_0]) \operatorname{ind}_{\Delta_0}^\Delta \varepsilon_{\Delta_0} \\ &= f(E/F) \operatorname{ind}_{\Delta_0}^\Delta \varepsilon_{\Delta_0}.\end{aligned}$$

Thus

$$\begin{aligned}\deg(\operatorname{nind}_\Delta^\Gamma \chi) &= f(E/F)\langle \chi, \operatorname{ind}_{\Delta_0}^\Delta \varepsilon_{\Delta_0}\rangle_\Delta \\ &= f(E/F)\langle \operatorname{res}_{\Delta_0}^\Delta \chi, \varepsilon_{\Delta_0}\rangle_{\Delta_0},\end{aligned}$$

which by comparison with (1.1) yields (i).

Applying (i) to (ii), we see that $\deg(\operatorname{nind}\chi) = 0$. But $\operatorname{nind}\chi$ is a genuine character or $= 0$. Hence $\operatorname{nind}\chi = 0$.

Next we come to (iii). If E/F is non-ramified then, by (i),

$$\deg(\operatorname{nind}\chi) = [E:F]\deg(\chi) = \deg(\operatorname{ind}\chi).$$

As $\operatorname{ind}\chi = \operatorname{nind}\chi + \psi$, ψ an actual character or $\psi = 0$, we get $\operatorname{ind}\chi = \operatorname{nind}\chi$.

Now suppose that E/F is totally ramified. Then $\operatorname{nind}\chi = \eta$ is an actual character, and by (i) is Abelian. It is an irreducible component of $\operatorname{ind}\chi$ and it is by definition non-ramified. Also

$$\langle \chi, \operatorname{res}_\Delta^\Gamma \eta\rangle_\Delta = \langle \operatorname{ind}\chi, \eta\rangle_\Gamma = 1,$$

i.e., $\operatorname{res}_\Delta^\Gamma \eta = \chi$. Uniqueness: If $\operatorname{res}_\Delta^\Gamma \eta' = \chi$ (whence η' is Abelian) then $\eta' = \phi\eta$, ϕ Abelian, $\operatorname{res}_\Delta^\Gamma \phi = \varepsilon_\Delta$. If η' is non-ramified, so is ϕ, hence as E/F is totally ramified, $\phi = \varepsilon_\Gamma$, i.e., $\eta' = \eta$. Thus (b) and (c) determine η uniquely. For (b) and (d) the same is true as by the preceding argument (b) and (d) imply that $\eta = \operatorname{nind}\chi$. □

Now we come to the real work of this section, dealing with resolvents. Here we are breaking new ground. In this context the coefficient field A for group representations will always be the algebraic closure of the fields involved in Galois extensions, i.e., $A = \mathbb{Q}^c$, if we are looking at Galois extensions of number fields, and $A = \mathbb{Q}_p^c$ (or $\mathbb{C}$) for Galois extensions of local fields (recall that for the study of L-functions, etc. of local fields, A had still to be $\mathbb{Q}^c$). Moreover we shall now restrict ourselves to tame Galois characters, i.e., the field $\tilde{F}$ in our original discussion is F^t.

Let first E/F be a tame Galois extension of local fields with Galois group Γ, $F \supset \mathbb{Q}_p$, p finite. If a and b are both free generators of $\mathfrak{o}_E$ over $\mathfrak{o}_F\Gamma$, then $b = a^\lambda$, $\lambda \in \mathfrak{o}_F\Gamma^*$. By the Corollary to Proposition I.4.2, we get, for all $\chi \in R_{E/F} = R_{E/F}(\mathbb{Q}_p^c)$, the equation

$$(b \mid \chi) = (a \mid \chi)\operatorname{Det}_\chi(\lambda).$$

As $\operatorname{Det}_\chi(\lambda) \in \mathfrak{U}(\mathbb{Q}_p^c)$, the class $P(E/F, \chi)$ of $(a \mid \chi) \bmod \mathfrak{U}(\mathbb{Q}_p^c)$, i.e., the fractional ideal generated by $(a \mid \chi)$ is independent of choices. We thus obtain a family

$$P(E/F, \cdot)\colon R_{E/F} \to \mathfrak{J}(\mathbb{Q}_p^c) \tag{1.2}$$

of homomorphisms. We define a further such homomorphism $\mathcal{N}_{F/\mathbb{Q}_p}P(E/F,\cdot)$ with values

$$\mathcal{N}_{F/\mathbb{Q}_p}P(E/F,\chi) = \prod_\sigma P(E/F,\chi^{\sigma^{-1}})^\sigma, \tag{1.3}$$

where $\{\sigma\}$ is a right transversal of Ω_F in $\Omega_{\mathbb{Q}_p}$. This is the fractional ideal in $\mathbb{Q}_p^c$ generated by the element

$$\mathcal{N}_{F/\mathbb{Q}_p}(a\,|\,\chi) = \prod_\sigma (a\,|\,\chi^{\sigma^{-1}})^\sigma, \tag{1.4}$$

with a as above. By Proposition I.4.4, the fractional ideal (1.3) is independent of the choice of σ. It need however not lie in $\mathfrak{J}(\mathbb{Q}_p)$.

We proceed analogously in the global situation. Here we consider a tame Galois extension N/K of number fields, with Galois group Γ. Let $\mathfrak{p}$ be a finite prime divisor of K. The resolvent $(a\,|\,\chi)$ $(\chi \in R_{N/K})$ of a free generator a of $\mathfrak{o}_{N,\mathfrak{p}}$ over $\mathfrak{o}_{K,\mathfrak{p}}\Gamma$ lies in $(\mathbb{Q}^c)_\mathfrak{p}^* \subset \mathfrak{J}(\mathbb{Q}^c)$, and its class $P_\mathfrak{p}(N/K,\chi)$ in $\mathfrak{J}(\mathbb{Q}^c)/\mathfrak{U}(\mathbb{Q}^c) = \mathfrak{J}(\mathbb{Q}^c)$ is independent of the choice of a. Moreover, by Proposition I.4.3, $P_\mathfrak{p}(N/K,\chi) = 1$, for almost all $\mathfrak{p}$, whence the product $P(N/K,\chi) = \prod_\mathfrak{p} P_\mathfrak{p}(N/K,\chi)$ is well defined. We thus get homomorphisms

$$\left.\begin{array}{l} P_\mathfrak{p}(N/K,\cdot): \\ P(N/K,\cdot): \end{array}\right\} R_{N/K} \to \mathfrak{J}(\mathbb{Q}^c), \tag{1.5}$$

and similarly

$$\left.\begin{array}{l} \mathcal{N}_{K/\mathbb{Q}}P_\mathfrak{p}(N/K,\cdot): \\ \mathcal{N}_{K/\mathbb{Q}}P(N/K,\cdot): \end{array}\right\} R_{N/K} \to \mathfrak{J}(\mathbb{Q}^c), \tag{1.5a}$$

with, for instance

$$\mathcal{N}_{K/\mathbb{Q}}P(N/K,\chi) = \prod_\sigma P(N/K,\chi^{\sigma^{-1}})^\sigma,$$

$\{\sigma\}$ being a right transversal of Ω_K in $\Omega_\mathbb{Q}$.

1.4. Proposition. *The families $P(E/F,\cdot)$, $\mathcal{N}_{F/\mathbb{Q}_p}P(E/F,\cdot)$ (local case) $P(N/K,\cdot)$, $P_\mathfrak{p}(N/K,\cdot)$, $\mathcal{N}_{K/\mathbb{Q}}P(N/K,\cdot)$, $\mathcal{N}_{K/\mathbb{Q}}P_\mathfrak{p}(N/K,\cdot)$ are inflation invariant for tame extensions and tame Galois characters.*

Accordingly we shall henceforth write $P(\chi)$, $P_\mathfrak{p}(\chi)$, $\mathcal{N}_{F/\mathbb{Q}_p}P(\chi)$ etc.

The proposition is an immediate consequence of the following Lemma stated here in the local case, and of its global analogue. □

1.5. Lemma. *Let E/F, L/F be tame Galois extensions of non-Archimedean local fields, with $L \supset E$. Let $\Gamma = \mathrm{Gal}(E/F)$, $\Sigma = \mathrm{Gal}(L/F)$. If a is a free generator of $\mathfrak{o}_L$ over $\mathfrak{o}_F\Sigma$ then $t_{L/E}a$ is a free generator of $\mathfrak{o}_E$ over $\mathfrak{o}_F\Gamma$, where $t_{L/E}$ is the trace. Also if $\chi \in R_{E/F}$ and χ' is its image under inflation in $R_{L/F}$, then $(a\,|\,\chi')_{L/F} = (t_{L/E}a\,|\,\chi)_{E/F}$.*

Proof. Immediate from the definitions. □

Remark. For the global analogue one considers a finite prime divisor $\mathfrak{p}$ of the base field K and free generators of, say, $\mathfrak{o}_{N,\mathfrak{p}}$ over the appropriate group ring $\mathfrak{o}_{K,\mathfrak{p}}\Gamma$ etc.

Resolvents also give rise to invariant signatures. The appropriate definitions, and the proofs that these lead to inflation-invariant objects will be contained in §4.

§2. Localization of Galois Gauss Sums and of Resolvents

Localization plays an essential role in the derivation of the arithmetic properties of Galois Gauss sums and of resolvents, and of their interrelations. In older versions of the proofs, as published in [F17], there was much more of a mixture of global and local aspects, while in our approach here they become more clearly separated, and this has the advantage of making the structure of the theory more transparent. It is also important for a proper understanding to realize, that we are actually considering simultaneously two distinct, but intertwined, localization procedures, with respect to two prime divisors, which – although lying above the same rational prime divisor – will in general be distinct. With some over-simplification we may say, that we are localizing global valued functions of global objects with respect both to their arguments, as well as to their values. The earlier section on localization, II §2, can serve as a useful introduction.

As we shall want to look simultaneously at representations over $\mathbb{Q}^c$ and over $\mathbb{Q}_p^c$, care will have to be taken now, to keep them notationally clearly apart. The following conventions and notations will be adopted. As in the preceding §1, representations of (infinite) Galois groups always have open kernel, and the additive group of virtual characters is to be understood in this sense. If F is a local field, or a number field, the symbol $R_{(F)}$ is now reserved for the group of virtual characters, as explained above, with respect to representations over $\mathbb{Q}^c$ of the group $\mathrm{Gal}(F^t/F)$, with F^t the maximal tame extension of F in F^c. If E/F is a Galois extension of finite degree, with $E \subset F^t$, the virtual characters for representations of $\mathrm{Gal}(E/F)$ over $\mathbb{Q}^c$ form a subgroup $R_{E/F}$ of $R_{(F)}$. If F is local, p is always the rational prime divisor with $\mathbb{Q}_p \subset F$. In this local case we also have to consider representations over $\mathbb{Q}_p^c$, and the respective groups of virtual characters will then be denoted by $R_{(F),p}$ and $R_{E/F,p}$.

Recall the local-global definitions of the conductor for the tame case (cf. (I.5.26), (I.5.27)) and of the root number of infinity (cf. (I.5.16), (I.5.20)). For both these, both localization procedures are transparent. The crux of the problem lies with the Galois Gauss sums and the resolvents. We shall first deal with the former.

Theorem 18. *Let F be a non-Archimedean local field. Then there exist homomorphisms $\chi \mapsto \tau(E,\chi)$ of $R_{(E)}$ into $\mathbb{Q}^{c*}$, for all $E \supset F$, E/F tame of finite degree, so that*

(i) *If $\chi \in R_{(E)}$ is Abelian, $\theta = \theta_\chi$ the corresponding multiplicative character of E, then $\tau(E,\chi) = \tau(\theta)$ is the Gauss sum of θ defined in* (I.5.5), (I.5.6).

(ii) *If* $E \supset E' \supset F, \chi \in R_{(E)}, \deg(\chi) = 0$ *and* $\operatorname{ind}\chi$ *denotes the induced character in* $R_{(E')}$ *then*

$$\tau(E, \chi) = \tau(E', \operatorname{ind}\chi) \qquad \textit{(Inductivity in degree zero)}.$$

No proof of this theorem will be given, except for an outline (in Chap. IV).
□

Remark 1. The theorem is actually true without assuming tame ramification, i.e., the field E may be any extension of finite degree and χ any Galois character. We have stated it in this restricted fashion in order to keep within our present framework and its conventions. In fact the proof we shall outline uses the tameness hypothesis, and is entirely in the spirit of our "tame" theory, fitting beautifully into the overall picture. See also Remark 5 below.

Remark 2. Tame Abelian characters $\chi \in R_{(E)}$ are either non-ramified, i.e., $\mathfrak{f}(\chi) = 1$ and

$$\tau(E, \chi) = \theta_\chi(\mathfrak{D}_E)^{-1}, \qquad \mathfrak{D}_E \text{ the different of } E/\mathbb{Q}_p,$$

or $\mathfrak{f}(\chi) = \mathfrak{p}_E$, and then

$$\tau(E, \chi) = \sum_{\substack{u \in \mathfrak{o}_E^* \\ u \bmod \mathfrak{p}_E}} \theta_\chi(uc^{-1})\psi_E(uc^{-1})$$

with $c \in E^*$, $(c) = \mathfrak{D}_E\mathfrak{p}_E$, and ψ_E the standard additive character, defined in I §5, prior to (5.5). (See (I.5.5) and (I.5.6).)

Remark 3. As, by Theorem 18 (i), $\tau(E, \varepsilon_E) = 1$ for the identity character ε_E (with $\varepsilon_E(\omega) = 1$ all ω), we get by Theorem 18 (ii), the induction formula

(2.1) $$\tau(E', \operatorname{ind}\chi) = \tau(E, \chi)\tau(E', \operatorname{ind}\varepsilon_E)^{\deg(\chi)}, \qquad \text{for all } \chi \in R_{(E)},$$

ind the induction map $R_{(E)} \to R_{(E')}$. By the strengthened version of Brauer's induction theorem, already quoted in the proof of Lemma 1.1 (cf. [Se2] (Ex. 2, p. 96)), the properties (i) and (ii) in Theorem 18 will determine the $\tau(E, \chi)$, all E, all χ, uniquely. Unless it is relevant to indicate the base field E we shall in the sequel write $\tau(E, \chi) = \tau(\chi)$. We shall refer to this as the *Galois Gauss sum.*

Remark 4. The Galois Gauss sum, the local root number (Langlands' constant) and the conductor for Galois characters χ of E are connected by the equation

(2.2) $$W(\chi) = \tau(\bar{\chi})/\mathbf{N}_E\mathfrak{f}(\chi)^{1/2},$$

in analogy to (I.5.8). Given the definition of the conductor, which is an easier problem, τ and W determine each other, i.e, the existence of the Galois Gauss sum with the properties (i) and (ii) of Theorem 18 is equivalent with the existence of a local root number $W(\chi) = W(E, \chi)$, $\chi \in R_{(E)}$, so that

(2.3) $$\begin{cases} \text{(i)} & W(E, \chi) = W(\theta_\chi) \text{ (as defined in (I.5.8)).} \\ \text{(ii)} & W(E', \operatorname{ind}\chi) = W(E, \chi) \text{ if } \deg(\chi) = 0 \text{ (notation as in the Theorem).} \end{cases}$$

The theorem is usually formulated in terms of the local constants $W(\chi)$. We prefer to use the Galois Gauss sum, for a number of reasons. Firstly, as we shall show later in this section,

(2.4) $$\mathbf{N}_E\mathfrak{f}(\chi)^{1/2} = |\tau(\chi)| = |\tau(\bar{\chi})|,$$

and thus $\tau(\bar{\chi})$ determines both $W(\chi)$ and $\mathbf{N}\mathfrak{f}(\chi)^{1/2}$, while the same is not true the other way round. Secondly, the basic arithmetic properties come naturally attached to the Galois Gauss sum, rather than the root number, and in the third place the deep connection with Galois module structure is formulated essentially in terms of $\tau(\chi)$ in the first place (Theorem 5).

Remark 5. The question of the existence of a local root number with the right properties was first raised in [Ha]. Dwork (cf. [Dw]) proved their existence modulo ± 1. The first complete, but unpublished proof is due to Langlands announced in [Ls], and this was entirely local, yielding in particular properties (2.3) (i) and (ii). Deligne (cf. [De1]) published a proof, using the existence of the inductive global root number and a version of this proof is contained in [Tt2]. In [Ma3] the local Galois Gauss sums are introduced via Eqs. (2.2), assuming the existence and properties of the local root numbers. This forms the basis for the essentially local method which we have now adopted. The proofs of Langlands and of Deligne work in full generality, without any tame hypothesis. A further proof, valid in the tame case only, and proceeding via Galois Gauss sums rather than root numbers, appears in [FT] as natural part of the wider arithmetic theory, and based on an internal characterization of Galois Gauss sums. We shall give an outline of this at the end of Chap. IV.

From Theorem 18 we now get the crucial local decomposition of global Galois Gauss sums and Artin root numbers. We consider number fields K together with their local completions $K_{\mathfrak{p}}$. The local Galois groups $\mathrm{Gal}(K^t_{\mathfrak{p}}/K_{\mathfrak{p}})$ map into $\mathrm{Gal}(K^t/K)$, uniquely to within conjugacy, giving rise to a unique homomorphism $R_{(K)} \to R_{(K_{\mathfrak{p}})}$; the image $\chi_{\mathfrak{p}}$ of χ is the local component. We then have the

Corollary to Theorem 18. *Let $\chi \in R_{(K)}$. For almost all $\mathfrak{p}$, $\tau(K_{\mathfrak{p}}, \chi_{\mathfrak{p}}) = 1$, and for the global Galois Gauss sum we have*

$$\tau(K,\chi) = \prod_{\mathfrak{p}} \tau(K_{\mathfrak{p}}, \chi_{\mathfrak{p}}) \qquad (\textit{product over all finite prime divisors of } K).$$

Similarly

$$W(K,\chi) = \prod_{\mathfrak{p}} W(K_{\mathfrak{p}}, \chi_{\mathfrak{p}}) \qquad (\textit{product over all prime divisors of } K).$$

Proof. Suppose $\chi \in R_{(L/K)}$. For almost all finite prime divisors $\mathfrak{p}$, the rational prime divisor p below $\mathfrak{p}$ is non-ramified in L. Hence $\chi_{\mathfrak{p}}$ is non-ramified, and $\mathfrak{D}_{K_{\mathfrak{p}}} = 1$, thus $\tau(K_{\mathfrak{p}}, \chi_{\mathfrak{p}}) = 1$. Now define for all number fields E, tame over K, and all $\chi \in R_{(E)}$, a function τ' by $\tau'(E,\chi) = \prod_{\mathfrak{p}} \tau(E_{\mathfrak{p}}, \chi_{\mathfrak{p}})$ (product over all finite prime divisors of E). Then τ' and the global Galois Gauss sum τ coincide on all pairs (E, χ)

with χ Abelian; they are both inductive in degree zero. Hence they coincide, by the uniqueness argument we have used before. Similarly for W. □

Next we need to establish the properties of the local conductor under induction. Although the result is quite general, we restrict ourselves to the tame case, using the definition (I.5.26):

$$\mathfrak{f}(E, \chi) = \mathfrak{p}_E^{\deg(\chi) - \deg(\mathrm{n}\chi)}, \qquad \chi \in R_{(E)}.$$

For the conductor of $\operatorname{ind} \chi \in R_{(F)}$ where $E \supset F$, we have by Proposition 1.3,

$$\mathfrak{f}(F, \operatorname{ind} \chi) = \mathfrak{p}_F^m, \qquad m = [E:F] \deg(\chi) - f \deg(\mathrm{n}\chi),$$

as $\deg(\operatorname{ind} \chi) = [E:F] \deg(\chi)$. Here $f = f(E/F)$ is the residue class degree. If $e = e(E/F)$ is the ramification index then $ef = [E:F]$, and so we have

$$m = f(\deg(\chi) - \deg(\mathrm{n}\chi)) + \deg(\chi)(f(e-1)).$$

Now

$$N_{E/F}\mathfrak{p}_E = \mathfrak{p}_F^f, \qquad \mathfrak{p}_F^{f(e-1)} = \mathfrak{d}(E/F) \qquad \text{(relative discriminant).}$$

Thus we now get

$$\mathfrak{f}(F, \operatorname{ind} \chi) = N_{E/F}\mathfrak{f}(E, \chi) \cdot \mathfrak{d}(E/F)^{\deg(\chi)}. \tag{2.5}$$

For the properties of local fields we have used, see [Se1] or [F5] – the only result which is not widely known is the stated formula for the discriminant of tame extensions. From (2.5) we conclude that

$$\mathbf{N}_F\mathfrak{f}(F, \operatorname{ind} \chi)^{1/2} = \mathbf{N}_E\mathfrak{f}(E, \chi)^{1/2}, \qquad \text{if} \qquad \deg(\chi) = 0. \tag{2.6}$$

Thus both $|\tau(E, \bar{\chi})|$ and $\mathbf{N}_E\mathfrak{f}(E, \chi)^{1/2}$ are inductive in degree zero. By a classical computation, one verifies that they coincide on Abelian characters (as shown in (I.5.7)). These two observations yield (2.4).

Now we shall turn to resolvents. For these the localization formula was derived in [F23]. We shall consider a tame Galois extension N/K of number fields with Galois group Γ, and a prime divisor $\mathfrak{p}$ of K. In the sequel the symbol p will denote the rational prime divisor below $\mathfrak{p}$. We choose a prime divisor $\mathfrak{q}$ of N, lying above $\mathfrak{p}$. We write

$$k: N \to N_{\mathfrak{q}}, \qquad x \mapsto x^k \qquad (N_{\mathfrak{q}} \subset \mathbb{Q}_p^c)$$

for the associated embedding. Let $\Gamma_{\mathfrak{q}} = \operatorname{Gal}(N_{\mathfrak{q}}/K_{\mathfrak{p}})$. We then get an embedding, again using the same symbol

$$k: \Gamma_{\mathfrak{q}} \to \Gamma, \qquad \gamma \mapsto k(\gamma),$$

the two maps being connected by the equation

$$(x^k)^\gamma = (x^{k(\gamma)})^k, \qquad x \in N, \qquad \gamma \in \Gamma_{\mathfrak{q}}. \tag{2.7}$$

In other words, we are viewing k as a morphism of "rings with group action".

We now consider all the prime divisors of N above $\mathfrak{p}$. Their completions will all be identified with $N_\mathfrak{q}$, with respect to different embeddings $i: N \to N_\mathfrak{q}$. More precisely, fixing $N_\mathfrak{q}$, and hence $K_\mathfrak{p}$, as subfields of $\mathbb{Q}_p^c$, the closure of the image of N in $\mathbb{Q}_p^c$ will coincide with $N_\mathfrak{q}$, for any embedding $N \to \mathbb{Q}_p^c$ which extends $K \to K_\mathfrak{p} \subset \mathbb{Q}_p^c$. Let I be a set of such embeddings, containing the "canonical" chosen k as above, and containing exactly one embedding corresponding to each prime divisor of N above $\mathfrak{p}$. For each $i \in I$, we get an algebra surjection

$$i \otimes 1: N \otimes_K K_\mathfrak{p} \to N_\mathfrak{q}.$$

The product map

(2.8) $$\prod_I (i \otimes 1): N \otimes_K K_\mathfrak{p} \to \prod_I N_\mathfrak{q}$$

(product of copies of $N_\mathfrak{q}$, indexed by I) is an isomorphism. Given an element $a \in N_\mathfrak{q}$, let

$$b = b(a) \in N \otimes_K K_\mathfrak{p}$$

be the unique element, with

(2.9) $$b^{k \otimes 1} = a, \qquad b^{i \otimes 1} = 0 \quad \text{if} \quad i \neq k, \qquad i \in I.$$

With the usual convention for infinite $\mathfrak{p}$, that $\mathfrak{o}_{K,\mathfrak{p}} = K_\mathfrak{p}$ etc. we now get

2.1. Proposition. *If a is a free generator of $\mathfrak{o}_{N,\mathfrak{q}}$ over $\mathfrak{o}_{K,\mathfrak{p}}\Gamma_\mathfrak{q}$, then $b(a)$ is a free generator of $\mathfrak{o}_{N,\mathfrak{p}} = \mathfrak{o}_N \otimes_{\mathfrak{o}_K} \mathfrak{o}_{K,\mathfrak{p}}$ over $\mathfrak{o}_{K,\mathfrak{p}}\Gamma$.*

Proof. (2.8) restricts to an isomorphism

(2.10) $$\mathfrak{o}_{N,\mathfrak{p}} \cong \prod_I \mathfrak{o}_{N,\mathfrak{q}}.$$

The set I corresponds to a left transversal $\{\delta\}$ of $k(\Gamma_\mathfrak{q})$ in Γ, where $x^i = x^{\delta k}$. Written in this way, the product $\prod_I \mathfrak{o}_{N,\mathfrak{q}}$ clearly has the structure of the $\mathfrak{o}_{K,\mathfrak{p}}$ Γ-module $\mathrm{Map}_{\Gamma_\mathfrak{q}}(\Gamma, \mathfrak{o}_{N,\mathfrak{q}})$, induced by the $\mathfrak{o}_{K,\mathfrak{p}}\Gamma_\mathfrak{q}$-module $\mathfrak{o}_{N,\mathfrak{q}}$, and (2.10) is an isomorphism of $\mathfrak{o}_{K,\mathfrak{p}}\Gamma$-modules. Reinterpreting $b(a)$ in these terms immediately gives the assertion. – Alternatively verify directly! □

Let, as before, p be the rational prime divisor below $\mathfrak{p}$, and let a and $b = b(a)$ be as in Proposition 2.1. Our aim is to compare norm resolvents of a over $\mathbb{Q}_p$ with the norm resolvents over $\mathbb{Q}$ we shall define with respect to the semi-local generator b. The latter is an idele and we shall have then to look at all its localizations above p. For this purpose we shall need a rather more careful description, than usual, of the maps connecting local multiplicative groups and ideles, in relation to Galois action. We shall work inside a big enough number field, to be denoted by E. We shall require all representations of Γ and of $\Gamma_\mathfrak{q}$ over $\mathbb{Q}^c$ to be realizable, to within equivalence, already over E, also $E/\mathbb{Q}$ to be Galois and E to contain N. Thus E will contain the values of the characters of Γ and of $\Gamma_\mathfrak{q}$. We shall write $\Sigma = \mathrm{Gal}(E/\mathbb{Q})$.

We also fix a "canonical" prime divisor $\mathfrak{P}$ of E, lying above $\mathfrak{q}$. (In the infinite case this may or may not be the one which comes from the originally given embedding of $\mathbb{Q}^c$ in $\mathbb{C}$.) We write $\Sigma_{\mathfrak{P}} = \mathrm{Gal}(E_{\mathfrak{P}}/\mathbb{Q}_p)$, where $E_{\mathfrak{P}}$ is the completion of E inside $\mathbb{Q}_p^c$. We extend the embedding k of fields, given above, to one of E in $E_{\mathfrak{P}}$, corresponding to $\mathfrak{P}$, and use the same symbol both for this and for the associated embedding $\Sigma_{\mathfrak{P}} \to \Sigma$, the two maps being again connected by (2.7), but now with $x \in E$ and $\gamma \in \Sigma_{\mathfrak{P}}$. For the reader's convenience we indicate the fields and Galois groups diagrammatically.

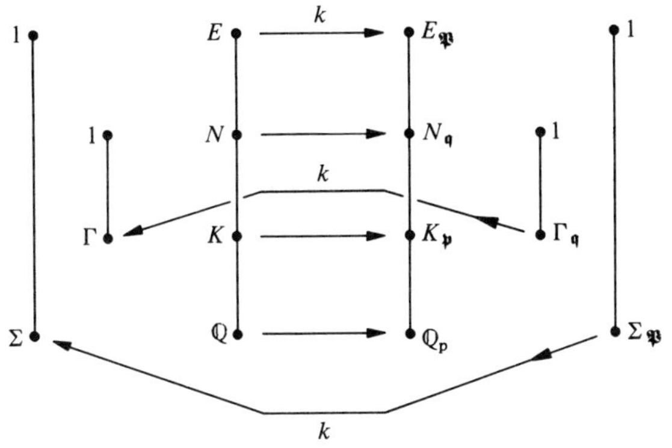

For any embedding $E \to \mathbb{Q}_p^c$, the closure of E coincides with $E_{\mathfrak{P}}$, i.e., factorizes through an embedding $j: E \to E_{\mathfrak{P}}$. We denote by $\mathfrak{P}_j$ the prime divisor of E associated with j. We now choose a set J of such embeddings in biunique correspondence with the prime divisors of E above p under the map $j \mapsto \mathfrak{P}_j$. We shall assume that $k \in J$, and that if $\mathfrak{P}_j$ lies above $\mathfrak{q}$ (respectively above $\mathfrak{p}$), then the restriction of j to N (respectively to K) coincides with that of k. We shall denote by J_0 the subset of J of embeddings $j: E \to E_{\mathfrak{P}}$, whose restriction to K is the same as that of k. Note that the distinct restrictions to N of the j in J_0 may be assumed to form the set I considered earlier. As in (2.8) we get an isomorphism

$$\prod_{J_0} (j \otimes 1): E \otimes_K K_{\mathfrak{p}} \cong \prod_{J_0} E_{\mathfrak{P}}.$$

Going over to multiplicative groups we get an isomorphism

$$r = \prod_{J_0} (j \otimes 1)^*: (E \otimes_K K_{\mathfrak{p}})^* \xrightarrow{\sim} \prod_{J_0} E_{\mathfrak{P}}^* \subset \mathfrak{J}(E), \tag{2.11}$$

and as indicated we shall view r at the same time as a map into the idele group $\mathfrak{J}(E)$. (Usually one will tend to view $(E \otimes_K K_{\mathfrak{p}})^*$ as a subgroup of $\mathfrak{J}(E)$, via any one of the many equally good injections – in the present context however the chosen map r enters into and affects the computations, and thus had to be introduced explicitly).

As usual the component of an idele x at a prime divisor $\mathfrak{R}$ will be denoted by $x_{\mathfrak{R}}$. Note that, for all j, the components $x_{\mathfrak{P}_j}$ lie in $E_{\mathfrak{P}}^*$. Moreover if $x \in E^*$ then $x_{\mathfrak{P}_j} = x^j$.

We shall now define the semi-local norm resolvent, with which the next theorem will be concerned. Let $\chi \in R_{N/K} \subset R_{(K)}$. Then, with $b \in N \otimes_K K_{\mathfrak{p}}$ as above,

$(b \mid \chi) \in (E \otimes_K K_{\mathfrak{p}})^*$, and $(b \mid \chi)^r \in \mathfrak{J}(E)$. Let $\{\omega\}$ be a right transversal of $\mathrm{Gal}(E/K)$ in Σ (or equivalently of Ω_K in $\Omega_{\mathbb{Q}}$). We define

(2.12) $$\mathcal{N}_{K/\mathbb{Q}}(b \mid \chi)^r = \prod_{\omega} (b \mid \chi^{\omega^{-1}})^{r\omega} \in \mathfrak{J}(E).$$

We shall be interested in the local components of this idele at the $\mathfrak{P}_j$. (Warning: $\mathcal{N}_{K/\mathbb{Q}}$ in (2.12) is not strictly a special case of the earlier definition in I §4, but is closely related to it.) Observe here also that, by its definition, b is attached to the particular prime divisor $\mathfrak{p}$ of K.

The original embedding $k: \Gamma_{\mathfrak{q}} \to \Gamma$ yields, by restricting χ, a character of $\Gamma_{\mathfrak{q}}$, which however only depends on $\mathfrak{p}$, and which we accordingly denote by $\chi_{\mathfrak{p}}$. Thus $\chi_{\mathfrak{p}} \in R_{N_{\mathfrak{q}}/K_{\mathfrak{p}}} \subset R_{(K_{\mathfrak{p}})}$, and

(2.13) $$\chi_{\mathfrak{p}}(\gamma) = \chi(k(\gamma)), \qquad \text{for} \quad \gamma \in \Gamma_{\mathfrak{q}}.$$

On the other hand if $j \in J$, then $\chi_{\mathfrak{p}}^j \in R_{(K_{\mathfrak{p}}),p}$. (Note that the values of any $\psi \in R_{N_{\mathfrak{q}}/K_{\mathfrak{p}}}$ will lie in E). With a as in Proposition 2.1, the resolvent $(a \mid \chi_{\mathfrak{p}}^j)$ is thus defined and lies in $E_{\mathfrak{P}}^*$. Via a right transversal $\{\sigma\}$ of $\mathrm{Gal}(E_{\mathfrak{P}}/K_{\mathfrak{p}})$ in $\Sigma_{\mathfrak{P}}$ we get the norm resolvent

(2.14) $$\mathcal{N}_{K_{\mathfrak{p}}/\mathbb{Q}_p}(a \mid \chi_{\mathfrak{p}}^j) = \prod_{\sigma} (a \mid \chi_{\mathfrak{p}}^{j\sigma^{-1}})^{\sigma} \in E_{\mathfrak{P}}^*.$$

For $\chi \in R_\Gamma$, the restriction of Det_χ to Γ, as a subset of group rings, is an Abelian character of Γ which, for distinction, we denote by $\det_\chi$ (see I §4).

On viewing $R_\Gamma = R_{N/K}$ as embedded in $R_{(K)}$, $\det_\chi$ is naturally defined within $R_{(K)}$.

Theorem 19 (cf. [F23]). *Let a be a free generator of $\mathfrak{o}_{N,\mathfrak{q}}$ over $\mathfrak{o}_{K,\mathfrak{p}}\Gamma_{\mathfrak{q}}$, and let $b = b(a)$ (cf. (2.9)). Let $j \in J, \chi \in R_{N/K}$. Then with definitions* (2.12)–(2.14),

$$(\mathcal{N}_{K/\mathbb{Q}}(b \mid \chi)^r)_{\mathfrak{P}_j} = \mathcal{N}_{K_{\mathfrak{p}}/\mathbb{Q}_p}(a \mid \chi_{\mathfrak{p}}^j) \det_\chi(\gamma)^j,$$

where $\gamma \in \Gamma$ is independent of χ (but dependent on j).

Note: This is an equation in $E_{\mathfrak{P}}^*$.

Proof. We begin by choosing a particular transversal W of $\mathrm{Gal}(E/K)$ in Σ, and subsequently show that in fact the choice of transversal is immaterial. We start with a given right transversal $W_{\mathfrak{P}}$ of $\mathrm{Gal}(E_{\mathfrak{P}}/K_{\mathfrak{p}})$ in $\Sigma_{\mathfrak{P}}$. Then $k(W_{\mathfrak{P}})$ is part of a right transversal of $\mathrm{Gal}(E/K)$ in Σ, namely that covering the product set $\mathrm{Gal}(E/K) \cdot k(\Sigma_{\mathfrak{P}})$, and we enlarge it to a right transversal W.

2.2. Lemma. *Let $x \in \mathrm{Im}\, r, \sigma \in \Sigma, (x^\sigma)_{\mathfrak{P}} \neq 1$ ($\mathfrak{P} = \mathfrak{P}_k$). Then $\sigma \in \mathrm{Gal}(E/K) \cdot k(\Sigma_{\mathfrak{P}})$.*

Proof of Lemma. We have

$$(x^\sigma)_{\mathfrak{P}} = (x_{\mathfrak{P}\sigma^{-1}})^\delta \qquad \text{for some} \quad \delta \in \Sigma_{\mathfrak{P}}.$$

Thus $x_{\mathfrak{P}\sigma^{-1}} \neq 1$. As ideles in $\operatorname{Im} r$ are non-trivial at most above $\mathfrak{p}$, we conclude that $\mathfrak{P}^{\sigma^{-1}}$ lies above $\mathfrak{p}$ i.e., is of form $\mathfrak{P}^{\rho}$ for some $\rho \in \operatorname{Gal}(E/K)$. Therefore $\mathfrak{P}^{\rho\sigma} = \mathfrak{P}$, i.e., $\rho\sigma \in k(\Sigma_{\mathfrak{P}})$, and $\sigma = \rho^{-1} \cdot \rho\sigma$ has the required form. □

To continue with the proof of Theorem 19, assume now that $W_{\mathfrak{P}}$ and W are used in the definition of the respective norm resolvents (cf. (2.12), (2.14)). We shall first of all show, specializing to the canonical embedding k, that

$$(2.15) \qquad (\mathcal{N}_{K/\mathbb{Q}}(b \mid \chi)^r)_{\mathfrak{P}} = \mathcal{N}_{K_{\mathfrak{p}}/\mathbb{Q}_p}(a \mid \chi_{\mathfrak{p}}^k).$$

By the Lemma, we have

$$(2.16) \qquad ((b \mid \chi^{\omega^{-1}})^{r\omega})_{\mathfrak{P}} = 1 \quad \text{if} \quad \omega \in W, \quad \omega \notin k(W_{\mathfrak{P}}).$$

On the other hand, if $\delta \in \Sigma_{\mathfrak{P}}$, then

$$(x^{k(\delta)})_{\mathfrak{P}} = (x_{\mathfrak{P}})^{\delta},$$

while, recalling the definition of r in (2.11),

$$(y^r)_{\mathfrak{P}} = y^{(k \otimes 1)} \qquad \text{for} \qquad y \in (E \otimes_K K_{\mathfrak{p}})^*.$$

Combining these equations we get, for $\omega \in W_{\mathfrak{P}}$,

$$(2.17) \qquad ((b \mid \chi^{k(\omega)^{-1}})^{rk(\omega)})_{\mathfrak{P}} = (b \mid \chi^{k(\omega)^{-1}})^{(k \otimes 1)\omega}.$$

We shall also show that, for $\omega \in W_{\mathfrak{P}}$,

$$(2.18) \qquad (b \mid \chi^{k(\omega)^{-1}})^{(k \otimes 1)} = (a \mid \chi_{\mathfrak{p}}^{k\omega^{-1}}).$$

From (2.16)–(2.18) we get (2.15).

To derive (2.18) we put $\chi^{k(\omega)^{-1}} = \phi$, taking this to be the character of a representation $T: \Gamma \to \mathrm{GL}_n(E)$. As remarked in the proof of Proposition 2.1, the chosen embeddings $i: N \to N_{\mathfrak{q}}$, forming the set I, are in biunique correspondence with a left transversal $\{\delta\}$ of $k(\Gamma_{\mathfrak{q}})$ in Γ. Summing over $\{\delta\}$ and over a variable $\gamma \in \Gamma_{\mathfrak{q}}$ we get,

$$\begin{aligned}
(b \mid \phi)^{(k \otimes 1)} &= \left(\operatorname{Det}\left(\sum_{\delta,\gamma} b^{\delta k(\gamma)} T(k(\gamma))^{-1} T(\delta)^{-1}\right)\right)^{k \otimes 1} \\
&= \operatorname{Det}\left(\left(\sum_{\delta,\gamma} b^{\delta k(\gamma)} T(k(\gamma))^{-1} T(\delta)^{-1}\right)^{k \otimes 1}\right) \\
&= \operatorname{Det}\left(\sum_{\delta,\gamma} b^{\delta(k \otimes 1)\gamma} (T(k(\gamma))^{-1} T(\delta)^{-1})^k\right) \\
&= \operatorname{Det}\left(\sum_{\gamma} a^{\gamma} T(k(\gamma)^{-1})^k\right) \qquad \text{(definition of } b \text{ in (2.9))} \\
&= (a \mid \phi_{\mathfrak{p}}^k),
\end{aligned}$$

which gives us (2.18), hence (2.15).

Next we shall prove that if we have a second right transversal W' of $\mathrm{Gal}(E/K)$ in Σ, and we use this to define a new norm resolvent, for the moment to be indicated by the symbol $\mathcal{N}'_{K/\mathbb{Q}}$, then

$$(\mathcal{N}'_{K/\mathbb{Q}}(b\,|\,\chi)^r)_{\mathfrak{P}} = (\mathcal{N}_{K/\mathbb{Q}}(b\,|\,\chi)^r)_{\mathfrak{P}} \operatorname{det}_\chi(\gamma')^k \tag{2.19}$$

for some $\gamma' \in \Gamma$, independent of χ.

We shall look at the factor in the product for the norm resolvent corresponding to the fixed coset $\mathrm{Gal}(E/K)\omega$, with $\omega \in W$. The corresponding representative in W' is of form $\rho\omega$, with $\rho \in \mathrm{Gal}(E/K)$. Write $\phi = \chi^{\omega^{-1}}$, to simplify notation. Then

$$(b\,|\,\phi^{\rho^{-1}})^{r\rho} = (b\,|\,\phi^{\rho^{-1}})^{\rho r} = (b\,|\,\phi)^r \operatorname{det}_\phi(\rho)^r,$$

by Proposition I.4.4. Thus

$$(b\,|\,\chi^{\omega^{-1}\rho^{-1}})^{r\rho\omega} = (b\,|\,\chi^{\omega^{-1}})^{r\omega} \operatorname{det}_\phi(\rho)^{r\omega}.$$

If first $\omega \notin k(\Sigma_{\mathfrak{P}})$, then by the Lemma $(\operatorname{det}_\phi(\rho)^{r\omega})_{\mathfrak{P}} = 1$. On the other hand, if $\omega = k(\sigma)$, $\sigma \in W_{\mathfrak{P}}$, then

$$(\operatorname{det}_\phi(\rho)^{r\omega})_{\mathfrak{P}} = (\operatorname{det}_\phi(\rho)^r)_{\mathfrak{P}}^\sigma = \operatorname{det}_\phi(\rho)^{k\sigma} = \operatorname{det}_\phi(\rho)^{\omega k} = \operatorname{det}_\chi(\rho)^k.$$

Replacing in the last expression, ρ by its image in Γ and collecting all factors we now obtain (2.19). Applying (2.19) twice we also see that it holds for any pair W, W' of transversals.

We have now proved the formula in the Theorem for the particular prime divisor $\mathfrak{P} = \mathfrak{P}_k$ (extending $\mathfrak{p}$). We need it however for an arbitrary prime divisor $\mathfrak{P}_j$ above p, where $j \in J$. Now observe that $j = \sigma^{-1} \circ k$ for some $\sigma \in \Sigma$, and $\mathfrak{P}_j = \mathfrak{P}^\sigma$, $x_{\mathfrak{P}_j} = (x^{\sigma^{-1}})_{\mathfrak{P}}$, for $x \in \mathfrak{J}(E)$. Thus if $\mathcal{N}_{K/\mathbb{Q}}$ is defined via some right transversal W, not necessarily the original one, then

$$\begin{aligned}(\mathcal{N}_{K/\mathbb{Q}}(b\,|\,\chi)^r)_{\mathfrak{P}_j} &= (\mathcal{N}_{K/\mathbb{Q}}(b\,|\,\chi)^{r\sigma^{-1}})_{\mathfrak{P}} \\ &= (\mathcal{N}'_{K/\mathbb{Q}}(b\,|\,\chi^{\sigma^{-1}})^r)_{\mathfrak{P}}\end{aligned} \tag{2.20}$$

with $\mathcal{N}'_{K/\mathbb{Q}}$ defined via $W\sigma^{-1}$. By (2.19) and the fact that

$$\operatorname{det}_{\chi^{\sigma^{-1}}}(\gamma)^k = \operatorname{det}_\chi(\gamma)^{\sigma^{-1}k} = \operatorname{det}_\chi(\gamma)^j,$$

we get

$$(\mathcal{N}_{K/\mathbb{Q}}(b\,|\,\chi)^r)_{\mathfrak{P}_j} = (\mathcal{N}_{K/\mathbb{Q}}(b\,|\,\chi^{\sigma^{-1}})^r)_{\mathfrak{P}} \operatorname{det}_\chi(\gamma)^j, \tag{2.21}$$

for some $\gamma \in \Gamma$. By (2.19) and (2.15) we now have

$$\begin{aligned}(\mathcal{N}_{K/\mathbb{Q}}(b\,|\,\chi^{\sigma^{-1}})^r)_{\mathfrak{P}} &= \mathcal{N}_{K_\mathfrak{p}/\mathbb{Q}_p}(a\,|\,\chi_\mathfrak{p}^{\sigma^{-1}k}) \operatorname{det}_{\chi^{\sigma^{-1}}}(\gamma_1)^k \\ &= \mathcal{N}_{K_\mathfrak{p}/\mathbb{Q}_p}(a\,|\,\chi_\mathfrak{p}^j) \operatorname{det}_\chi(\gamma_1)^j\end{aligned}$$

for some $\gamma_1 \in \Gamma$, and in conjunction with (2.21) this gives the required equation. □

Remark. It is clear that the last theorem is still valid if just the prime divisor $\mathfrak{p}$ of K is at most tamely ramified in N, and in particular if $\mathfrak{p}$ is infinite – irrespective of what happens elsewhere.

Corollary 1. *If the Abelian character* $\det_\chi$ *of* Γ *is trivial (i.e.,* $=\varepsilon_\Gamma$*), and in particular if* χ *is symplectic then, in the notation of Theorem 19,*

$$(\mathcal{N}_{K/\mathbb{Q}}(b\,|\,\chi)^r)_{\mathfrak{P}_j} = \mathcal{N}_{K_\mathfrak{p}/\mathbb{Q}_p}(a\,|\,\chi_\mathfrak{p}^j). \qquad \square$$

Next suppose that $\mathfrak{p}$ is a finite prime divisor. With $\chi \in R_{N/K}$ and $j \in J$, we consider the fractional ideal $\mathcal{N}_{K_\mathfrak{p}/\mathbb{Q}_p} P(N_\mathfrak{q}/K_\mathfrak{p}, \chi_\mathfrak{p}^j)$ in $\mathfrak{J}(\mathbb{Q}_p^c)$ – actually already in $\mathfrak{J}(E_\mathfrak{P})$, and the fractional ideal $\mathcal{N}_{K/\mathbb{Q}} P_\mathfrak{p}(N/K, \chi)$ in $\mathfrak{J}(\mathbb{Q}^c)$ – actually already in $\mathfrak{J}(E)$ (see (1.3) and (1.5a)). We get the crucial

Corollary 2.

$$(\mathcal{N}_{K/\mathbb{Q}} P_\mathfrak{p}(N/K, \chi))_{\mathfrak{P}_j} = \mathcal{N}_{K_\mathfrak{p}/\mathbb{Q}_p} P(N_\mathfrak{q}/K_\mathfrak{p}, \chi_\mathfrak{p}^j). \qquad \square$$

Recall that $\mathfrak{P}_j$ can be any prime divisor above p, and the $\mathfrak{P}_j$-component of a fractional ideal in $\mathfrak{J}(E)$ is viewed as lying in $\mathfrak{J}(E_{\mathfrak{P}_j}) = \mathfrak{J}(E_\mathfrak{P})$.

For completeness we note also the equation

(2.22) $\quad (\mathcal{N}_{K/\mathbb{Q}} P_\mathfrak{p}(N/K, \chi))_\mathfrak{R} = (1) \qquad$ if $\mathfrak{R}$ is a prime divisor not above p.

This is obvious, as in fact we have $(\mathcal{N}_{K/\mathbb{Q}}(b\,|\,\chi)^r)_\mathfrak{R} = 1$.

§3. Galois Action

In this section we determine the action of Galois groups on resolvents and on Galois Gauss sums. This turns out – globally – to be exactly the "same" in both situations, and thus we will obtain a proof of Theorem 5B. We shall see that what we get is a genuinely global result, although reached via localization. As far as the Galois Gauss sums are concerned, the simplicity of the Galois action formula is deceptive. This is one place, where global class field theory enters directly and significantly into our proofs, firstly via the local decomposition of global characters, secondly via the explicit idelic description of the rational Artin map and thirdly via the class field theoretic interpretation of the transfer map.

It is worth pointing out that the Theorem on Galois action (Theorem 20) is valid without any tameness restriction.

We shall also derive some rationality properties of symplectic Galois Gauss sums, which are however valid only in the tame case. These are of some importance in themselves, and also will imply the Galois invariance of the Artin root number, i.e., Proposition I.6.2.

By Proposition I.4.4, we already know how the Galois groups act on resolvents, and we shall extend this result to norm resolvents. Let E/F be a Galois extension of finite degree (of fields of characteristic zero) with Galois group Γ, and let F be an extension of finite degree of a field F_0, with B_0 a commutative F_0-algebra. If a is a free generator of $E \otimes_{F_0} B_0$ over $(F \otimes_{F_0} B_0)(\Gamma)$ we define the *norm resolvent* by

$$(3.1) \qquad \mathcal{N}_{F/F_0}(a \mid \chi) = \prod_{\sigma} (a \mid \chi^{\sigma^{-1}})^{\sigma},$$

where $\{\sigma\}$ is a right transversal of Ω_F in Ω_{F_0}. In various special cases, in particular for $F_0 = \mathbb{Q}$, or $F_0 = \mathbb{Q}_p$, this has already occurred regularly. We shall write v_{F/F_0} for the cotransfer map from Abelian characters of Ω_F to Abelian characters of Ω_{F_0}, i.e., the dual of the transfer $\Omega_{F_0}^{ab} \to \Omega_F^{ab}$. If $\chi \in R_\Gamma$, then $\det_\chi$ is an Abelian character, which we shall view as lifted to an Abelian character of Ω_F, via the surjection $\Omega_F \to \Gamma$ of Galois groups. Thus $\mathrm{v}_{F/F_0} \det_\chi$ is an Abelian character of Ω_{F_0}.

We shall state and prove the next theorem in two separate parts.

Theorem 20A (cf. [F13], [F17]). *Let a be a free generator of $E \otimes_{F_0} B_0$ over $(F \otimes_{F_0} B_0)(\Gamma)$, let $\chi \in R_\Gamma$ and $\omega \in \Omega_{F_0}$. Then*

$$[\mathcal{N}_{F/F_0}(a \mid \chi^{\omega^{-1}})]^{\omega} = \mathcal{N}_{F/F_0}(a \mid \chi)(\mathrm{v}_{F/F_0} \det_\chi)(\omega).$$

Proof. Let S be the right transversal of Ω_F in Ω_{F_0}, used to define $\mathcal{N}_{F/F_0}$. For $\omega \in \Omega_{F_0}$ and $\sigma \in S$ we can then write

$$\sigma\omega = \delta(\sigma, \omega)\overline{\sigma\omega}, \qquad \delta(\sigma, \omega) \in \Omega_F, \qquad \overline{\sigma\omega} \in S.$$

Thus

$$\begin{aligned}
\mathcal{N}_{F/F_0}(a \mid \chi^{\omega^{-1}})^{\omega} &= \prod_S (a \mid \chi^{\omega^{-1}\sigma^{-1}})^{\sigma\omega} \\
&= \prod_S (a \mid \chi^{\overline{\sigma\omega}^{-1}\delta(\sigma,\omega)^{-1}})^{\delta(\sigma,\omega)\overline{\sigma\omega}} \\
&= \prod_S (a \mid \chi^{\sigma^{-1}})^{\sigma} \cdot \prod_S \det{}_{\chi^{\overline{\sigma\omega}^{-1}}}(\delta(\sigma, \omega))^{\overline{\sigma\omega}} \qquad \text{(by Proposition I.4.4)} \\
&= \mathcal{N}_{F/F_0}(a \mid \chi) \cdot \prod_S \det{}_\chi(\delta(\sigma, \omega)).
\end{aligned}$$

As indeed $\prod_S \det_\chi(\delta(\sigma, \omega)) = (\mathrm{v}_{F/F_0} \det_\chi)(\omega)$, the result now follows. □

To discuss the problem for Galois Gauss sum we shall have to introduce some further notation. We let μ_{p^∞} be the group of p^n-th roots of unity, for all n, in $\mathbb{Q}^c$. The action of $\Omega_{\mathbb{Q}}$ on μ_{p^∞} defines a homomorphism

$$u_p \colon \Omega_{\mathbb{Q}} \to \mathbb{Z}_p^* = \mathfrak{U}(\mathbb{Q}_p) = \mathfrak{U}_p(\mathbb{Q}).$$

Explicitly

$$y = y^{\omega \cdot u_p(\omega)}, \qquad \text{for} \qquad \omega \in \Omega_{\mathbb{Q}}, \qquad y \in \mu_{p^\infty},$$

where the action of $\mathbb{Z}_p^* = \mathrm{Aut}(\mu_{p^\infty})$ on μ_{p^∞} is written exponentially. We shall also denote the multiplicative character of a local field F, corresponding to the Abelian character $\det_\chi$ of the Galois group Ω_F, by $\theta_{\det_\chi}$ rather than use iterated subscripts.

Theorem 20B. (i) *Let F be a non-Archimedean local field, $\chi \in R_{(F)}$ a character of Ω_F in $\mathbb{Q}^c$ and $\omega \in \Omega_{\mathbb{Q}}$. Then*

$$\tau(\chi^{\omega^{-1}})^{\omega} = \tau(\chi)\theta_{\det\chi}(u_p(\omega)).$$

(ii) *Let K be a number field, $\chi \in R_{(K)}$ and $\omega \in \Omega_{\mathbb{Q}}$. Then*

$$\tau(\chi^{\omega^{-1}})^{\omega} = \tau(\chi)(\mathrm{v}_{K/\mathbb{Q}}\det{}_{\chi})(\omega).$$

Remark. The global result, which is what we really require, was first established in [F13], see also [F17] of course. The first stage in the original proof was that of the formula for local Abelian Gauss sums and thence for global Abelian Gauss sums. This was carried out in [F7], see also [F9]. The general formula, as based on the Abelian results and on global inductivity properties was then suggested by Serre. The realization of the parallel to the behaviour of norm resolvents and the resulting Theorem 5B came to the author at a somewhat later stage.

The local part of the theorem is due to Martinet (cf. [Ma3]) (see also [FT]) and this of course opens a second path to the global formula, via Theorem 18 and its Corollary. This is the way we shall do it here.

Proof of Theorem 20B. We first consider Gauss sums for a multiplicative character θ of a non-Archimedean local field F (cf. (I.5.5), (I.5.6)). Suppose θ is ramified. If $x \in F$, $\omega \in \Omega_{\mathbb{Q}}$ then $\psi_F(x)^{\omega} = \psi_p(t_{F/\mathbb{Q}_p}x)^{\omega} = (e^{2\pi i m/p^r})^{\omega}$, where $t_{F/\mathbb{Q}_p}x \equiv m/p^r \pmod{\mathbb{Z}_p}$. But

$$(e^{2\pi i m/p^r})^{\omega} = (e^{2\pi i m/p^r})^{u_p(\omega)^{-1}} = \psi_p(u_p(\omega)^{-1}t_{F/\mathbb{Q}_p}x) = \psi_F(u_p(\omega)^{-1}x).$$

Therefore

$$\begin{aligned}\tau(\theta^{\omega^{-1}})^{\omega} &= \sum_u \theta(uc^{-1})(\psi_F(uc^{-1}))^{\omega}\\ &= \sum_u \theta(uc^{-1})\psi_F(u_p(\omega)^{-1}uc^{-1})\\ &= \theta(u_p(\omega))\sum_u \theta(u_p(\omega)^{-1}uc^{-1})\psi_F(u_p(\omega)^{-1}uc^{-1})\\ &= \theta(u_p(\omega))\tau(\theta).\end{aligned}$$

Via the map $\chi \mapsto \theta_\chi$ of class field theory, we have thus proved the formula in Theorem 20b (i) for any non-Archimedean local field and any ramified Abelian character χ. For the non-ramified characters the formula is in any case trivial. We shall show that both sides of the equation in Theorem 20b (i) are inductive in degree zero, and this implies the full validity of the equation. For the maps $\chi \mapsto \tau(\chi^{\omega^{-1}})^{\omega}$ and $\chi \mapsto \tau(\chi)$ we already have the required property. Let then $\chi \in R_{(E)}$, $F \subset E$, $\deg(\chi) = 0$. Then if $\operatorname{ind}\chi$ is the image under the induction map $R_{(E)} \to R_{(F)}$, we know that $\det_{\operatorname{ind}\chi} = \mathrm{v}_{E/F}\det_{\chi}$, as Abelian characters of Ω_F. On the other hand, one knows from class field theory, that if ϕ is an Abelian character of Ω_E, then

(3.2) $$\operatorname{res}_F^E \theta_\phi = \theta_{\mathrm{v}_{E/F}\phi},$$

where res_F^E is restriction from E^* to F^*. Thus

$$\theta_{\det \mathrm{ind}\,\chi}(u_p(\omega)) = \theta_{\mathrm{v}_{E/F}\det\chi}(u_p(\omega)) = \mathrm{res}_F^E\,\theta_{\det\chi}(u_p(\omega)) = \theta_{\det\chi}(u_p(\omega)),$$

which is the required result.

We now turn to the global case. The homomorphisms $u_p\colon \Omega_{\mathbb{Q}} \to \mathfrak{U}_p(\mathbb{Q})$ give rise to a product homomorphism onto the finite rational unit ideles,

$$u\colon \Omega_{\mathbb{Q}} \to \mathfrak{U}_{\mathrm{fin}}(\mathbb{Q}) = \prod_{p\,\mathrm{finite}} \mathfrak{U}_p(\mathbb{Q}),$$

so that the composition with the Artin map

$$A\colon \mathfrak{U}_{\mathrm{fin}}(\mathbb{Q}) \to \Omega_{\mathbb{Q}}^{ab}$$

is the canonical surjection $\Omega_{\mathbb{Q}} \to \Omega_{\mathbb{Q}}^{ab}$. In particular if $\phi \in R_{(\mathbb{Q})}$, ϕ Abelian then

$$\theta_\phi(u(\omega)) = \phi(\omega), \tag{3.3}$$

where θ_ϕ is now the idele class character, which corresponds to ϕ under global class field theory.

Now let $\chi \in R_{(K)}$ and apply the Corollary to Theorem 18. In what follows products run over finite prime divisors only. We get

$$\begin{aligned}\tau(\chi^{\omega^{-1}})^\omega \cdot \tau(\chi)^{-1} &= \prod_{\mathfrak{p}} \tau(\chi_{\mathfrak{p}}^{\omega^{-1}})^\omega \cdot \tau(\chi_{\mathfrak{p}})^{-1}\\ &= \prod \theta_{\det\chi_{\mathfrak{p}}}(u_p(\omega)) \quad \text{(always for } \mathfrak{p}\,|\,p\text{), by the local theorem,}\\ &= \prod (\theta_{\det\chi})_{\mathfrak{p}}(u_p(\omega)),\end{aligned}$$

as taking local components commutes with the maps $\chi \mapsto \det_\chi$ and $\phi \mapsto \theta_\phi$. Moreover the $\mathfrak{p}$-component $u(\omega)_{\mathfrak{p}}$ of $u(\omega)$ is $u_p(\omega)$. Thus $(\theta_{\det\chi})_{\mathfrak{p}}(u_p(\omega)) = (\theta_{\det\chi})_{\mathfrak{p}}(u(\omega))$. As the components of $u(\omega)$ at infinity are trivial, we get

$$\begin{aligned}\prod_{\mathfrak{p}} (\theta_{\det\chi})_{\mathfrak{p}}(u(\omega)) &= \theta_{\det\chi}(u(\omega))\\ &= (\mathrm{res}_{\mathbb{Q}}^K\,\theta_{\det\chi})(u(\omega))\\ &= (\theta_{\mathrm{v}_{K/\mathbb{Q}}\det\chi})(u(\omega)),\end{aligned}$$

by the global analogue to (3.2). By (3.3), we now get

$$\tau(\chi^{\omega^{-1}})^\omega \cdot \tau(\chi)^{-1} = (\mathrm{v}_{K/\mathbb{Q}}\det{}_\chi)(\omega)$$

as we had to show. □

From now on we consider symplectic characters.

3.1. Proposition. (i) *Let* $F = \mathbb{R}$ *or* $\mathbb{C}$, $\chi \in R_{(F)}^s$, $\omega \in \Omega_{\mathbb{Q}}$. *Then*

$$W(\chi) = W(\chi^\omega) = W(\chi)^\omega = \pm 1.$$

(ii) *Let K be a number field,* $\chi \in R^s_{(K)}$, $\omega \in \Omega_{\mathbb{Q}}$. *Then*

$$W_\infty(\chi) = W_\infty(\chi^\omega) = W_\infty(\chi)^\omega = \pm 1.$$

Proof. (i) This follows immediately from the definition (cf. (I.5.16)). If $\chi \in R^s_{(F)}$ then $\deg(\chi_-) \equiv 0 \pmod 2$. Thus, $W(\chi) = \pm 1$, and so $W(\chi)^\omega = W(\chi)$. Also $\deg((\chi^\omega)_-) = \deg(\chi_-)$, so $W(\chi^\omega) = W(\chi)$.

(ii) follows now from the definition of $W_\infty(\chi)$ (cf. (I.5.20)). □

Theorem 21. *Let F be a non-Archimedean local field,* $\chi \in R^s_{(F)}$ *a tame, symplectic character. Then*

(i) $\mathfrak{f}(\chi)$ *is the square of a fractional ideal in F.*

(ii) $\tau(\chi) \in \mathbb{Q}^*$.

(iii) $W(\chi) = \pm 1$, *and* $W(\chi^\omega) = W(\chi)$, *for all* $\omega \in \Omega_{\mathbb{Q}}$.

Remark. For χ symplectic and tame, we see that $\mathbf{N}_F\mathfrak{f}(\chi)$ is an even power of p, the residue class characteristic of F, and hence $\mathbf{N}_F\mathfrak{f}(\chi)^{1/2} = p^m$ and $\tau(\chi) = \pm p^m$, with $W(\chi) = \operatorname{sign} \tau(\chi)$. We may thus view $\tau(\chi)$ as an element, in fact a unit, in any ring in which p is a unit.

Proof of Theorem 21. View χ as a character of $\mathrm{Gal}(E/F) = \Gamma$, for some tame Galois extension E/F of finite degree. Let σ be a generator of the inertia group Γ_0. We may take χ as the character of an actual representation T. Then $\deg(\mathrm{n}\chi)$ is the number of eigenvalues $= 1$ of $T(\sigma)$, and we write m for the number of its eigenvalues $= -1$. As χ is real valued, non-real eigenvalues occur in complex conjugate pairs, so in the first place their number $\deg(\chi) - \deg(\mathrm{n}\chi) - m$ is even, and in the second place their product contributes 1 to $\det_\chi(\sigma)$. Thus $\det_\chi(\sigma) = (-1)^m$. But $\det_\chi = \varepsilon$, hence m is even. Thus $\deg(\chi) - \deg(\mathrm{n}\chi)$ is even, i.e., $\mathfrak{f}(\chi)$ is a square, by its definition (cf. (I.5.26)). It follows that

$$\text{(3.4)} \quad \mathbf{N}_F\mathfrak{f}(\chi)^{1/2} \in \mathbb{Q}^*, \quad (\mathbf{N}_F\mathfrak{f}(\chi)^{1/2})^\omega = \mathbf{N}_F\mathfrak{f}(\chi)^{1/2} = \mathbf{N}_F\mathfrak{f}(\chi^\omega)^{1/2}, \quad (\omega \in \Omega_{\mathbb{Q}})$$

the last equation holding, as clearly $\mathfrak{f}(\chi^\omega) = \mathfrak{f}(\chi)$. By Theorem 20B, and as $\det_\chi = \varepsilon$,

$$\text{(3.5)} \qquad \tau(\chi^{\omega^{-1}})^\omega = \tau(\chi).$$

Choose now an element $\sigma \in \Omega_{\mathbb{Q}}$ with $y^\sigma = y^{-1}$ for all roots of unity, e.g. the restriction of complex conjugation to $\mathbb{Q}^c$. By the definition (2.2) of $W(\chi)$ and as χ is real valued, i.e., $\chi^\sigma = \chi$, we conclude that $W(\chi)^\sigma = W(\chi)$, i.e., $W(\chi)$ is totally real. By (2.2) and (2.4), $|W(\chi)| = 1$. Therefore

$$W(\chi) = \pm 1, \qquad W(\chi)^\omega = W(\chi) \qquad \text{for all } \omega.$$

By (3.4) and (2.2), $\tau(\chi) = \mathbf{N}_F\mathfrak{f}(\chi)^{1/2} \cdot W(\chi) \in \mathbb{Q}^*$. Finally, by (2.2),

$$\begin{aligned} W(\chi^{\omega^{-1}})/W(\chi) &= W(\chi^{\omega^{-1}})^\omega/W(\chi) \\ &= [\tau(\chi^{\omega^{-1}})^\omega/\tau(\chi)][(\mathbf{N}_F\mathfrak{f}(\chi^{\omega^{-1}})^{1/2})^\omega/\mathbf{N}_F\mathfrak{f}(\chi)^{1/2}] \end{aligned}$$

and this is $= 1$, by (3.4) and (3.5).

Corollary. *Let K be a number field,* $\chi \in R^s_{(K)}$, χ *tame. Then*
(i) $\mathfrak{f}(\chi)$ *is the square of a fractional ideal in K.*
(ii) $\tau(\chi) \in \mathbb{Q}^*$.
(iii) $W(\chi) = W(\chi^\omega)$ $(= \pm 1)$, *for all* $\omega \in \Omega_{\mathbb{Q}}$.

Remark. (iii) is precisely Proposition I.6.2.

Proof. (i) and (ii) follow from Theorem 20B, in conjunction with the appropriate localization formulae, viz. the Corollary to Theorem 18 for $\tau(\chi)$, the definition (I.5.27) for $\mathfrak{f}(\chi)$. For (iii) we need Theorem 20B, the Corollary to Theorem 18, as well as Proposition 3.1. □

Remark. The original proofs (cf. [F17]) were not so entirely local, as those given here, where we have followed in the main the exposition in [Ma3]. The first proof – via explicit computation for quaternion characters – was published in [F7].

Proof of Theorem 5B. With N/K a Galois extension of number fields, of finite degree, with Galois group Γ, and $\chi \in R_\Gamma$, let

$$u(\chi) = \mathcal{N}_{K/\mathbb{Q}}(a \mid \chi) \cdot \tau(\chi)^{-1} W'(\chi),$$

as in (I.5.24). Here W' is the adjusted root number (cf. (I.5.25)), and for each prime divisor $\mathfrak{p}$ in K, $a_\mathfrak{p}$ is a free generator of $\mathfrak{o}_{N,\mathfrak{p}}$ over $\mathfrak{o}_{K,\mathfrak{p}}\Gamma$. We have to show that for all $\omega \in \Omega_{\mathbb{Q}}$

(3.6) $$u(\chi^{\omega^{-1}})^\omega = u(\chi).$$

We have to take again a certain amount of care, in distinguishing between determinants with different rings. The virtual characters χ are defined via representations T of Γ in $\mathrm{GL}_m(\mathbb{Q}^c)$, for some m. We take χ as the actual character of such a T. Then $(\mathrm{v}_{K/\mathbb{Q}} \det_\chi)(\omega)$ will lie in $\mathbb{Q}^{c*}$. Now let $\mathfrak{p}$ be a prime divisor of K. The quotient

$$\mathcal{N}_{K/\mathbb{Q}}(a_\mathfrak{p} \mid \chi^{\omega^{-1}})^\omega / \mathcal{N}_{K/\mathbb{Q}}(a_\mathfrak{p} \mid \chi)$$

will lie in $(\mathbb{Q}^c \otimes_K K_\mathfrak{p})^*$. Theorem 20A tells us that this quotient is the image $((\mathrm{v}_{K/\mathbb{Q}} \det_\chi)(\omega))_\mathfrak{p}$ of $(\mathrm{v}_{K/\mathbb{Q}} \det_\chi)(\omega)$ in $(\mathbb{Q}^c \otimes_K K_\mathfrak{p})^*$. Thus, for all $\mathfrak{p}$,

$$[\mathcal{N}_{K/\mathbb{Q}}(a \mid \chi^{\omega^{-1}})^\omega / \mathcal{N}_{K/\mathbb{Q}}(a \mid \chi)]_\mathfrak{p} = (\mathrm{v}_{K/\mathbb{Q}} \det_\chi)(\omega)_\mathfrak{p},$$

whence

(3.7) $$\mathcal{N}_{K/\mathbb{Q}}(a \mid \chi^{\omega^{-1}})^\omega = \mathcal{N}_{K/\mathbb{Q}}(a \mid \chi) \mathrm{v}_{K/\mathbb{Q}} \det_\chi(\omega).$$

By the Corollary of Theorem 21

$$W'(\chi^{\omega^{-1}})^\omega = W'(\chi).$$

This together with (3.7) and Theorem 20B (ii) implies (3.6) and thus also Theorem 5B. □

Remark. It is clear from the proof of Theorem 5B, that the result remains true if $a_\mathfrak{p}$ is a free generator of $N_\mathfrak{p}$ over $K_\mathfrak{p}\Gamma$, for all $\mathfrak{p}$, and a free generator of $\mathfrak{o}_{N,\mathfrak{p}}$ over $\mathfrak{o}_{K,\mathfrak{p}}\Gamma$ for almost all $\mathfrak{p}$.

§4. Signatures

In the present section we shall look at invariants at infinity for symplectic characters. The aim is to prove Theorem 5A. We shall incidentally also restate our result, so that it appears instead as an analogue at infinity of our Theorem 7 on conductors. Throughout, the restriction of tameness is irrelevant, but we shall nevertheless just state the results within our given framework. The localization theorems of §2 are now crucial. The inflation invariance of the signatures we are looking at is obvious.

We shall start off with the local result. If $a \in \mathbb{R}^*$, by definition, $\operatorname{sign} a = a/|a| = \pm 1$. Recall (cf. I. §5 prior to (5.16)) the definition of χ_- for a character of $\operatorname{Gal}(\mathbb{C}/\mathbb{R})$. If χ is symplectic, then $\deg(\chi_-)$ is even, as the number of eigenvalues -1 of the generator of $\operatorname{Gal}(\mathbb{C}/\mathbb{R})$ determines $\det_\chi$.

4.1. Proposition (cf. [F13], [F17]). *Let F be a local Archimedean field, $\chi \in R^s_{(F),\infty}$, a a free generator of $\mathbb{C}$ over $F\Gamma$, $\Gamma = \operatorname{Gal}(\mathbb{C}/F)$. Then $\mathcal{N}_{F/\mathbb{R}}(a \mid \chi)$ is real, and*

$$\operatorname{sign}(\mathcal{N}_{F/\mathbb{R}}(a \mid \chi)) = (-1)^{\deg(\chi_-)/2},$$

where for $F = \mathbb{C}$ we make the convention that $\deg(\chi_-) = 0$.

Proof. If $F = \mathbb{C}$, then $\chi = 2m\varepsilon$, $(a \mid \chi) = a^{2m}$, $\mathcal{N}_{F/\mathbb{R}}(a \mid \chi) = |a|^{4m} > 0$. If $F = \mathbb{R}$, and η is the non-trivial character of Γ, then $\chi = 2m\varepsilon + 2q\eta$, $a = c + id$, c, d real, $cd \neq 0$. Therefore

$$\mathcal{N}_{F/\mathbb{R}}(a \mid \chi) = (a \mid \chi) = (a \mid \varepsilon)^{2m}(a \mid \eta)^{2q} = (2c)^{2m}(2di)^{2q} = (-1)^q x^2, \qquad x \in \mathbb{R}.$$

As $q = \deg(\chi_-)/2$, the result follows. □

In the sequel N/K is again a Galois extension of number fields with Galois group Γ, and E is a big enough number field, i.e., $N \subset E$, $E/\mathbb{Q}$ is Galois, and all characters correspond to representations over E.

Theorem 22 (cf. [F13], [F17]) (see note [2]). *Let $\mathfrak{p}$ be an infinite prime divisor of K, b a free generator of $N_\mathfrak{p}$ over $K_\mathfrak{p}\Gamma$, $\mathfrak{P}'$ any infinite prime divisor of E, $\chi \in R^s_\Gamma$. Then $(\mathcal{N}_{K/\mathbb{Q}}((b \mid \chi)^r))_{\mathfrak{P}'}$ is real and*

$$\mathfrak{s}_\mathfrak{p}(\chi) = \operatorname{sign}_{\mathfrak{P}'}(\mathcal{N}_{K/\mathbb{Q}}((b \mid \chi)^r))$$

is independent of the particular choice of b, or of $\mathfrak{P}'$, and

$$\mathfrak{s}_\mathfrak{p}(\chi) = W(\chi_\mathfrak{p}).$$

Here r is an embedding $(E \otimes_K K_\mathfrak{p})^* \to \mathfrak{J}(E)$, as in (2.11), with $\operatorname{Im} r = \prod_{J_0} E_{\mathfrak{P}}^*$.

Proof. We shall first of all show that if both b and b^λ $(\lambda \in K_\mathfrak{p}\Gamma^*)$ are free generators of $N_\mathfrak{p}$ over $K_\mathfrak{p}\Gamma$ then for the quotient we get

$$(4.1) \qquad (\mathscr{N}_{K/\mathbb{Q}}((b \mid \chi)^r))_{\mathfrak{P}'}(\mathscr{N}_{K/\mathbb{Q}}((b^\lambda \mid \chi)^r))_{\mathfrak{P}'}^{-1} > 0.$$

We shall prove this for a specially chosen right transversal V of $\operatorname{Gal}(E/K)$ in $\operatorname{Gal}(E/\mathbb{Q})$. As $\det_\chi = \varepsilon$, the norm resolvents for all choices of V are in fact the same (see (2.19), (2.20)).

We shall use again the notation introduced in §2 for the statement and proof of Theorem 19. Let $\mathfrak{P}' = \mathfrak{P}_{j_1}$, $j_1 \in J$. We may suppose that $E_{\mathfrak{P}'} = \mathbb{C}$, and we denote complex conjugation by ν. Then for some unique $\rho \in \Sigma = \operatorname{Gal}(E/\mathbb{Q})$, $\rho j_1 = j_1 \nu$. For any $\omega \in \Sigma$, each of the following statements implies the other, the subscript denoting restriction to K,

$$(4.2) \qquad \begin{cases} \text{(a)} & (\omega j_1)_{|K} = (\omega j_1 \nu)_{|K} \\ \text{(b)} & \operatorname{Gal}(E/K)\omega = \operatorname{Gal}(E/K)\omega\rho. \end{cases}$$

If these conditions hold we say that ω is real, and otherwise imaginary. We now choose V as the union of three disjoint sets $V_\mathbb{R}$, V', $V'\rho$, where $V_\mathbb{R}$ consists of real ω only, V' of imaginary ones no two lying in the same double coset $\operatorname{Gal}(E/K)\omega\langle\rho\rangle$, and $V'\rho = [\omega\rho \mid \omega \in V']$.

By the Corollary to Proposition I.4.2,

$$(b \mid \chi)^{-1}(b^\lambda \mid \chi) = \operatorname{Det}_\chi(\lambda), \qquad \lambda \in K_\mathfrak{p}\Gamma^*.$$

If, say, χ is the character of a representation $T: \Gamma \to \operatorname{GL}_m(E)$, the determinant Det_χ above is defined via the extended homomorphism $T: K_\mathfrak{p}\Gamma^* \to \operatorname{GL}_m(E \otimes_K K_\mathfrak{p})$. View $K_\mathfrak{p}\Gamma^*$ as embedded in $\prod_{J_0}(E_\mathfrak{P}\Gamma)^*$ diagonally. Then

$$\operatorname{Det}_\chi(\lambda)^{(j \otimes 1)} = \operatorname{Det} T^j(\lambda) = \operatorname{Det}_{\chi^j}(\lambda), \qquad \text{if} \qquad j \in J_0.$$

Thus, for all χ,

$$(4.3) \qquad \begin{cases} \text{writing} \quad (b \mid \chi)^r((b^\lambda \mid \chi)^r)^{-1} = x(\chi) \in \mathfrak{J}(E), \quad \text{we get} \\ x(\chi)_{\mathfrak{P}_j} = \operatorname{Det}_{\chi^j}(\lambda) \text{ if } j \in J_0; \quad x(\chi)_\mathfrak{R} = 1, \text{ otherwise.} \end{cases}$$

The left hand side of (4.1) is thus

$$(4.4) \qquad \prod_{V_\mathbb{R}} (x(\chi^{\omega^{-1}})^\omega)_{\mathfrak{P}'} \cdot \prod_{V'} (x(\chi^{\omega^{-1}})^\omega \cdot x(\chi^{\rho^{-1}\omega^{-1}})^{\omega\rho})_{\mathfrak{P}'}.$$

Let $\sigma \in \operatorname{Gal}(E/\mathbb{Q})$, $\mathfrak{P}'' = \mathfrak{P}'^{\sigma^{-1}}$. Then $\exists \delta \in \operatorname{Gal}(\mathbb{C}/\mathbb{R})$, i.e., $\delta = \nu$ or $\delta = 1$, so that

$$(4.5) \qquad \text{for all} \quad x \in \mathfrak{J}(E), \qquad (x^\sigma)_{\mathfrak{P}'} = (x_{\mathfrak{P}''})^\delta.$$

Specifically if $\sigma = \rho$ then $\mathfrak{P}'' = \mathfrak{P}'$ and indeed $(y^\rho)_{\mathfrak{P}'} = (y_{\mathfrak{P}'})^\nu$. Combining this with (4.5), we have in addition

(4.5a) $$(x^{\sigma\rho})_{\mathfrak{P}'} = (x_{\mathfrak{P}''})^{\delta v}.$$

From (4.3), (4.5), (4.5a) we now deduce that

(4.6) $$(x(\chi^{\omega^{-1}})^{\omega})_{\mathfrak{P}'} = (x(\chi^{\rho^{-1}\omega^{-1}})^{\omega\rho})_{\mathfrak{P}'} = 1,$$

if $\mathfrak{P}'^{\omega^{-1}} \neq \mathfrak{P}_j$, i.e., $\mathfrak{P}'^{\rho^{-1}\omega^{-1}} \neq \mathfrak{P}_j$ for every $j \in J_0$.

So from now on we assume that

$$\mathfrak{P}' = \mathfrak{P}_j^{\omega} = \mathfrak{P}_j^{\omega\rho}, \qquad \text{for some} \quad j \in J_0,$$

and, without loss of generality $j = \omega j_1$. If first ω is real, then $jv_{|K} = \omega j_1 v_{|K} = \omega\rho j_{1|K} = \omega j_{1|K} = j_{|K}$, i.e., the restriction of j to K is real, in other words $K_{\mathfrak{p}} = \mathbb{R}$. Now we get

$$\begin{aligned}(x(\chi^{\omega^{-1}})^{\omega})_{\mathfrak{P}'} &= ((x(\chi^{\omega^{-1}}))_{\mathfrak{P}_j})^{\delta}, &&\text{by (4.5)}\\ &= \mathrm{Det}_{\chi^{\omega^{-1}j}}(\lambda)^{\delta}, &&\text{by (4.3), replacing } \chi \text{ by } \chi^{\omega^{-1}}.\end{aligned}$$

As $K_{\mathfrak{p}} = \mathbb{R}$, $\mathrm{Det}_{\chi^{\omega^{-1}j}}(\lambda)^{\delta} = \mathrm{Det}_{\chi^{\omega^{-1}j}}(\lambda) > 0$, $\chi^{\omega^{-1}j}$ being a symplectic $\mathbb{C}$-character. Thus

(4.7) $$(x(\chi^{\omega^{-1}})^{\omega})_{\mathfrak{P}'} > 0 \qquad \text{if} \qquad \omega \in V_{\mathbb{R}}.$$

Next take $\omega \in V'$. Then, as χ is symplectic, i.e., χ^{j_1} is real valued, we have $\chi^{\rho^{-1}j_1} = \chi^{j_1}$, i.e., $\chi^{\rho^{-1}} = \chi$. Thus

$$\begin{aligned}(x(\chi^{\omega^{-1}})^{\omega} \cdot x(\chi^{\rho^{-1}\omega^{-1}})^{\omega\rho})_{\mathfrak{P}'} &= ((x(\chi^{\omega^{-1}}))_{\mathfrak{P}_j})^{\delta}((x(\chi^{\omega^{-1}}))_{\mathfrak{P}_j})^{\delta v}\\ &= ((x(\chi^{\omega^{-1}}))_{\mathfrak{P}_j})^{1+v} > 0.\end{aligned}$$

Thus, collecting all factors in (4.4), by the last equation, and by (4.6) and (4.7), we get (4.1), and thus the sign of the norm resolvent is independent of the choice of b.

Now let $\mathfrak{q}$ be a prime divisor of N above $\mathfrak{p}$, and let a be a free generator of $N_{\mathfrak{q}}$ over $K_{\mathfrak{p}}\Gamma_{\mathfrak{q}}$, $\Gamma_{\mathfrak{q}}$ the local Galois group. Let $b = b(a)$ be given as in (2.9), so that Theorem 19 becomes applicable.

As $\det_{\chi} = \varepsilon$, we deduce that

$$(\mathcal{N}_{K/\mathbb{Q}}(b \mid \chi)^r)_{\mathfrak{P}_j} = \mathcal{N}_{K_{\mathfrak{p}}/\mathbb{R}}(a \mid \chi_{\mathfrak{p}}^j).$$

By Proposition 4.1, this is real, and moreover

(4.8) $$\mathrm{sign}_{\mathfrak{P}_j}(\mathcal{N}_{K/\mathbb{Q}}(b \mid \chi)^r) = \mathrm{sign}(\mathcal{N}_{K_{\mathfrak{p}}/\mathbb{R}}(a \mid \chi_{\mathfrak{p}}^j)).$$

Now $\deg(\chi_{\mathfrak{p}^-}^j) = \deg(\chi_{\mathfrak{p}^-})$ only depends on $\chi_{\mathfrak{p}}$ and not on j. Thus, by Proposition 4.1, and (4.8),

$$\mathrm{sign}_{\mathfrak{P}_j}(\mathcal{N}_{K/\mathbb{Q}}(b \mid \chi)^r) = (-1)^{\deg(\chi_{\mathfrak{p}^-})/2}.$$

By the definition of $W(\chi_{\mathfrak{p}})$ (cf. (I.5.16)) the theorem now follows. □

Remark. In the proofs of Theorems 19 and 22 we had to distinguish formally between $(E \otimes_K K_\mathfrak{p})^*$ and its image in $\mathfrak{J}(E)$. This is now no longer necessary and we return to our former convention of viewing $(E \otimes_K K_\mathfrak{p})^*$ as a subgroup of $\mathfrak{J}(E)$.

Proof of Theorem 5A. We now choose for every prime divisor $\mathfrak{p}$ of K, whether finite or infinite, a free generator $b_\mathfrak{p}$ of $N_\mathfrak{p}$ over $K_\mathfrak{p}\Gamma$. ($\Gamma = \mathrm{Gal}(N/K)$.) Then we shall show that for any infinite prime divisor $\mathfrak{P}'$,

$$\mathrm{sign}_{\mathfrak{P}'}(\mathcal{N}_{K/\mathbb{Q}}(b \mid \chi)) = W_\infty(\chi) \qquad (\chi \in R_\Gamma^s), \tag{4.9}$$

where $(b \mid \chi)_\mathfrak{p} = (b_\mathfrak{p} \mid \chi)$. The left hand side in (4.9) clearly coincides with

$$\prod_\mathfrak{p} \mathrm{sign}_{\mathfrak{P}'}(\mathcal{N}_{K/\mathbb{Q}}(b_\mathfrak{p} \mid \chi)) \qquad \text{(product over the infinite prime divisors of } K)$$

and, by Theorem 22 and the definition of $W_\infty(\chi)$ (cf. (I.5.20)), this yields (4.9).

Now we shall use the global formula (I.5.22), which in the case of symplectic characters χ can be restated as

$$\tau(\chi) = W(\chi) W_\infty(\chi) \mathbf{N}_K \mathfrak{f}(\chi)^{1/2}. \tag{4.10}$$

As $W(\chi)W_\infty(\chi) = \pm 1$ (see for instance Proposition 3.1 and Theorem 21), and $\mathbf{N}_K\mathfrak{f}(\chi)^{1/2} > 0$, we deduce that, for any infinite prime $\mathfrak{P}'$,

$$\mathrm{sign}_{\mathfrak{P}'}(\tau(\chi)) = W(\chi) W_\infty(\chi). \tag{4.11}$$

By (4.9), (4.11), the definition of $W'(\chi)$ (cf. (I.5.25)) and of $u(\chi)$ (cf. (I.5.24)) it follows that, for $\chi \in R_\Gamma^s$, $u(\chi)$ is positive at all infinite prime divisors. This is Theorem 5A. □

Remark 1. Equations (I.5.22) were originally purely formal, viewed as defining relations for both $\tau(\chi)$ and $\mathbf{N}_K\mathfrak{f}(\chi)^{1/2}$. But subsequently the positive number $\mathbf{N}_K\mathfrak{f}(\chi)^{1/2}$ was shown to be indeed the square root of the norm conductor. To get (4.11) one does however not need this; it suffices to have the fact that $\mathbf{N}_K\mathfrak{f}(\chi)^{1/2}$ is positive – which is already part of (I.5.22). An alternative way to (4.10) is via the corresponding local formulae at the finite prime divisors (cf. (2.2), (2.4)) as well as the infinite ones (definition of $W_\infty(\chi)$).

Remark 2. The theorem is true also in the wild case, and even the individual steps of the proof are essentially the same.

We now will reinterpret our results in terms of a local analogue at infinity to Theorem 7 (cf. [F13]). We take $F = \mathbb{R}$ as basefield, with $\phi \in R_{(F),\infty}$. Then $\phi + \bar{\phi} = \chi$ is symplectic, $\deg(\chi_-)/2 = \deg(\phi_-) = \deg(\phi) - \deg(\phi_+)$. Thus, if a is a generator of $\mathbb{C}$ over $\mathbb{R}\Gamma$, $\Gamma = \mathrm{Gal}(\mathbb{C}/\mathbb{R})$ then

$$\mathrm{sign}((a \mid \phi)(a \mid \bar{\phi})) = (-1)^{\deg(\phi) - \deg(\phi_+)} \tag{4.12}$$

and indeed ϕ_+ is the obvious infinite analogue to the finite $\mathrm{n}\phi$.

§5. The Local Main Theorems

We shall state here an entirely local theorem, which lies behind the important global Theorem 5C, and which incorporates the deepest and most essential aspects of the Galois Gauss sum – norm resolvent relation. Such a local result is of course implicit in the previous formulations of the proof of Theorem 5 (cf. [F17]), but this can now be separated out. The global result is then obtained from the local one via the local-global transition procedure, developed in §2. We shall also state the local version of the less deep Theorem 7 (note [3]).

Let F be a non-Archimedean local field, $F \supset \mathbb{Q}_p$, and let $j: \mathbb{Q}^c \to \mathbb{Q}_p^c$ be an embedding. We consider characters $\chi \in R_{(F)}$. Here it is essential that these are tame and are $\mathbb{Q}^c$-characters. Then the local Gauss sum $\tau(\chi)$ is defined, and lies in $\mathbb{Q}^{c*}$. Thus $\tau(\chi)^j \in \mathbb{Q}_p^{c*}$. On the other hand χ^j is a $\mathbb{Q}_p^c$-character. If say $\chi^j \in R_{L/F,p}$, where L/F is tame with Galois group Γ, and if a is a free generator of $\mathfrak{o}_L$ over $\mathfrak{o}_F\Gamma$, then $\mathcal{N}_{F/\mathbb{Q}_p}(a \mid \chi^j)$ is defined, and, for any such a, generates the ideal $\mathcal{N}_{F/\mathbb{Q}_p}P(\chi^j)$, defined in (1.3).

Theorem 23 (cf. [FT]).

$$\mathcal{N}_{F/\mathbb{Q}_p}P(\chi^j) = (\tau(\chi)^j). \qquad \square$$

We shall show that this theorem implies Theorem 5C.

So from now on K is a number field, and χ is a tame Galois character in $R_{(K)}$. We take E to be again a large enough number field. Specifically $E/\mathbb{Q}$ is Galois, $N \subset E$ where $\chi \in R_{N/K}$, and the values of χ and of the local components $\chi_\mathfrak{p}$ lie in E. By Theorem 20, the norm resolvents and Galois Gauss sums are ideles, and numbers respectively, in E. We then have to prove that for all finite prime divisors $\mathfrak{P}'$ of E

(5.1) $\quad (\mathcal{N}_{K/\mathbb{Q}}P(\chi))_{\mathfrak{P}'} = (\tau(\chi))_{\mathfrak{P}'}, \qquad$ (the $\mathfrak{P}'$-component of the ideal $(\tau(\chi))$).

Now, by the definition of $\mathcal{N}_{K/\mathbb{Q}}P$ we have

$$\mathcal{N}_{K/\mathbb{Q}}P(\chi) = \prod_\mathfrak{p} \mathcal{N}_{K/\mathbb{Q}}P_\mathfrak{p}(\chi)$$

(cf. §1), and by the Corollary to Theorem 18,

$$(\tau(\chi)) = \prod_\mathfrak{p} (\tau(\chi_\mathfrak{p})),$$

where both products run over the finite prime divisors $\mathfrak{p}$ of K, and both products are actually finite. Thus (5.1) will follow provided we can show that

(5.2) $$(\mathcal{N}_{K/\mathbb{Q}}P_\mathfrak{p}(\chi))_{\mathfrak{P}'} = (\tau(\chi_\mathfrak{p}))_{\mathfrak{P}'},$$

for every finite prime divisor $\mathfrak{p}$ of K and every finite prime divisor $\mathfrak{P}'$ of E. Warning: At this stage we cannot assume any relation between $\mathfrak{p}$ and $\mathfrak{P}'$.

By (2.22), we have

(5.3) $\quad (\mathcal{N}_{K/\mathbb{Q}}P_\mathfrak{p}(\chi))_{\mathfrak{P}'} = 1 \quad$ if $\mathfrak{p}$ and $\mathfrak{P}'$ do not lie above the same rational p.

We shall prove analogously that

(5.4) $(\tau(\chi_{\mathfrak{p}}))_{\mathfrak{P}'} = 1$ if $\mathfrak{p}$ and $\mathfrak{P}'$ do not lie above the same rational prime p.

By (5.3), (5.4), we get (5.2) for any such pair $\mathfrak{p}, \mathfrak{P}'$.

We fix again a finite rational prime p. Let θ be a multiplicative character of a local field $F \supset \mathbb{Q}_p$. By its definition, $\tau(\theta)$ is an algebraic integer. The formula $\tau(\theta)\overline{\tau(\theta)} = \mathbf{N}\mathfrak{f}(\theta)$ $(= p^m)$ shows that it divides a power of p. Let now $\mathfrak{P}'$ be a finite prime divisor not above p. Then we conclude that $\tau(\theta)_{\mathfrak{P}'}$ is a unit. Therefore $\tau(\phi)_{\mathfrak{P}'}$ is a unit, if ϕ is an Abelian character of a local Galois group of fields in $\mathbb{Q}_p^c$. By additivity and inductivity this extends to arbitrary characters of such Galois groups, i.e., we get (5.4).

From now on we assume $\mathfrak{p}$ and $\mathfrak{P}'$ to lie above the same rational p. In the notation of Theorem 19, there is an embedding j so that $\mathfrak{P}' = \mathfrak{P}_j$. By Corollary 2 to that theorem,

(5.5) $$(\mathcal{N}_{K/\mathbb{Q}} P_{\mathfrak{p}}(\chi))_{\mathfrak{P}_j} = \mathcal{N}_{K_{\mathfrak{p}}/\mathbb{Q}_p} P(\chi_{\mathfrak{p}}^j) \qquad (\mathfrak{P}_j = \mathfrak{P}').$$

Trivially

(5.6) $$(\tau(\chi_{\mathfrak{p}}))_{\mathfrak{P}_j} = (\tau(\chi_{\mathfrak{p}})^j).$$

By Theorem 23, (5.5), (5.6), we now get (5.2). Thus (5.2) holds in all cases, and this implies Theorem 5C. □

Next we state the local analogue to Theorem 7. This was tentatively conjectured by E. Noether (cf. [No]).

Theorem 24. *Let F be a non-Archimedean local field, $\phi \in R_{(F),p}$ a tame Galois character. Then*

$$P(\phi)P(\bar{\phi}) = \mathfrak{f}(\phi).$$

(Remark: $\bar{\phi}$ is the contragredient to ϕ). □

We shall show that Theorem 24 implies Theorem 7. Let then N/K be a tame Galois extension of number fields with Galois group Γ, $\chi \in R_\Gamma$. We have

$$\mathfrak{f}(\chi) = \prod_{\mathfrak{p}} \mathfrak{f}(\chi_{\mathfrak{p}}), \qquad P(\chi) = \prod_{\mathfrak{p}} P_{\mathfrak{p}}(\chi)$$

(product over all finite prime divisors $\mathfrak{p}$ of K).

Let $\mathfrak{P}'$ be a finite prime divisor in E, as before. As $\mathfrak{f}(\chi_{\mathfrak{p}})$ is a power of $\mathfrak{p}$, we have

(5.7) $$\mathfrak{f}(\chi_{\mathfrak{p}})_{\mathfrak{P}'} = 1, \qquad \text{unless } \mathfrak{P}' \mid \mathfrak{p}.$$

If b is a free generator of $\mathfrak{o}_{N,\mathfrak{p}}$ over $\mathfrak{o}_{K,\mathfrak{p}}\Gamma$, then $(b \mid \chi)$ as an idele is trivial at prime divisors not above $\mathfrak{p}$. Thus

(5.8) $$(P_{\mathfrak{p}}(\chi))_{\mathfrak{P}'} = 1, \qquad \text{unless } \mathfrak{P}' \mid \mathfrak{p}.$$

Thus in order to prove Theorem 7, we have to show that

(5.9) $$(\mathfrak{f}(\chi_\mathfrak{p}))_{\mathfrak{P}'} = (P_\mathfrak{p}(\chi)P_\mathfrak{p}(\bar{\chi}))_{\mathfrak{P}'}, \qquad \text{if } \mathfrak{P}' \mid \mathfrak{p}.$$

We may now assume that $\mathfrak{P}' = \mathfrak{P} = \mathfrak{P}_k$, as in the proof of Theorem 19. By (2.18), with $\omega = 1$, we get

(5.10) $$P_\mathfrak{p}(\chi)_{\mathfrak{P}'} = P(\chi_\mathfrak{p}^k).$$

By Theorem 24,

(5.11) $$P(\chi_\mathfrak{p}^k)P(\bar{\chi}_\mathfrak{p}^k) = \mathfrak{f}(\chi_\mathfrak{p}^k).$$

But trivially $\mathfrak{f}(\chi_\mathfrak{p}^k) = \mathfrak{f}(\chi_\mathfrak{p})$, and thus (5.9) will follow from (5.11), and from (5.10) with respect to both χ and $\bar{\chi}$. □

§6. Non-Ramified Base Field Extension

In the present section we shall describe the transformation properties of resolvents, Galois Gauss sums and conductors under a non-ramified extension of the basefield. This will be an essential reduction step in the proofs of Theorems 23 and 24. It is the striking similarity of functorial behaviour of resolvents on the one hand, and of Galois Gauss sums and Artin conductors on the other, which has turned out to be a central element in the whole theory. This initially purely formal similarity was commented on by the author in the early stages (cf. [F14]), as indicating deeper arithmetic connections, and has now in turn become instrumental in establishing such connections (see here note [4]).

Throughout this section F is a non-Archimedean local field, containing $\mathbb{Q}_p$, L is a tame Galois extension of F of finite degree with Galois group Γ, and B is an intermediary field, the fixed field of the subgroup Δ of Γ, with B/F non-ramified. We shall compare functions of virtual characters χ in R_Γ, or in $R_{\Gamma,p}$ with the corresponding functions of their restrictions $\operatorname{res}_\Delta^\Gamma \chi = \operatorname{res} \chi$.

Theorem 25 (cf. [F17], [F18]). (i) *Let $\chi \in R_\Gamma$. Then*

$$\mathfrak{f}(\operatorname{res} \chi) = \mathfrak{f}(\chi)$$

and

$$\tau(\operatorname{res} \chi) = \tau(\chi)^{[B:F]} u(\chi), \qquad u(\chi) \textit{ a root of unity.}$$

(ii) *Let a be a free generator of $\mathfrak{o}_L$ over $\mathfrak{o}_F\Gamma$, b a free generator of $\mathfrak{o}_L$ over $\mathfrak{o}_B\Delta$. Then there exists $\lambda \in (\mathfrak{o}_B\Gamma)^*$, so that for all $\chi \in R_{\Gamma,p}$*

$$(b \mid \operatorname{res} \chi)_{L/B} = (a \mid \chi)_{L/F} \operatorname{Det}_\chi(\lambda).$$

In particular

$$\mathcal{N}_{B/\mathbb{Q}_p}(b \mid \operatorname{res}\chi)_{L/B} = \mathcal{N}_{F/\mathbb{Q}_p}(a \mid \chi)_{L/F}^{[B:F]} v(\chi),$$

$v(\chi)$ *a unit.*

Remark. A more careful evaluation of the factors $u(\chi)$ and $v(\chi)$, turning up in Theorem 25, would give

$$u(\chi) = \operatorname{Det}_\chi(t) = t^{\deg(\chi)},$$

for some fixed t $(= \pm 1)$, independent of χ, and

$$v(\chi) = \operatorname{Det}_\chi(\lambda^1),$$

for some $\lambda^1 \in \mathbb{Z}_p\Gamma^*$, λ^1 independent of χ.

Corollary 1. (i) *If* $\chi \in R_\Gamma$, *we have the equation*

$$(\tau(\operatorname{res}\chi)) = (\tau(\chi))^{[B:F]}$$

for ideals.

(ii) *If* $\chi \in R_{\Gamma,p}$, *we have*

$$P(\operatorname{res}\chi) = P(\chi),$$
$$\mathcal{N}_{B/\mathbb{Q}_p} P(\operatorname{res}\chi) = \mathcal{N}_{F/\mathbb{Q}_p} P(\chi)^{[B:F]}.$$

Corollary 2. *If* χ and χ' *have the same restriction to the inertia group, then*

$$\mathfrak{f}(\chi) = \mathfrak{f}(\chi'), \qquad (\tau(\chi)) = (\tau(\chi')) \qquad (\chi, \chi' \in R_\Gamma),$$
$$P(\chi) = P(\chi'), \qquad \mathcal{N}_{F/\mathbb{Q}_p} P(\chi) = \mathcal{N}_{F/\mathbb{Q}_p} P(\chi') \qquad (\chi, \chi' \in R_{\Gamma,p}).$$

Proof of Theorem 25. We first look at the conductors. By hypothesis on B, Δ contains the inertia group Γ_0 and hence

$$\deg(\mathrm{n}\operatorname{res}\chi) = \deg(\mathrm{n}\chi).$$

As B/F is non-ramified, the prime ideals $\mathfrak{p}_B$ and $\mathfrak{p}_F$ are the same in our identification of ideal groups. Hence indeed $\mathfrak{f}(\operatorname{res}\chi) = \mathfrak{f}(\chi)$ – recalling the definition $\mathfrak{f}(\chi) = \mathfrak{p}_F^{\deg(\chi) - \deg(\mathrm{n}\chi)}$.

For the Galois Gauss sums we need a Lemma.

6.1. Lemma. *Let* χ *be a virtual character and* α *a non-ramified Abelian character of* Ω_F. *Then*

$$\tau(\alpha\chi) = \tau(\chi) w(\chi, \alpha), \qquad w(\chi, \alpha) \text{ a root of unity.}$$

Proof of 6.1. This is well known (see e.g., [FT]). If first χ is Abelian, then by going back to the multiplicative characters $\theta_\chi, \theta_\alpha$ one has

$$\begin{aligned}\tau(\alpha\chi) = \tau(\theta_\alpha\theta_\chi) &= \theta_\alpha(\mathfrak{f}(\chi)\mathfrak{D}_F)^{-1}\tau(\theta_\chi)\\ &= w(\chi, \alpha)\tau(\chi), \qquad \text{as required.}\end{aligned}$$

Thus the equation of the Lemma holds for all local fields, and Abelian characters χ.

Next if χ is induced from an Abelian character ϕ of Ω_E, where $E \supset F$, then $\chi\alpha$ is induced from $\phi\beta$, β the restriction of α to Ω_E. Hence, as $\deg(\alpha\chi - \chi) = 0$, we get $\tau(\alpha\chi - \chi) = \tau(\beta\phi - \phi) = w(\phi, \beta)$, a root of unity. Additivity of $w(\chi, \alpha)$ in χ now does the rest. □

Now we go back to the proof of the formula of Theorem 25 for Galois Gauss sums. By Frobenius reciprocity we have

$$\operatorname{ind}_\Delta^\Gamma(\operatorname{res}_\Delta^\Gamma \chi) = \operatorname{ind}_\Delta^\Gamma(\varepsilon_\Delta \cdot \operatorname{res}_\Delta^\Gamma \chi) = \operatorname{ind}_\Delta^\Gamma \varepsilon_\Delta \cdot \chi.$$

As $\Delta \supset \Gamma_0$, and Γ/Γ_0 is Abelian, we see that $\operatorname{ind}_\Delta^\Gamma \varepsilon_\Delta = \sum \alpha_i$, α_i running through the Abelian characters of Γ/Δ lifted to Γ. These are all non-ramified and their number is $[B:F]$. Thus, by the Lemma,

$$\tau(\operatorname{ind}_\Delta^\Gamma(\operatorname{res}_\Delta^\Gamma \chi)) = \prod_i \tau(\alpha_i\chi) = \tau(\chi)^{[B:F]} u_1(\chi), \tag{6.1}$$

$u_1(\chi)$ a root of unity – namely $u_1(\chi) = \prod_i w(\chi, \alpha_i)$.

On the other hand, by the induction formula for Galois Gauss sums,

$$\tau(\operatorname{ind}_\Delta^\Gamma(\operatorname{res}_\Delta^\Gamma \chi)) = \tau(\operatorname{res}_\Delta^\Gamma \chi)\tau(\operatorname{ind}_\Delta^\Gamma \varepsilon_\Delta)^{\deg(\chi)}. \tag{6.2}$$

As $\operatorname{ind}_\Delta^\Gamma \varepsilon_\Delta = \sum \alpha_i$, we have

$$\tau(\operatorname{ind}_\Delta^\Gamma \varepsilon_\Delta)^{\deg(\chi)} = \prod \tau(\alpha_i)^{\deg(\chi)} = u_2(\chi), \tag{6.3}$$

a root of unity. Comparing (6.1), (6.2), (6.3) we get the required result.

Now we turn to resolvents, and thus have to consider representations over $\mathbb{Q}_p^c$ and their characters. We shall first show that the equation for norm resolvents follows from that for resolvents. We shall express everything in terms of congruences modulo units. The resolvent relation is then

$$(b \mid \operatorname{res} \chi) \equiv (a \mid \chi).$$

Let $\{\sigma\}$ be a right transversal of Ω_F in $\Omega_{\mathbb{Q}_p}$, $\{\gamma\}$ one of Ω_B in Ω_F. The Galois action formula on resolvents tells us that

$$(a \mid \chi^{\gamma^{-1}})^\gamma \equiv (a \mid \chi), \qquad \text{for all } \chi.$$

Therefore

$$\begin{aligned}
\mathcal{N}_{B/\mathbb{Q}_p}(b \mid \operatorname{res} \chi) &= \prod_{\gamma,\sigma} (b \mid \operatorname{res} \chi^{\sigma^{-1}\gamma^{-1}})^{\gamma\sigma} \\
&\equiv \prod_{\gamma,\sigma} (a \mid \chi^{\sigma^{-1}\gamma^{-1}})^{\gamma\sigma} \\
&\equiv \prod_{\sigma} (a \mid \chi^{\sigma^{-1}})^{\sigma[B:F]} \\
&= \mathcal{N}_{F/\mathbb{Q}_p}(a \mid \chi)^{[B:F]}.
\end{aligned}$$

The remaining part of the proof requires some rather more extensive technical tools, which we shall now introduce.

We consider the set $\mathrm{Map}_\Delta(\Gamma, L)$ of maps of Δ-sets $\Gamma \to L$, Δ acting on Γ by right multiplication. This is a ring, addition and multiplication being defined via that in L, so that, for instance, $f_1 \cdot f_2(\gamma) = f_1(\gamma)f_2(\gamma)$. It is a B-algebra, where for $x \in B$, $f \in \mathrm{Map}_\Delta(\Gamma, L)$, we define xf by $(xf)(\gamma) = x(f(\gamma))$. It is a Γ-module, where for $\gamma \in \Gamma$, $f \in \mathrm{Map}_\Delta(\Gamma, L)$, $f^\gamma(\gamma_1) = f(\gamma\gamma_1)$. Γ acts then by B-algebra automorphisms.

6.2. Lemma. *There is a commutative diagram*

$$\begin{array}{ccc} B \otimes_F L & \xrightarrow{j} & \mathrm{Map}_\Delta(\Gamma, L) \\ & {\scriptstyle e}\searrow \quad \swarrow{\scriptstyle e'} & \\ & L & \end{array} \tag{6.4}$$

of B-algebra homomorphisms, where $e'(f) = f(1)$, and for $x \in B$, $y \in L$, $e(x \otimes y) = xy$, $(j(x \otimes y))(\gamma) = xy^\gamma$. e and e' are surjective and j is an isomorphism of Γ-modules.

(6.4) *restricts to a commutative diagram of $\mathfrak{o}_B$-algebras*

$$\begin{array}{ccc} \mathfrak{o}_B \otimes_{\mathfrak{o}_F} \mathfrak{o}_L & \xrightarrow{j_1} & \mathrm{Map}_\Delta(\Gamma, \mathfrak{o}_L) \\ & {\scriptstyle e_1}\searrow \quad \swarrow{\scriptstyle e'_1} & \\ & \mathfrak{o}_L & \end{array} \tag{6.5}$$

where e_1, e'_1 are surjective and j_1 is an isomorphism.

Proof. All assertions as regards (6.4) are easily checked, except the bijectivity of j. As $B \otimes_F L$ and $\mathrm{Map}_\Delta(\Gamma, L)$ have the same dimensions $[B:F][L:F]$ as F-vector spaces, it suffices to show that j is injective. Let $\{y_k\}$ be an F-basis of L. Any element of $\mathrm{Ker}\, j$ is then of form

$$u = \sum_k x_k \otimes y_k, \qquad x_k \in B, \qquad \text{with} \qquad \sum_k x_k y_k^\gamma = 0, \qquad \text{all } \gamma.$$

As $\mathrm{Det}(y_k^\gamma)_{k,\gamma} \neq 0$, we must have $x_k = 0$, all k, i.e., $u = 0$.

Again all statements as regards (6.5) are obvious, except those relating to j_1. Now j preserves integrality, hence indeed it maps $\mathfrak{o}_B \otimes_{\mathfrak{o}_F} \mathfrak{o}_L$ into the maximal order $\mathrm{Map}_\Delta(\Gamma, \mathfrak{o}_L)$ of $\mathrm{Map}_\Delta(\Gamma, L)$. The resulting restriction j_1 is injective, together with j. As, however, B/F is non-ramified, $\mathfrak{o}_B \otimes_{\mathfrak{o}_F} \mathfrak{o}_L$ is the maximal order of $B \otimes_F L$, and this implies that j_1 is surjective. □

Now let $T: \Gamma \to \mathrm{GL}_n(E)$ be a representation, with $L \subset E \subset \mathbb{Q}_p^c$, and E/F Galois of finite degree. Modulo equivalence, every representation of Γ over $\mathbb{Q}_p^c$ is of this form, for some E. As usual, we extend T to algebra maps

$$T: L(\Gamma) \to \mathrm{M}_n(E),$$

$$B \otimes T: (B \otimes_F L)(\Gamma) \to \mathrm{M}_n(B \otimes_F E),$$

where we now have to be somewhat careful with our notations. We obtain a diagram

$$
(6.6)\qquad
\begin{array}{ccccccc}
 & & (B\otimes_F L)(\Gamma)^* & \xrightarrow{B\otimes T} & \mathrm{GL}_n(B\otimes_F E) & \xrightarrow{\mathrm{Det}} & (B\otimes_F E)^* \\
 & \nearrow & & & & & \\
B(\Gamma)^* & & \downarrow^{e} & & \downarrow^{e} & & \downarrow^{e} \\
 & \searrow & & & & & \\
 & & L(\Gamma)^* & \xrightarrow{T} & \mathrm{GL}_n(E) & \xrightarrow{\mathrm{Det}} & E^*
\end{array}
$$

The horizontal maps are as clearly indicated, the vertical maps all come from the multiplication $B\otimes_F E\to E$ $(x\otimes y\mapsto xy)$ which extends the map e of (6.4). The lower arrow in the triangle is given by the embedding $B\subset L$, the upper one by mapping $B\to B\otimes_F 1\subset B\otimes_F L$.

6.3. Lemma. *Diagram* (6.6) *commutes.*

Proof. For the triangle on the left this is obvious. For the remainder one only needs to observe that e is a ring homomorphism, which thus maps group rings into group rings, matrix rings into matrix rings, multiplicative groups into multiplicative groups, and commutes with determinant formation and with maps induced from group representations. □

Now we return to the proof of Theorem 25 (ii). Associated with the element b, define $g\colon \Gamma\to L$ by

$$
(6.7)\qquad g(\gamma)=\begin{cases} b^{\gamma}, & \text{if } \gamma\in\Delta,\\ 0, & \text{if } \gamma\in\Gamma,\quad \gamma\notin\Delta.\end{cases}
$$

One verifies easily that $g\in\mathrm{Map}_\Delta(\Gamma,\mathfrak{o}_L)$. We shall show that g is a free generator of this module over $\mathfrak{o}_B\Gamma$. For, let $\{\gamma_j\}$ be a left transversal of Δ in Γ. An element h of $\mathrm{Map}_\Delta(\Gamma,\mathfrak{o}_L)$ is uniquely determined by the values $h(\gamma_j)\in\mathfrak{o}_L$, and those can be prescribed arbitrarily. Write

$$h(\gamma_j)=\sum_{\delta\in\Delta} c_{j,\delta}b^{\delta},\qquad c_{j,\delta}\in\mathfrak{o}_B.$$

Then

$$\sum_{k,\delta} c_{k,\delta}g^{\delta\gamma_k^{-1}}(\gamma_j)=h(\gamma_j),$$

whence

$$h=\sum_{k,\delta} c_{k,\delta}g^{\delta\gamma_k^{-1}}.$$

Now apply j^{-1}. By Lemma 6.2,

$$(6.8)\qquad j^{-1}g \text{ is a free generator of } \mathfrak{o}_B\otimes_{\mathfrak{o}_F}\mathfrak{o}_L \text{ over } \mathfrak{o}_B\Gamma.$$

Next, with a as given in the theorem, $1\otimes a$ is a free generator of $\mathfrak{o}_B\otimes_{\mathfrak{o}_F}\mathfrak{o}_L$ over $\mathfrak{o}_B\Gamma$ as well. By (6.8) there exists an element $\lambda\in(\mathfrak{o}_B\Gamma)^*$, so that $j^{-1}g=(1\otimes a)^{\lambda}$.

Therefore in $(\mathfrak{o}_B \otimes_{\mathfrak{o}_F} \mathfrak{o}_L)(\Gamma)$,

(6.9) $$\sum_\gamma (j^{-1}g)^\gamma \gamma^{-1} = \left(\sum_\gamma (1 \otimes a)^\gamma \gamma^{-1}\right)(\lambda \otimes 1).$$

Now let χ be the character of the representation T, considered above. We apply the composite map $e \circ \mathrm{Det} \circ (B \otimes T)$ of the diagram (6.6) to Eq. (6.9), and apply Lemma 6.3 to each of the three factors occurring in (6.9). Firstly we have $e(\lambda \otimes 1) = \lambda$, and hence

$$(e \circ \mathrm{Det} \circ (B \otimes T))(\lambda \otimes 1) = \mathrm{Det}\, T(e(\lambda \otimes 1)),$$

i.e.,

(6.10) $$(e \circ \mathrm{Det} \circ (B \otimes T))(\lambda \otimes 1) = \mathrm{Det}_\chi(\lambda).$$

Next,

$$(e \circ \mathrm{Det} \circ (B \otimes T))\left(\sum_\gamma (1 \otimes a)^\gamma \gamma^{-1}\right) = \mathrm{Det}\, T(e(1 \otimes \textstyle\sum a^\gamma \gamma^{-1})) = \mathrm{Det}\, T(\textstyle\sum a^\gamma \gamma^{-1})$$
$$= \mathrm{Det}_\chi(\textstyle\sum a^\gamma \gamma^{-1})$$

i.e.,

(6.11) $$(e \circ \mathrm{Det} \circ (B \otimes T))\left(\sum_\gamma (1 \otimes a)^\gamma \gamma^{-1}\right) = (a \mid \chi).$$

By Lemma 6.2,

$$e\left(\sum_\gamma (j^{-1}g)^\gamma \gamma^{-1}\right) = e\left(\sum_\gamma j^{-1}(g^\gamma) \cdot \gamma^{-1}\right)$$
$$= \sum_\gamma e'(g^\gamma)\gamma^{-1} = \sum_\gamma g^\gamma(1)\gamma^{-1}$$
$$= \sum_\gamma g(\gamma)\gamma^{-1} = \sum_\delta b^\delta \delta^{-1}$$

(definition of g). Therefore, by Lemma 6.3,

(6.12) $$(e \circ \mathrm{Det} \circ (B \otimes T))\left(\sum_\gamma (j^{-1}g)^\gamma \gamma^{-1}\right) = \mathrm{Det}_\chi(\textstyle\sum b^\delta \delta^{-1}) = (b \mid \mathrm{res}\, \chi).$$

Comparison of (6.9)–(6.12) finally gives the required resolvent equation of Theorem 25. □

§7. Abelian Characters, Completion of Proofs

In this section we compute the relevant ideals for tame Abelian characters. This is of some interest in itself. On this basis we then complete the proofs of Theorems 23 and 24.

Throughout F is a non-Archimedean local field of residue class characteristic p. We shall view any homomorphism $\eta\colon \mathfrak{o}_F^* \to \mathbb{Q}_p^{c*}$ with $\operatorname{Ker}\eta \supset 1 + \mathfrak{p}_F$ (a harmless abuse of notation!) also as a homomorphism $(\mathfrak{o}_F/\mathfrak{p}_F)^* \to \mathbb{Q}_p^{c*}$, and vice versa. η will have order dividing $\mathbf{N}_F\mathfrak{p}_F - 1$, and thus will take values in the group μ_F^* of $\mathbf{N}_F\mathfrak{p}_F - 1$-st roots of unity, which is contained in F^*. The composite map

$$(\mathfrak{o}_F/\mathfrak{p}_F)^* \xrightarrow{\eta} \mu_F^* \xrightarrow{\text{residue class}} (\mathfrak{o}_F/\mathfrak{p}_F)^*$$

is an endomorphism of the cyclic group $(\mathfrak{o}_F/\mathfrak{p}_F)^*$. There thus exists a unique

$$\text{(7.1)} \qquad \begin{cases} s(\eta) = s \in \mathbb{Q}^*,\ 0 \leqslant s < 1 \\ \text{with} \quad \eta(u) \equiv u^{-s(\mathbf{N}_F\mathfrak{p}_F - 1)} \pmod{\mathfrak{p}_F}, \quad \text{for all} \quad u \in \mathfrak{o}_F^*. \end{cases}$$

Let now ϕ be a tame, Abelian character in $R_{(F),p}$. Then we have

Theorem 26. (i) $P(\phi) = \mathfrak{p}_F^{s(\eta)}$, *where η is the restriction to $\mathfrak{o}_F^*$ of the multiplicative character θ_ϕ of F, which corresponds to ϕ under class field theory.*

(ii) *Suppose F contains the primitive m-th roots of unity, where $m =$ order (ϕ). Let*

$$U_\phi = [x \in \mathbb{Q}_p^c \mid x^\omega = x\phi(\omega) \text{ for all } \omega \in \Omega_F].$$

Then $P(\phi)$ is the ideal generated by $U_\phi \cap \mathbb{Z}_p^c$. □

Remark 1. As we are viewing fractional ideals as elements of the uniquely divisible group $\mathfrak{I}(\mathbb{Q}_p^c)$, raising an ideal to a rational power $s(\eta)$ makes sense.

Remark 2. We shall only need the case (order $(\phi), p) = 1$, but for completeness we shall prove the result as stated.

Remark 3. Theorem 26 (ii) deals with an aspect of Galois module structure of Kummer extensions. For a detailed early treatment of this topic, including the wild case, see [F2].

Proof of Theorem 26. We first prove (ii). Let $x_0 \in U_\phi$, $x_0 \neq 0$, $E = F(x_0)$. Then $U_\phi \subset E$. E is a tame Galois extension of F, whose Galois group Δ is cyclic. Let $\mathfrak{b}_\phi$ be the fractional ideal generated by $U_\phi \cap \mathbb{Z}_p^c = U_\phi \cap \mathfrak{o}_E$, and let a be a free generator of $\mathfrak{o}_E$ over $\mathfrak{o}_F\Delta$. Then clearly $(a \mid \phi) \in U_\phi \cap \mathfrak{o}_E$, whence $P(\phi) \subset \mathfrak{b}_\phi$. Let Y_ϕ be the ideal of $\mathfrak{o}_F\Delta$, of elements y with $y\delta = y\phi(\delta)$. $U_\phi \cap \mathfrak{o}_E$ is the image of Y_ϕ under the isomorphism $\mathfrak{o}_F\Delta \cong \mathfrak{o}_E$ given by $\lambda \mapsto a^\lambda$. But Y_ϕ is generated over $\mathfrak{o}_F$ by $\sum_\delta \phi(\delta)\delta^{-1}$, and so every element in $U_\phi \cap \mathfrak{o}_E$ is of form $c(a \mid \phi), c \in \mathfrak{o}_F$, i.e., lies in $P(\phi)$. Thus $\mathfrak{b}_\phi = P(\phi)$.

For (i) we may suppose that the order m of ϕ is prime to p. For, otherwise $\phi = \phi'\phi''$ where (order $(\phi'), p) = 1$, $\operatorname{ord}(\phi'') = p^n$. But then ϕ'' is non-ramified, and thus $\theta_{\phi|\mathfrak{o}_F^*} = \theta_{\phi'|\mathfrak{o}_F^*} = \eta$, i.e., ϕ and ϕ' yield the same $s(\eta)$. On the other hand, by Corollary 2 to Theorem 25, $P(\phi) = P(\phi')$.

Next we show that we may assume that F contains the primitive m-th roots of unity, $m =$ order (ϕ). Suppose then that the result is true in this case, and let $L = F(m)$ be the field of m-th roots of unity over F and ϕ^* be the restriction of ϕ to Ω_L. Then order $(\phi^*) \mid m$. Let $\eta = \theta_{\phi \mid \mathfrak{o}_F^*}$, and suppose that s satisfies (7.1). By local class field theory,

$$\theta_{\phi^*} = \theta_\phi \circ N_{L/F}, \qquad N_{L/F} \text{ the norm } L^* \to F^*. \tag{7.2}$$

But, as L/F is non-ramified, m being prime to p,

$$N_{L/F} v \equiv v^{(\mathrm{N}_L \mathfrak{p}_L - 1)/(\mathrm{N}_F \mathfrak{p}_F - 1)} \pmod{\mathfrak{p}_L}, \qquad \text{for all} \quad v \in \mathfrak{o}_L^*. \tag{7.3}$$

Thus if $\eta^* = \theta_{\phi^* \mid \mathfrak{o}_L^*}$, then by (7.2) and (7.3),

$$\eta^*(v) \equiv \eta(v)^{(\mathrm{N}_L \mathfrak{p}_L - 1)/(\mathrm{N}_F \mathfrak{p}_F - 1)} \pmod{\mathfrak{p}_L},$$

and, in conjunction with (7.1), we get

$$\eta^*(v) \equiv v^{-s(\mathrm{N}_L \mathfrak{p}_L - 1)} \pmod{\mathfrak{p}_L}, \qquad \text{for all} \quad v \in \mathfrak{o}_L^*.$$

Therefore $P(\phi^*) = \mathfrak{p}_L^s$. As L/F is non-ramified, we have in our identification of $\mathfrak{I}(F)$ and $\mathfrak{I}(L)$ as subgroups of $\mathfrak{I}(\mathbb{Q}_p^c)$, that $\mathfrak{p}_F = \mathfrak{p}_L$. Hence

$$P(\phi^*) = \mathfrak{p}_F^s. \tag{7.4}$$

By Theorem 25,

$$P(\phi) = P(\phi^*),$$

and so finally, by (7.4), $P(\phi) = \mathfrak{p}_F^s$.

From now on we assume that F contains the primitive m-th roots of unity. We suppose that $P(\phi) = \mathfrak{p}_F^s$ and show that s satisfies (7.1), i.e., that $s = s(\eta)$, with η as given in the theorem. Let $A \colon F^* \to \Omega_F^{ab}$ be the Artin map, let x be a generator of $U_\phi \cap \mathbb{Z}_p^c$ and let $\pi \in F^*$, $(\pi) = \mathfrak{p}_F$. Then $x^m = \pi^r w$, $w \in \mathfrak{o}_F^*$, and $s = r/m$. Using the properties of the local norm residue symbol (,) we get, for $u \in \mathfrak{o}_F^*$, $\sqrt[m]{u} = y$,

$$\begin{aligned}
\eta(u) = \theta_\phi(u) = x\phi(Au) \cdot x^{-1} = x^{Au} x^{-1} \\
&= (\pi^r w, u)_m \qquad \text{(definition of (,))} \\
&= (\pi, u)_m^r (w, u)_m \qquad \text{(multiplicativity)} \\
&= (\pi, u)_m^r \qquad (\text{as } (m, p) = 1, \text{ hence } (w, u)_m = 1) \\
&= (u, \pi)_m^{-r} \qquad \text{(skew symmetry)} \\
&= (y^{A(\pi)} y^{-1})_m^{-r} \qquad \text{(definition)} \\
&\equiv y^{-r(\mathrm{N}_F \mathfrak{p}_F - 1)} \pmod{\mathfrak{p}_F} \qquad (\text{as } A(\pi) = \text{Frobenius}) \\
&\equiv u^{-s(\mathrm{N}_F \mathfrak{p}_F - 1)} \pmod{\mathfrak{p}_F} \qquad (\text{as } sm = r).
\end{aligned}$$

This then gives the result. □

Reduction step for Theorems 23 and 24. By Theorem 25, one can restrict oneself to characters of the Galois group $\Gamma = \mathrm{Gal}(E/F)$ of a totally, tamely ramified extension. As such an extension is cyclic of order prime to p, the irreducible characters of Γ are Abelian of order prime to p, and by additivity it then suffices to consider such characters only. □

Proof of Theorem 24. Let ϕ be a tame Abelian character of Ω_F, and write $s(\eta) = s(\phi)$ if $\eta = \theta_{\phi \mid \mathfrak{o}_F^*}$. One verifies immediately that either $\eta = \eta^{-1} = \varepsilon$, i.e., θ_ϕ is non-ramified and in this case $s(\phi) = s(\phi^{-1}) = 0$, i.e.,

$$P(\phi)P(\phi^{-1}) = 1,$$

or $\eta \neq \varepsilon \neq \eta^{-1}$, θ_ϕ is ramified, and $s(\phi^{-1}) = 1 - s(\phi)$, i.e.,

$$P(\phi)P(\phi^{-1}) = \mathfrak{p}_F.$$

In the first case, by local class field theory, ϕ is non-ramified, i.e., $\phi = \mathrm{n}\phi$, hence $\mathfrak{f}(\phi) = 1$, in the second case $\mathrm{n}\phi = 0$, i.e., $\mathfrak{f}(\phi) = \mathfrak{p}_F$. Comparison now gives the required equation. □

Our final aim in this section is the proof of Theorem 23. The exponent of (p), which turns up in the evaluation of the Gauss sums and the norm resolvents, is of some interest and it seems worthwhile to spend a little extra time to give several explicit formulae, and also an axiomatic characterization. The exponent for the Gauss sum is of course known. Our determination of it follows essentially that in [DH] (Appendix II), but seems simpler – we need only some of their equations, and so some of the standard computations can be omitted. For a rigorous derivation one should keep in mind, that we are looking at the image of the Gauss sum under a specific local embedding j – this is implicit in all the usual proofs, but is made quite explicit in the form Theorem 23 was stated. The result for the norm resolvent was first published in [F17]. The purely local formulation was given in [FT].

We shall first define a variant of local Gauss sums. Let, as before, η be a homomorphism $(\mathfrak{o}_F/\mathfrak{p}_F)^* \to \mathbb{Q}_p^{c*}$, and let ψ_* be the non-null homomorphism from the additive group $(\mathfrak{o}_F/\mathfrak{p}_F)^+$ into $\mathbb{Q}_p^{c*}$, i.e., into the group of p-th roots of unity, given as follows.

Let $1 + \pi$ be a primitive p-th root of unity in $\mathbb{Q}_p^c$. If $z \in \mathbb{F}_p$, let $(1 + \pi)^z = (1 + \pi)^y$, where y is any integer with residue class z. Then $\psi_*(x) = (1 + \pi)^{Tx}$, T the trace $\mathfrak{o}_F/\mathfrak{p}_F \to \mathbb{F}_p$.

We define

$$G(\eta) = \begin{cases} 1, & \text{if} \quad \eta = \varepsilon, \\ \sum\limits_{x \in (\mathfrak{o}_F/\mathfrak{p}_F)^*} \eta(x)\psi_*(x), & \text{if} \quad \eta \neq \varepsilon. \end{cases} \tag{7.5}$$

Thus $G(\eta) \in \mathbb{Q}_p^c$.

7.1. Lemma. *Let $k\colon \mathbb{Q}^c \to \mathbb{Q}_p^c$ be an embedding, and let χ be a tame Abelian character of Ω_F. Let η be the restriction of θ_χ^k to $\mathfrak{o}_F^*$, viewed as a character of $(\mathfrak{o}_F/\mathfrak{p}_F)^*$.*

Then

(7.6) $$(\tau(\chi)^k) = (G(\eta)).$$

Proof. We know that $\tau(\chi) = \tau(\theta_\chi)$, as defined in (I.5.5), (I.5.6). If χ is non-ramified then $\tau(\theta_\chi)$ is a root of unity in $\mathbb{Q}^c$, so $\tau(\chi)^k$ is a root of unity in $\mathbb{Q}_p^c$, and $\eta = \varepsilon$. By (7.5), we get (7.6) in this case. If $\mathfrak{f}(\chi) = \mathfrak{p}_F$, then $\varepsilon \neq \eta$ and the map $x \mapsto \psi_F(c^{-1}x)^k = \psi'(x)$, with c as in (I.5.5), defines a non-null homomorphism $(\mathfrak{o}_F/\mathfrak{p}_F)^+ \to \mathbb{Q}_p^{c*}$. With ψ_* as in (7.5), we have $\psi'(x) = \psi_*(ax)$ for some fixed a, $a \neq 0$ in $\mathfrak{o}_F/\mathfrak{p}_F$. Thus

$$\begin{aligned}\tau(\chi)^k &= \left(\sum_{x \in (\mathfrak{o}_F/\mathfrak{p}_F)^*} \eta(x)\psi_*(ax))\theta_\chi^k(c^{-1})\right)\\ &= G(\eta)\eta(a^{-1})\theta_\chi^k(c^{-1}),\end{aligned}$$

and as $\eta(a^{-1})\theta_\chi^k(c^{-1})$ is a root of unity we again get (7.6). □

For simplicity's sake we shall write from now on

(7.7) $$\mathbf{N}_F\mathfrak{p}_F = q = p^f$$

and

(7.8) $$s(\eta)(q-1) = s(q-1) = h = h(\eta).$$

Thus

(7.9) $$h \in \mathbb{Z}, \qquad 0 \leqslant h < q-1.$$

The residue class map defines an isomorphism from the group of $q-1$-st roots of unity in $\mathbb{Q}_p^c$ onto $(\mathfrak{o}_F/\mathfrak{p}_F)^*$. Therefore $\eta \mapsto h(\eta)$ is a bijection from the set of homomorphisms $(\mathfrak{o}_F/\mathfrak{p}_F)^* \to \mathbb{Q}_p^{c*}$ onto the range given in (7.9).

We define, with h as in (7.9),

(7.10) $$\begin{cases} t(h) = \sum_{j=0}^{f-1} h_j \quad \text{if} \quad h = \sum_{j=0}^{f-1} h_j p^j, \quad 0 \leqslant h_j \leqslant p-1,\\ r(h) = \sum_{j=0}^{f-1} \left\{\dfrac{p^j h}{q-1}\right\},\end{cases}$$

where $\{c\}$ stands for the fractional part of a rational number c, i.e., $\{c\} \in \mathbb{Q}$, $\{c\} \equiv c \pmod{\mathbb{Z}}$, $0 \leqslant \{c\} < 1$.

Theorem 27. *Let $\phi \in R_{(F),p}$ be a tame Abelian character, let $\eta = \theta_{\phi|\mathfrak{o}_F^*}$, and view η also as a homomorphism $\mathbb{F}_q^* \to \mathbb{Q}_p^{c*}$, and let $h = h(\eta)$. Then*

$$\mathcal{N}_{F/\mathbb{Q}_p}P(\phi) = (p)^{r(h)} = (p)^{t(h)/p-1} = (G(\eta)).$$ □

The proof is based on a number of intermediate statements. We consider functions $g\colon \mathbb{Z} \to \mathbb{Q}$ of period $q-1$, i.e., so that

(7.11) $$g(h) = g(h'), \quad \text{if} \quad h \equiv h' \pmod{q-1}$$

and we agree to extend t and r, as defined in (7.10) on the domain (7.9) by such periodicity to the whole of $\mathbb{Z}$. We consider in particular functions g satisfying the following further conditions:

$$(7.12)\quad \begin{cases} \text{(a)} & g(0) = 0 \\ \text{(b)} & g(1) = 1 \\ \text{(c)} & g(h + h') \leqslant g(h) + g(h'), & 0 \leqslant h,\quad h' < q - 1, \\ \text{(d)} & g(p^i h) = g(h), & 0 \leqslant h < q - 1, \\ \text{(e)} & g(h) + g(-h) = (p - 1)f, & 0 < h < q - 1. \end{cases}$$

We shall then prove:

A. g satisfies (7.12) if and only if $g = t$.
B. $(p - 1)r$ satisfies (7.12).
C. If $(G(\eta)) = (p)^{t_1(h)/p-1}$ for $h = h(\eta)$, then t_1 satisfies (7.12).
D. $\mathcal{N}_{F/\mathbb{Q}_p} P(\phi) = (p)^{r(h)}$.

It is clear that A to D imply Theorem 27. □

Proof of A. Clearly t satisfies (7.12). Conversely suppose g satisfies (7.12). By (b) and (c),

$$(7.13)\qquad g(h) \leqslant h \qquad \text{for} \qquad 0 < h \leqslant p - 1.$$

If now $h = \sum_{j=0}^{f-1} h_j p^j$, $0 \leqslant h_j \leqslant p - 1$ then it follows from 7.12 (a) and (d), and from (7.13) that $g(h_j p^j) \leqslant h_j$. Applying (7.12) (c) again we get $g(h) \leqslant t(h)$; for $h \neq 0$ equality $g(h) = t(h)$ follows now from (e), and for $h = 0$ from (a). □

Proof of B. (a) is trivial. (b) follows from $\{p^j/(q - 1)\} = p^j/(q - 1)$ and summing, (c) from the corresponding inequality for $\{p^j h/(q - 1)\}$. For (d) we observe that multiplication by p^i simply permutes the terms in the sum for $r(h)$. Finally we get (e) from the relation $\{c\} + \{-c\} = 1$ if $c \neq 0$. □

Proof of C. The case $h(\eta) = 0$, i.e., $\eta = \varepsilon$ is obvious.

Now assume $h(\eta) = 1$. Let π be the element used in defining ψ_*, whence $(\pi)^{p-1} = (p)$, and let for the moment $\mathfrak{P}$ be the maximal ideal of $\mathbb{Z}_p^c$. With y running through the $q - 1$-st roots of unity in $\mathbb{Q}_p^c$, we see that $\eta(y) = y^{-1}$ and get

$$G(\eta) = \sum_y y^{-1}(1 + \pi)^{[Ty]}$$

where $[Ty] \in \mathbb{Z}$ represents the trace of the residue class of y. As $\pi^2 \in \pi\mathfrak{P}$ and $[Ty] \equiv \sum_{j=0}^{f-1} y^{p^j} \pmod{\mathfrak{P}}$ we get

$$G(\eta) \equiv \sum_y y^{-1}(1 + [Ty]\pi) \equiv \sum_y y^{-1}\left(1 + \sum_{j=0}^{f-1} y^{p^j}\pi\right) \pmod{\pi\mathfrak{P}}.$$

But $\sum_y y^{-1} = 0 = \sum_y y^{p^j-1}$ for $j \neq 0$, whence

$$G(\eta) \equiv \left(\sum_y 1\right)\pi \equiv (q-1)\pi \equiv -\pi \pmod{\pi\mathfrak{P}}.$$

Thus $(G(\eta)) = (\pi) = (p)^{1/p-1}$, and this establishes (7.12) (b) for t_1.

(7.12) (c) in the case $h = 0$ or $h' = 0$ is trivial, and in the case $h + h' \equiv 0 \pmod{q-1}$ follows from (a) and (e). In the remaining case, when the homomorphisms η, η' are not inverses and both distinct from ε we have

$$G(\eta)G(\eta') = \sum_{c \in \mathbb{F}_q} S_c \psi_*(c)$$

$$S_c = \sum_{x,y} \eta(x)\eta'(y) \qquad \text{(where } x, y \in \mathbb{F}_q^*, x + y = c).$$

For $c = 0$ this is $\eta'(-1)\sum_x \eta\eta'(x) = 0$. For $c \neq 0$ we write $x = c(c^{-1}x)$, $y = c(c^{-1}y)$, and replace x, y by $c^{-1}x, c^{-1}y$. Then we get

$$S_c = \eta\eta'(c)S_1,$$

hence

$$G(\eta)G(\eta') = S_1 G(\eta\eta').$$

As S_1 is integral, (7.12) (c) now follows.

For (7.12) (d) observe that $x \mapsto x^{p^i}$ is an automorphism of $\mathbb{F}_q$, and $T(x) = T(x^{p^i})$. Thus

$$\begin{aligned} G(\eta^{p^i}) &= \sum \eta(x^{p^i})\psi_*(x) \\ &= \sum \eta(x^{p^i})\psi_*(x^{p^i}) \\ &= G(\eta). \end{aligned}$$

Finally (7.12) (e) comes from the equation

$$G(\eta)G(\eta^{-1}) = \eta(-1)q \qquad \text{for} \qquad \eta \neq \varepsilon,$$

which is standard. (See I.5.7.a.) □

Proof of D. Let M be the maximal non-ramified subfield of F, and choose a right transversal $\{\sigma\}$ of Ω_F in Ω_M. We extend this to a right transversal $\{\sigma\omega^{-j}\}$ ($j = 0, \ldots, f-1$), where ω acts as the Frobenius over $\mathbb{Q}_p$ on M. ϕ and η are now again as in Theorem 27. As the values of η lie in M, $\eta^{\omega^j\sigma^{-1}} = \eta^{\omega^j} = \eta^{p^j}$. Thus $\phi^{\omega^j\sigma^{-1}}$ differs from ϕ^{p^j} by a non-ramified character, and so $P(\phi^{\omega^j\sigma^{-1}}) = P(\phi^{p^j})$. If $h = h(\eta)$, then by (7.8) and Theorem 26,

$$P(\phi^{p^j}) = \mathfrak{p}_F^{\{p^jh/q-1\}}$$

and as $\Omega_{\mathbb{Q}_p}$ acts trivially on ideals, we get

$$\prod_\sigma P(\phi^{\omega^j\sigma^{-1}})^{\sigma\omega^{-j}} = \mathfrak{p}_F^{\{p^jh/q-1\}e},$$

where $e = [F:M]$ is a ramification index, i.e., $\mathfrak{p}_F^e = (p)$. Therefore finally

$$\mathcal{N}_{F/\mathbb{Q}_p} P(\phi) = \prod_{j,\sigma} P(\phi^{\omega^j \sigma^{-1}})^{\sigma\omega^{-j}} = (p)^{r(h)}. \qquad \square$$

Proof of Theorem 23. Immediate, by the reduction step given earlier (see Corollary 1 to Theorem 25), by Theorem 27 and by Lemma 7.1). □

§8. Module Conductors and Module Resolvents

We shall only present a short survey. Module conductors and module resolvents are defined without any restriction on ramification, and have important applications for wild extensions. This lies outside the scope of the present volume. They do however also throw new light on some aspects of the "tame" theory. They have, as indicated in Chap. I, §1, played a role in its origins. We do not propose to give an exhaustive treatment, but merely illustrate the new connections which arise. We therefore keep to the original formulation in [F3], [F14], where the underlying integral representations were restricted to be realizable over the base ring. In recent work, necessarily of more technical complexity, this restriction has been removed (cf. [Ne1]). We shall also keep entirely within the framework of local fields.

Throughout F is a non-Archimedean local field, $\mathfrak{o} = \mathfrak{o}_F$, L/F is a Galois extension of finite degree with Galois group Γ, and initially no condition of tameness is imposed. Let X be an $\mathfrak{o}\Gamma$-lattice, spanning the $F\Gamma$-module V – here an "$\mathfrak{o}\Gamma$-lattice" is an $\mathfrak{o}\Gamma$-module, free of finite rank over $\mathfrak{o}$. Let $D_{\mathfrak{o}}(X)$ be the $\mathfrak{o}$-dual of X, i.e., $D_{\mathfrak{o}}(X) = \mathrm{Hom}_{\mathfrak{o}}(X, \mathfrak{o})$, with the contragredient structure of right $\mathfrak{o}\Gamma$-module, by the rule

$$\langle y\gamma, x\rangle = \langle y, x\gamma^{-1}\rangle, \tag{8.1}$$

$y \in D_{\mathfrak{o}}(X)$, $x \in X$, $\gamma \in \Gamma$. Here

$$\langle\ ,\ \rangle : D_{\mathfrak{o}}(X) \times X \to \mathfrak{o}$$

is the standard pairing. $D_{\mathfrak{o}}(X)$ spans the F-dual $D_F(V) = \mathrm{Hom}_F(V, F)$ with its contragredient structure as $F\Gamma$-module. We now define a pairing

$$\beta\colon \mathrm{Hom}_{F\Gamma}(V, L) \times \mathrm{Hom}_{F\Gamma}(D_F(V), L) \to F \tag{8.2}$$

as follows: Choose an F-basis $\{v_i\}$ of V and a dual F-basis $\{u_i\}$ of $D_F(V)$ – i.e., so that under the pairing $\langle\ ,\ \rangle$ extended to $D_F(V) \times V \to F$, we have $\langle u_i, v_j\rangle = \delta_{ij}$ (Kronecker delta). Then for $f \in \mathrm{Hom}_{F\Gamma}(V, L)$, $g \in \mathrm{Hom}_{F\Gamma}(D_F(V), L)$ we define

$$\beta(f, g) = \sum_i f(v_i) g(u_i). \tag{8.3}$$

$\beta(f, g)$ lies indeed in F, and is independent of the choice of the pair $\{v_i\}$, $\{u_i\}$ of dual bases. This can be verified directly, or it can be derived via a less down to earth definition of β, which we shall now state. Note that there is an isomorphism

$$[\ ]\colon V \otimes_F D_F(V) \to \mathrm{End}_F(V) \tag{8.4}$$

of $F\Gamma$-modules, Γ acting on the tensor product diagonally and on End by conjugation. This is given by

$$[v \otimes u](v') = v\langle u, v'\rangle, \qquad v, v' \in V, \qquad u \in D_F(V).$$

Moreover we have a map

$$\otimes : \mathrm{Hom}_{F\Gamma}(V, L) \times \mathrm{Hom}_{F\Gamma}(D_F(V), L) \to \mathrm{Hom}_{F\Gamma}(V \otimes_F D_F(V), L)$$

with $(f \otimes g)(v \otimes u) = f(v)g(u)$. Under the map [ ] we have, for $\{v_i\}, \{u_i\}$ as above,

$$\left[\sum_i v_i \otimes u_i\right] = 1_V = \text{identity of the ring } \mathrm{End}_F(V).$$

We thus get

(8.5) $$\beta(f, g) = (f \otimes g)([\ ]^{-1}(1_V)).$$

This shows that the original definition (8.1) is independent of the choice of dual pairs of bases, and, as $1_V^\gamma = 1_V$, that $\beta(f, g) \in F$.

Now let $\{f_i\}$, $\{g_i\}$ be $\mathfrak{o}$-bases of the two $\mathfrak{o}$-lattices $\mathrm{Hom}_{\mathfrak{o}\Gamma}(X, \mathfrak{o}_L)$, and $\mathrm{Hom}_{\mathfrak{o}\Gamma}(D_{\mathfrak{o}}(X), \mathfrak{o}_L)$ respectively. These modules span $\mathrm{Hom}_{F\Gamma}(V, L)$ and $\mathrm{Hom}_{F\Gamma}(D_F(V), L)$ respectively, and the latter both have F-dimension $= \dim_F(V)$. By studying the effect of a change of basis, we see immediately that the fractional ideal, in F,

(8.6) $$\mathfrak{c}(L/F, X) = \mathrm{Det}(\beta(f_i, g_j))_{i,j}\mathfrak{o}$$

only depends on X. One can prove in general that it is non-zero, but we shall not do this here, as in the tame case this will be a consequence of another result, which we shall derive. $\mathfrak{c}(L/F, X)$ is the *module conductor*. (Actually in [F14] two conductors $\mathfrak{c}^*$, $\mathfrak{c}_*$ were defined, but here we shall stick to the one $\mathfrak{c}$, which is the $\mathfrak{c}^*$ of [F14]).

In the same notation as before, and in particular with $\{f_i\}$ still an $\mathfrak{o}$-basis of $\mathrm{Hom}_{\mathfrak{o}\Gamma}(X, \mathfrak{o}_L)$, and now $\{v_i\}$ an $\mathfrak{o}$-basis of X, we see that

(8.7) $$\mathfrak{r}(L/F, X) = \mathrm{Det}(f_i(v_j))_{i,j}\mathfrak{o}$$

only depends on X. This is an $\mathfrak{o}$-submodule of L. It is non-zero, i.e., of rank one, and again we shall not prove this at this stage. $\mathfrak{r}(L/F, X)$ is the *module resolvent*.

8.1. Proposition (cf. [F14]).

$$\mathfrak{r}(L/F, X)\mathfrak{r}(L/F, D_{\mathfrak{o}}(X)) = \mathfrak{c}(L/F, X).$$

Proof. We use a pair of dual bases $\{v_i\}$ of X, $\{u_i\}$ of $D_{\mathfrak{o}}(X)$. Then, by (8.3),

$$\mathrm{Det}(f_i(v_j)) \cdot \mathrm{Det}(g_i(u_j)) = \mathrm{Det}\, \beta(f_i, g_j),$$

with $\{f_i\}, \{g_i\}$ as before. (Transpose the second determinant on the left!) □

Module conductors and module resolvents are multiplicative with respect to direct sums $X \oplus X'$ of $\mathfrak{o}\Gamma$-lattices. They conform in full generality to the standard pattern of behaviour under induction, or non-ramified base field extension, but we shall not go into this here. From now on we shall assume L/F to be tame. Then we have

Theorem 28 (cf. [F13]) (note [5] and [6]). *Let a be a free generator of $\mathfrak{o}_L$ over $\mathfrak{o}_F\Gamma$, and let X be an $\mathfrak{o}\Gamma$-lattice so that χ is the character of the representation of Γ on $X \otimes_{\mathfrak{o}} F^c$. Then*

$$\mathfrak{r}(L/F, X) = (a \mid \chi)\mathfrak{o}.$$

Corollary 1 (cf. [F14]). *$\mathfrak{c}(L/F, X)$ and $\mathfrak{r}(L/F, X)$ are non-zero, and only depend on χ, not on X itself.*

Corollary 2 (cf. [F14]). $\mathfrak{c}(L/F, \chi) = \mathfrak{f}(\chi)$. (By Theorem 24.)

Remark 1. Neither the theorem, nor the Corollaries are true in the wild case.

Remark 2. The last theorem was first suggested by *S.* Ullom.

Proof of Theorem 28. We first define an isomorphism

$$\kappa\colon \mathrm{Hom}_{\mathfrak{o}}(X, \mathfrak{o}) = D_{\mathfrak{o}}(X) \to \mathrm{Hom}_{\mathfrak{o}\Gamma}(X, \mathfrak{o}_L) \tag{8.8}$$

in terms of the given free generator a of $\mathfrak{o}_L$ over $\mathfrak{o}\Gamma$. For $u \in D_{\mathfrak{o}}(X)$, define κu by

$$\kappu(v) = \sum_\gamma a^\gamma \langle u\gamma, v\rangle$$

One verifies immediately that we get a homomorphism κ as in (8.8). Any $g \in \mathrm{Hom}_{\mathfrak{o}\Gamma}(X, \mathfrak{o}_L)$ can be written uniquely in the form

$$g(v) = \sum_\gamma a^\gamma \langle u_\gamma, v\rangle,$$

and one verifies that $u_\gamma = u_1\gamma$, for all γ. Thus the composite maps $g \mapsto u_1 = u_1(g) \mapsto \kappa(u_1) = g$, and $u \mapsto \kappa(u) \mapsto u_1(\kappa(u)) = u$ are identities, i.e., κ is an isomorphism.

Now represent Γ on the given module X via a basis v_i, and denote by $T_{ij}(\gamma)$ the matrix entries of $T(\gamma)$. In other words,

$$v_i\gamma = \sum_j T_{ij}(\gamma)v_j.$$

Hence, if $\{u_k\}$ is the dual basis of $\{v_i\}$, then $T_{ij}(\gamma) = \langle u_j\gamma^{-1}, v_i\rangle$ and we get

$$\begin{aligned}(a \mid \chi) &= \mathrm{Det}\Big(\sum_\gamma a^\gamma T_{ij}(\gamma^{-1})\Big)_{i,j}\\ &= \mathrm{Det}\Big(\Big\langle \sum_\gamma a^\gamma u_j\gamma, v_i\Big\rangle\Big)_{i,j}\\ &= \mathrm{Det}((\kappa u_j)(v_i))_{i,j},\end{aligned}$$

as required. □

Example. We give an example of an application of the last theorem (cf. [F21]). Let X, Y be $\mathfrak{o}\Gamma$-lattices, χ, ψ the corresponding characters. Thus $X \otimes_{\mathfrak{o}} Y$, with diagonal action of Γ, is an $\mathfrak{o}\Gamma$-lattice, and $\chi\psi$ is its associated character. As χ and ψ are fixed under Ω_F, it follows from Proposition I.4.4, that – with a a free generator of $\mathfrak{o}_L$ over $\mathfrak{o}\Gamma$ –

$$(a \mid \chi)^{\deg(\psi)}(a \mid \psi)^{\deg(\chi)}(a \mid \chi\psi)^{-1} \in F^*. \tag{8.9}$$

Now the map $f, g \mapsto f \otimes g$, $(f \otimes g)(u \otimes v) = f(u)g(v)$ gives rise to an injective homomorphism

$$\mathrm{Hom}_{\mathfrak{o}\Gamma}(X, \mathfrak{o}_L) \otimes_{\mathfrak{o}} \mathrm{Hom}_{\mathfrak{o}\Gamma}(Y, \mathfrak{o}_L) \to \mathrm{Hom}_{\mathfrak{o}\Gamma}(X \otimes_{\mathfrak{o}} Y, \mathfrak{o}_L), \tag{8.10}$$

with a cokernel which is a finite $\mathfrak{o}$-module. As such it has an "order" $\mathfrak{s}(X, Y)$ which is an $\mathfrak{o}$-ideal, and one easily sees that

$$\mathfrak{s}(X, Y) = \mathfrak{r}(L/F, X)^{\deg(\psi)}\mathfrak{r}(L/F, Y)^{\deg(\chi)}\mathfrak{r}(L/F, X \otimes_{\mathfrak{o}} Y)^{-1}.$$

By Theorem 28 we conclude

$$((a \mid \chi)^{\deg(\psi)}(a \mid \psi)^{\deg(\chi)}(a \mid \chi\psi)^{-1}) = \mathfrak{s}(X, Y). \tag{8.11}$$

In particular the element (8.9) is integral. Using Theorem 23 one gets a similar interpretation for the integral ideals

$$(\tau(\chi)^{\deg(\psi)}\tau(\psi)^{\deg(\chi)}\tau(\chi\psi)^{-1}) \qquad (\text{"Galois Jacobi sum"}).$$

Notes to Chapter III

[1] The inductivity of the L-function for non-Archimedean local fields is easily deduced from Proposition 1.3. We shall use the notation of its proof. Clearly χ may be assumed to be irreducible. If first $\mathrm{n}\chi = 0$, then by (ii) $\tilde{L}(s, \mathrm{ind}\,\chi) = \tilde{L}(s, \chi) = 1$. Now assume χ to be non-ramified. If E/F is totally ramified, then $\mathbf{N}_E\mathfrak{p}_E = \mathbf{N}_F\mathfrak{p}_F$, and $\theta_\chi = \theta_\eta \circ N_{E/F}$, in the notation of Proposition 1.3. Thus

$$\begin{aligned}\tilde{L}(s, \chi) &= (1 - \theta_\chi(\mathfrak{p}_E)(\mathbf{N}_E\mathfrak{p}_E)^{-s})^{-1}\\ &= (1 - \theta_\eta(\mathfrak{p}_F)(\mathbf{N}_F\mathfrak{p}_F)^{-s})^{-1} = \tilde{L}(s, \eta).\end{aligned}$$

Finally if E/F is non-ramified, then on writing $f = f(E/F)$, we have $\mathbf{N}_E\mathfrak{p}_E = \mathbf{N}_F\mathfrak{p}_F^f$ and $\mathrm{n}\,\mathrm{ind}\,\chi = \sum \beta_i$, where the β_i are the distinct extensions of χ to a character of Ω_F. Writing $\theta_{\beta_i} = \theta_i$, we get $\theta_\chi(\mathfrak{p}_E) = \theta_i(\mathfrak{p}_F)^f$, for each i, and

$$\begin{aligned}\tilde{L}(s, \mathrm{ind}\,\chi) &= \prod_i (1 - \theta_i(\mathfrak{p}_F)(\mathbf{N}_F\mathfrak{p}_F)^{-s})^{-1}\\ &= (1 - \theta_\chi(\mathfrak{p}_E)(\mathbf{N}_E\mathfrak{p}_E)^{-s})^{-1} = \tilde{L}(s, \chi).\end{aligned}$$

[2] The root numbers at infinity, and their relation to the resolvents, can be used to determine certain invariants of the orthogonal representation of a Galois group $\Gamma = \mathrm{Gal}(N/K)$ on N, viewed as a quadratic module via the trace form. If, say, $\mathfrak{p}$ is a real prime divisor of K, then $N_\mathfrak{p}$ is the direct sum of a positive definite quadratic $K_\mathfrak{p}\Gamma$-module $N_\mathfrak{p}^+$, with character ρ^+, and a negative definite quadratic $K_\mathfrak{p}\Gamma$-module $N_\mathfrak{p}^-$, with character ρ^-, and these two characters determine $N_\mathfrak{p}$ as a quadratic $K_\mathfrak{p}\Gamma$-module (cf. [FM]). If χ is an orthogonal character of Γ, then the scalar product $\langle \chi, \rho^- \rangle$ is closely related to the value of $W(\chi_\mathfrak{p})$ – for details see [F13].

[3] There are some indications that the global invariant $U_{N/K}$ may in some sense – possibly under some further conditions – have a local decomposition. So of course certainly has $W_{N/K}$. Thus the global equation $U_{N/K} = tW_{N/K}$ would then reflect corresponding local equations. The fact that considerable parts of the proof of Theorem 5 are local points to this. The Hermitian theory (cf. [F20], [F24]) referred to in an earlier remark (at the end of II §4) should be relevant here; on the level of the general algebraic formalism developed in [F24] it does connect local and global class groups. In the present context of Galois module structure, the original theorem on H_8 was in a purely formal manner restated as a collection of local results in [Ma4]. New light is thrown on this problem in a recent paper (cf. [CN–T2]).

[4] One of the formal procedures, with respect to which resolvents behave "like" Galois Gauss sums is induction of characters. In fact the original proof of Theorems 5 and 7, or locally of Theorems 23 and 24 (cf. [F17], [F18] and [FT]), was based on reduction to the Abelian case via Brauer's induction theorem. Following a suggestion of Queyrut (cf. [Q2]) this step can now be omitted. It is, however, of some interest in itself and also useful for other applications. We shall state the result purely in terms of local fields, although it is of course more general.

We consider a tower $\mathbb{Q}_p \subset F \subset M \subset L$ of local fields, with L/F tame and Galois with Galois group Γ. Let $\mathrm{Gal}(L/M) = \Delta$. Write $\mathrm{ind}_\Delta^\Gamma = \mathrm{ind}$. We have (cf. [F17], [F18])

I. *Suppose a is a free generator of* $\mathfrak{o}_L$ *over* $\mathfrak{o}_F\Gamma$, *b a free generator of* $\mathfrak{o}_L$ *over* $\mathfrak{o}_M\Delta$. *Let* $\{c_i\}$ *be a free basis of* $\mathfrak{o}_M$ *over* $\mathfrak{o}_F$, *and let* $\{\sigma\}$ *be a right transversal of* Ω_M *in* Ω_F, *used to define the norm resolvent*

$$\mathcal{N}_{M/F}(b \mid \chi) = \prod_\sigma (b \mid \chi^{\sigma^{-1}})^\sigma.$$

Then $\exists \lambda \in \mathfrak{o}_F\Delta^*$, *so that for all* $\chi \in R_{\Delta,p}$,

$$(\mathrm{Det}\, c_i^\sigma)^{\deg(\chi)} \mathcal{N}_{M/F}(b \mid \chi) = (a \mid \mathrm{ind}\, \chi) \mathrm{Det}_\chi(\lambda).$$

It follows that $\mathcal{N}_{M/\mathbb{Q}_p} P(\chi)$ is inductive in degree zero. Moreover, the extra factor $(\mathrm{Det}\, c_i^\sigma)$ as an ideal is the same as that appearing in the induction formula for Galois Gauss sums. Indeed if ε is the identity character of Δ (over $\mathbb{Q}^c$) then

II. $(\tau(\mathrm{ind}_\Delta^\Gamma \varepsilon)) = (\mathrm{Det}\, c_i^\sigma)$ *and* $(\tau(\mathrm{ind}_\Delta^\Gamma \varepsilon))^2 = \mathfrak{d}(M/F)$, *the relative discriminant.*

In terms of the module $(\mathfrak{o}:\chi)$, introduced in I, note [4], the induction theorem can be stated in the form

$$\mathfrak{e}(M/F)^{\deg(\chi)}\mathcal{N}_{M/F}(\mathfrak{o}_M:\chi) = (\mathfrak{o}_F:\operatorname{ind}\chi),$$

where $\mathfrak{e}(M/F)$ is the $\mathfrak{o}_F$-module, generated by the $\operatorname{Det}(c_i^\sigma)$, as above, i.e., $\mathfrak{e}(M/F)^2 = \mathfrak{d}(M/F)$.

[5] In terms of the modules $(\mathfrak{o}:\chi)$ of I note [4], we have of course $\mathfrak{r}(L/F, X) = (\mathfrak{o}:\chi)$ in the notation of the theorem. Thus in the wild case the module resolvents $\mathfrak{r}(L/F, X)$ are generalizations of $(\mathfrak{o}:\chi)$. Note also that, in the tame case, $P(\chi)$ is the ideal generated by $\mathfrak{r}(L/F, X)$, i.e., by $(\mathfrak{o}:\chi)$.

[6] There is yet another, related module theoretic interpretation of resolvents, due to S. Chase (cf. [Chs]). The set $\operatorname{Map}(\Gamma, L)$ is the cartesian product of copies of L, and as such has the structure of an L-algebra. It moreover admits Γ as a group of automorphisms, by $f^\gamma(\gamma_1) = f(\gamma\gamma_1)$, for $f \in \operatorname{Map}(\Gamma, L), \gamma, \gamma_1 \in \Gamma$. (This and similar constructions are appearing repeatedly in this book). For x, $y \in L$, define $[x, y] \in \operatorname{Map}(\Gamma, L)$ by

(a) $$[x, y](\gamma) = x \cdot y^\gamma.$$

Give the tensor product $L \otimes_F L$ the structure of an L-algebra via the left tensor factor, and of a Γ-module via the right tensor factor: $(x \otimes y)^\gamma = x \otimes y^\gamma$. The map $x \otimes y \mapsto [x, y]$ then yields an isomorphism

(b) $$L \otimes_F L \xrightarrow{\sim} \operatorname{Map}(\Gamma, L)$$

of L-algebras and of Γ-modules, giving rise to an injection

(c) $$\mathfrak{o}_L \otimes_{\mathfrak{o}} \mathfrak{o}_L \to \operatorname{Map}(\Gamma, \mathfrak{o}_L)$$

with torsion (hence finite) cokernel. Identify the two sides of (b) via the given isomorphism. We end up with an $\mathfrak{o}_L\Gamma$-module

$$\operatorname{Map}(\Gamma, \mathfrak{o}_L)/\mathfrak{o}_L \otimes_{\mathfrak{o}} \mathfrak{o}_L$$

which is an $\mathfrak{o}$-torsion module and the quotient of two free $\mathfrak{o}_L\Gamma$-modules of rank one. (We are in the local case – globally we would have two locally free $\mathfrak{o}_L\Gamma$-modules). This module defines an element of the Grothendieck group $\mathfrak{K}_0T(\mathfrak{o}_L\Gamma)$ (cf. (II.1.2), (II.1.4)), and this in turn is represented by an element of $\operatorname{Hom}_{\Omega_L}(R_\Gamma, (\mathbb{Q}_p^c)^*)$ (cf. (II.2.3)). Chase shows:

I. *Under the isomorphism* (II.2.3), *the class of the module* $\operatorname{Map}(\Gamma, \mathfrak{o}_L)/\mathfrak{o}_L \otimes_{\mathfrak{o}} \mathfrak{o}_L$ *corresponds to the class of the resolvent map* $\chi \mapsto (a \mid \chi)$, *where* $a\mathfrak{o}\Gamma = \mathfrak{o}_L$.

Proof. The map $f \mapsto \sum_{\gamma \in \Gamma} f(\gamma)\gamma^{-1}$ is an isomorphism

$$\operatorname{Map}(\Gamma, L) \cong L\Gamma$$

UI CREDIT UNION

niversity of Illinois Credit Union
201 S. First Street
hampaign, IL 61820
17-278-7700

UIECU
llini Union Branch 1005125156
icah James 10/05/2006
12:51 PM

ccount: 102073 Micah James
hecking Sub: 1 (Share Draft)
Joint: Jennifer James

eposit		516.75
ew Balance	$	2,128.53

ccount: 102073 Micah James

hecks	20.00
hecks	5.75
hecks	511.00

- Member Service Representative -
Rose

Thank-You

Cash Back 20.00

yielding an isomorphism

(d) $$\mathrm{Map}(\Gamma, \mathfrak{o}_L) \cong \mathfrak{o}_L\Gamma.$$

A generator of $\mathfrak{o}_L \otimes_{\mathfrak{o}} \mathfrak{o}_L$ as an $\mathfrak{o}_L\Gamma$-submodule of $\mathrm{Map}(\Gamma, \mathfrak{o}_L)$ is $[1, a]: \gamma \mapsto a^\gamma$, while a generator of $\mathrm{Map}(\Gamma, \mathfrak{o}_L)$ is the map $e: 1 \mapsto 1$, and $\gamma \mapsto 0$ if $\gamma \in \Gamma$, $\gamma \neq 1$. The image of the latter under (d) is the identity of $\mathfrak{o}_L\Gamma$, that of the former is $\sum_\gamma a^\gamma \gamma^{-1}$. This gives the assertion.

Analogously Chase describes the $\mathfrak{o}_L\Gamma$-torsion module $\mathfrak{D}_{L/F}/\mathfrak{o}_L$ in terms of a character function given by the conductors. The resolvent-conductor formula (cf. Theorem 24) arises in this context as a relation between the given modules, as above. From this point of view there is yet another $\mathfrak{o}_L\Gamma$-module, which corresponds to the Galois Gauss sums and it is of some interest to describe this explicitly. The precise relationship of Chase's approach with the developments outlined in the next chapter and leading ultimately to the proof of Theorem 9, is as yet unclear.

IV. Congruences and Logarithmic Values

§1. The Non-Ramified Characteristic

The central Theorem 5 reveals Galois Gauss sums as the basic arithmetic ingredient of the Galois module structure theory for rings of integers. Once this theorem had been established, progress in the latter theory had to be bound up with the theory of Galois Gauss sums. Here the new developments were based on the method of congruences and of logarithmic evaluation, in close parallel with the application of the same method to group determinants. It is the main purpose of the present chapter to describe the principal results – although for determinants some of these had already been anticipated in Chap. II, §4–6. The guiding aim is the proof of the main conjecture, recently obtained by M. Taylor (cf. [Ty7]), i.e., the proof of Theorem 9, or rather of Theorem 11, to which it has already been reduced in Chap. I, §6. We shall follow essentially the strategy of [Ty7], except for a stronger local emphasis. In addition, there is a final section, which is devoted to a systematic survey of the theory of tame local Galois Gauss sums, mainly coming from [FT]. This will, among other things, contain an outline of the proof of two theorems, which are of independent interest. The first contains an internal characterization of the tame Galois Gauss sums over a given local field, by certain arithmetic properties within that field. The second, related one, is the existence theorem quoted earlier, i.e., Theorem 18.

In the present section we shall first of all introduce a simple, but nevertheless fundamental function of local Galois characters, which we call the *non-ramified characteristic*. It first appeared in another context in [De1], then "in embryo" in some of the early work on Galois module structure, referred to in note [7] of Chap. II. It was studied systematically in [FT], and was used there as one of the ingredients in the local characterization of Galois Gauss sums. It was also considered independently in [Q2]. It finally turned out to be of fundamental importance in the proof of Theorem 9. To indicate its role in this briefly, recall the presentation of $D(\mathbb{Z}\Gamma)$, following Theorem 1. As a first step in proving that the class of $U_{N/K} \cdot tW_{N/K}^{-1}$ is 1, we had to find a convenient representative for it in $\mathrm{Hom}^+_{\Omega_\mathbb{Q}}(R_\Gamma, \mathfrak{U}(\mathbb{Q}^c))$. Having done this we then wish to write this representative as a product of one factor from $\mathrm{Hom}^+_{\Omega_\mathbb{Q}}(R_\Gamma, \mathfrak{Y}(\mathbb{Q}^c))$, and another one from $\mathrm{Det}\,\mathfrak{U}\mathbb{Z}\Gamma$. The non-ramified characteristic essentially provides such a first factor canonically. The main effort then goes into showing, that what is left lies in $\mathrm{Det}\,\mathfrak{U}\mathbb{Z}\Gamma$. For the purpose of proving Theorem 11, it will however be convenient to introduce a non-canonical variant of the non-ramified characteristic, which is easier to handle, and which will enter the formulation of two entirely local theorems. These, as we shall show at the end of this section, in turn imply Theorem 11. The remainder of the present chapter is exclusively local.

Throughout this section F is any non-Archimedean local field of residue class characteristic p. Characters in this section always have values in $\mathbb{Q}^c$; thus R_Γ has its usual meaning.

Recall now the definition of the non-ramified part $\mathrm{n}\chi$ of a Galois character χ of F (cf. I §5, in particular (5.17)). This has already been used in our definition of the local Artin L-function (cf. (I.5.18)), and of the local Artin conductor for tame characters (cf. (I.5.26)). See in this context Proposition III.1.3 and its application (note [1] in III). Here we shall introduce yet a third object in terms of $\mathrm{n}\chi$. For any Galois character χ of F, not necessarily tame, the *non-ramified characteristic*

$$y(F,\chi) = y(\chi)$$

is defined by

$$y(\chi) = (-1)^{\deg(\mathrm{n}\chi)}\theta_{\det(\mathrm{n}\chi)}(\mathfrak{p}_F). \tag{1.1}$$

(As before we are writing $\det(\mathrm{n}\chi)$ in place of $\det_{\mathrm{n}\chi}$, to avoid double subscripts). As $\det(\mathrm{n}\chi)$ is non-ramified, $\theta_{\det(\mathrm{n}\chi)}(\mathfrak{p}_F)$ is uniquely defined as $\theta_{\det(\mathrm{n}\chi)}(a)$, for any $a \in F^*$, with $a\mathfrak{o}_F = \mathfrak{p}_F$.

If l is a rational prime, let $\mathfrak{L}_l$ be the radical of the ideal (l) in the ring of all algebraic integers – recall the explanation of congruences mod $\mathfrak{L}_l$ given in II §4, prior to Theorem 15A – and from the same place, in particular (II.4.2), also recall the definition of $\operatorname{Ker} d_l$.

Theorem 29. (i) *Let L/F be a Galois extension of local fields, with Galois group Γ. $y(F,\cdot)$ restricted to R_Γ lies in* $\operatorname{Hom}^+_{\Omega_{\mathbb{Q}}}(R_\Gamma, \mu(\mathbb{Q}^c))$, *where $\mu(\mathbb{Q}^c)$ is the group of roots of unity in $\mathbb{Q}^c$. In particular $y(R_\Gamma^s) = 1$.*

(ii) *y is fully inductive, i.e., if F' is an extension of F of finite degree, and* ind: $R_{(F')} \to R_{(F)}$ *is the induction map, then*

$$y(F',\chi) = y(F, \operatorname{ind}\chi).$$

(iii) *With L/F as in* (i), *but now restricted to be tame, let $\chi \in \operatorname{Ker} d_{\Gamma,l}$, for some prime number l. Then*

$$y(F,\chi) \equiv \tau(F,\chi) \pmod{\mathfrak{L}_l}.$$

Remark. This theorem comes from [FT], where however only tame characters were considered.

Proof. (i) It is clear that y is a homomorphism of $\Omega_{\mathbb{Q}}$-modules into $\mu(\mathbb{Q}^c)$. This already implies that $y(\operatorname{Tr}(\phi)) = 1$, where we recall that $\operatorname{Tr}(\phi) = \phi + \bar{\phi}$. On the other hand if χ is irreducible symplectic, then $\mathrm{n}\chi = 0$, i.e. $y(\chi) = 1$. By additivity, $y(\chi) = 1$ for all $\chi \in R_\Gamma^s$.

(ii) Let $\chi \in R_{(F')}$. To prove inductivity, we may suppose that χ is irreducible, and further that F'/F is non-ramified, or is totally ramified. In both cases, $\mathrm{n}\chi = 0$

implies $\mathrm{n}\,\mathrm{ind}\,\chi = 0$, by Proposition III.1.3; thus $y(F', \chi) = y(F, \mathrm{ind}\,\chi)$ if $\mathrm{n}\chi = 0$. So we now suppose that χ is non-ramified, hence Abelian.

First take F'/F to be totally ramified. Write res for the restriction map $R_{(F)} \to R_{(F')}$, and apply Proposition III.1.3. This tells us that for some Abelian character η, with $\mathrm{res}\,\eta = \chi$, we have $\mathrm{n}\,\mathrm{ind}\,\chi = \eta$. But

$$\theta_{\mathrm{res}\,\eta} = \theta_\eta \circ N_{F'/F}. \tag{1.2}$$

Hence $\theta_\chi(\mathfrak{p}_{F'}) = \theta_\eta(N_{F'/F}\mathfrak{p}_{F'}) = \theta_\eta(\mathfrak{p}_F)$. Trivially $\deg(\mathrm{n}\chi) = \deg(\eta) = 1$. Thus again $y(F', \chi) = y(F, \mathrm{ind}\,\chi)$.

Finally, with χ still non-ramified Abelian, assume now that F'/F is non-ramified of degree m. By Proposition III.1.3,

$$\mathrm{n}\,\mathrm{ind}\,\chi = \mathrm{ind}\,\chi$$

i.e.,

$$\mathrm{n}\,\mathrm{ind}\,\chi = \sum_{j=1}^{m} \eta\alpha_j$$

where $\mathrm{res}\,\eta = \chi$, and the α_j are the distinct Abelian characters of $\mathrm{Gal}(F'/F)$ inflated to Ω_F, or to Γ. One now computes the determinant, and gets

$$\det(\mathrm{n}\,\mathrm{ind}\,\chi) = \begin{cases} \eta^m, & \text{if } m \text{ is odd,} \\ \eta^m\alpha_1, \quad \alpha_1 \text{ the non-ramified quadratic character,} & \text{if } m \text{ is even.} \end{cases}$$

Then $\theta_{\alpha_1}(\mathfrak{p}_F) = -1$. As $m = \deg(\mathrm{ind}\,\chi)$ we get

$$\begin{cases} y(\mathrm{ind}\,\chi) = (-1)^m\theta_\eta(\mathfrak{p}_F)^m = -\theta_\eta(\mathfrak{p}_F)^m, & m \text{ odd,} \\ y(\mathrm{ind}\,\chi) = (-1)^m\theta_{\alpha_1}(\mathfrak{p}_F)\theta_\eta(\mathfrak{p}_F)^m = -\theta_\eta(\mathfrak{p}_F)^m, & m \text{ even.} \end{cases} \tag{1.3}$$

Let $a\mathfrak{o}_F = \mathfrak{p}_F$, so $a\mathfrak{o}_{F'} = \mathfrak{p}_{F'}$. Recall that $\mathrm{res}\,\eta = \chi$. As in (1.2) we get

$$\theta_\chi(\mathfrak{p}_{F'}) - \theta_\chi(a) = \theta_\eta(N_{F'/F}a) = \theta_\eta(a^m) = \theta_\eta(\mathfrak{p}_F)^m.$$

Thus indeed (1.3) gives the required equation.

For (iii) we use inductivity. By a refinement of Brauer's induction theorem (cf. [Del]), $\mathrm{Ker}\,d_l$ is generated by images under induction of virtual characters $\phi - \phi'$, where ϕ, ϕ' are Abelian, with ϕ' of order prime to l and $\phi'\phi^{-1}$ of l-power order.

By (ii) and the inductivity of τ in degree zero, it remains for us to show that for such a pair ϕ, ϕ',

$$\tau(\phi - \phi') \equiv y(\phi - \phi') \pmod{\mathfrak{L}_l}.$$

Writing $\theta = \theta_\phi$, $\theta' = \theta_{\phi'}$, $\theta = \theta'\theta''$, we get multiplicative characters θ', θ'' of orders prime to l, and a power of l, respectively, and $\tau(\phi - \phi') = \tau(\theta)/\tau(\theta')$. Evaluating

$y(\phi - \phi')$, we are now reduced to proving

$$(1.4)\quad \tau(\theta)/\tau(\theta') \equiv \begin{cases} 1(\text{mod}\,\mathfrak{L}_l), & \text{if } \theta, \theta' \text{ are both ramified,} \\ 1(\text{mod}\,\mathfrak{L}_l), & \text{if } \theta \text{ (hence } \theta'\text{) is non-ramified,} \\ -\theta'(\mathfrak{p}_F)^{-1}(\text{mod}\,\mathfrak{L}_l), & \text{if } \theta \text{ is ramified, but } \theta' \text{ is not.} \end{cases}$$

(Recall that l power roots of unity are $\equiv 1 \pmod{\mathfrak{L}_l}$.)

If θ is non-ramified the result is obvious. Now suppose that θ is ramified. As all values of θ'' are $\equiv 1 \pmod{\mathfrak{L}_l}$, we get from (I.5.5), and in the notation of (I.5.5),

$$(1.5)\qquad \tau(\theta) \equiv \sum_u \theta'(uc^{-1})\psi_F(uc^{-1}) \pmod{\mathfrak{L}_l}.$$

Assume also that $l \neq p$. If θ' is ramified, the right hand side of (1.5) is $\tau(\theta')$. By (III.5.4) we know this is a unit at all prime divisors above l, therefore we do get (1.4). If θ' is non-ramified, the right hand side of (1.5) is $\theta'(c)^{-1}\sum_u \psi_F(uc^{-1})$. Now $x \mapsto \psi_F(xc^{-1})$ defines a non-trivial character of the additive group of $\mathfrak{o}_F/\mathfrak{p}_F$. Thus

$$\begin{aligned} 0 &= \sum \psi_F(sc^{-1}) \qquad (\text{sum over } x \in \mathfrak{o}_F \bmod \mathfrak{p}_F) \\ &= 1 + \sum_u \psi_F(uc^{-1}). \end{aligned}$$

Therefore $\sum_u \psi_F(uc^{-1}) = -1$. (We shall use this observation subsequently, without repeating the proof). As $c\mathfrak{o}_F = \mathfrak{p}_F\mathfrak{D}_F$ we now deduce from (1.5) that

$$\tau(\theta) \equiv -\theta'(\mathfrak{p}_F\mathfrak{D}_F)^{-1} \pmod{\mathfrak{L}_l}.$$

As $\tau(\theta')$ is now $\theta'(\mathfrak{D}_F)^{-1}$ we again get (1.4).

Finally assume $l = p$, θ ramified. Then θ'' is non-ramified, and hence $\theta(u) = \theta'(u)$ for all units u. Hence $\tau(\theta)/\tau(\theta') = \theta''(c)^{-1} \equiv 1 \pmod{\mathfrak{L}_l}$, as required. □

The last theorem suggests that the function $\tau(\chi)y(\chi)^{-1}$ should have good determinantal properties – as indeed it has. At this stage we shall however introduce a twisted version of y, or rather of τy^{-1} (cf. [Ty7]). This is no longer "canonical", in that it depends on choices, and has a more restricted domain. It has however a very useful invariance property with respect to multiplication of characters by non-ramified Abelian characters.

We fix once and for all an element c of F^* with $c\mathfrak{o}_F = \mathfrak{D}_F\mathfrak{p}_F$. If F'/F is of finite degree, and is tame then we have the equation $\mathfrak{D}_{F'/F} = \mathfrak{p}_{F'}^{e-1}$ for the relative different, where $e = e(F'/F)$ is the ramification index (cf. [F5], 5, Theorem 2). Therefore also $c\mathfrak{o}_{F'} = \mathfrak{D}_{F'}\mathfrak{p}_{F'}$. For any such field F' and for every Galois character χ of F' we define

$$(1.6)\qquad z(F', \chi) = \theta_{\det(\chi)}(c),$$

$$(1.7)\qquad \tau^*(F', \chi) = \tau(F', \chi)y(F', \chi)^{-1}z(F', \chi).$$

(We omit the reference to F' if there is no danger of confusion.) We are thus replacing y by yz^{-1} and τy^{-1} by τ^*.

We shall state the basic properties of τ^* in the next proposition. It will be convenient in particular at the proof stage to think in terms of a tower of fields $L \supset F'' \supset F' \supset F$ with L/F of finite degree and tame, L/F' Galois with Galois group Γ, $F'' = L^{\Delta}$, $\Delta \subset \Gamma$. Thus the characters are assumed to be tame. As usual the symbol Γ_0 stands for the inertia subgroup of Γ. u_p is the homomorphism $\Omega_{\mathbb{Q}} \to \mathbb{Z}_p^*$ which occurred in Theorem 20B.

1.1. Proposition (cf. [Ty7]). (i) *The $\tau^*(\chi)$ are units at all prime divisors, not above p.*

(ii) $$\tau^*(\chi^{\omega^{-1}})^{\omega} = \tau^*(\chi)\theta_{\det(\chi)}(u_p(\omega)), \qquad \text{for} \qquad \omega \in \Omega_{\mathbb{Q}}.$$

In particular,

$$\tau^*(\chi^{\omega^{-1}})^{\omega} = \tau^*(\chi) \qquad \text{if} \qquad \omega \in \Omega_{\mathbb{Q}(p)}.$$

(iii) $\tau^*(\chi) \equiv 1 \pmod{\mathfrak{L}_l}$, *if l is a prime and* $\chi \in \operatorname{Ker} d_l$.

(iv) τ^* *is inductive in degree zero, i.e., if* $\chi \in R_{\Delta}$, $\deg(\chi) = 0$ *then*

$$\tau^*(F', \operatorname{ind}_{\Delta}^{\Gamma} \chi) = \tau^*(F'', \chi).$$

(v) *If χ is irreducible, then $\tau^*(\chi)$ lies in the field $\mathbb{Q}(p)(\operatorname{res}_{\Gamma_0}^{\Gamma} \chi)$ of values of $\operatorname{res}_{\Gamma_0}^{\Gamma} \chi$ over $\mathbb{Q}(p)$.*

(vi) *If α is non-ramified Abelian then*

$$\tau^*(\alpha\chi) = \tau^*(\chi).$$

Remark 1. Compare the congruence (iii), or for that matter the analogous congruence which, by Theorem 29, holds for τy^{-1}, with Theorem 15A. We see for the first time the close and obvious parallel between adjusted Galois Gauss sums and group determinants, as regards their congruence behaviour (see also note [1]). This will become even more marked later in this chapter, and will eventually lead to the proof that the adjusted Galois Gauss sum actually is a group determinant.

Remark 2. Properties (i)–(iv) of τ^* are shared by τy^{-1}, but (v) and (vi) are new, and are in a sense crucial for much that follows.

Remark 3. Parts of Proposition 1.1 remain valid for Galois characters which are not necessarily tame, but we shall not go into this here.

Remark 4. If β is a tame Abelian Galois character of F', then always

$$\tau^*(\beta) = \sum \theta_{\beta}(u)\psi_{F'}(uc^{-1}),$$

(sum over $u \in \mathfrak{o}_{F'}^* \pmod{\mathfrak{p}_{F'}}$). For β ramified this follows directly from the definition of τ^*. For β non-ramified verify that each expression is -1.

Proof of Proposition 1.1. (i) follows from the corresponding result for τ (see e.g. (III.5.4)) and the fact that y and z take root of unity values.

For (ii) recall that y is a homomorphism of $\Omega_{\mathbb{Q}}$-modules, by Theorem 29. The same is trivially true for z. The Galois action formula for τ of Theorem 20B thus implies that for τ^*. If now $\omega \in \Omega_{\mathbb{Q}(p)}$, then by the definition of u_p, we have $u_p(\omega) \equiv 1 \pmod{p}$. But $\theta_{\det(\chi)}$, being tame, is trivial on $1 + \mathfrak{p}_{F'}$, i.e., $\theta_{\det(\chi)}(u_p(\omega)) = 1$.

For (iii) use Theorem 29 (iii). We thus only have to show that $z(\chi) \equiv 1 \pmod{\mathfrak{L}_l}$, for $\chi \in \operatorname{Ker} d_l$. But for such a χ, $\det_\chi$ takes only values $\equiv 1 \pmod{\mathfrak{L}_l}$, as can be seen e.g., by Theorem 15A, or by an easy direct argument. Therefore in particular $\theta_{\det(\chi)}(c) \equiv 1 \pmod{\mathfrak{L}_l}$.

(iv) We already know that τ and y are inductive in degree zero (cf. Theorems 18 and 29). If now $\chi \in R_\Delta$, $\deg(\chi) = 0$, then $\det(\operatorname{ind}(\chi)) = \mathrm{v}_{\Delta/\Gamma} \det_\chi$, ($\operatorname{ind} = \operatorname{ind}_\Delta^\Gamma$), where $\mathrm{v}_{\Delta/\Gamma}$ is the cotransfer. In terms of the associated multiplicative characters of F'' and of F', this means that

$$\theta_{\det(\operatorname{ind}(\chi))} = \operatorname{res}_{F'}^{F''} \theta_{\det(\chi)}.$$

Hence indeed $\theta_{\det(\operatorname{ind}(\chi))}(c) = \theta_{\det(\chi)}(c)$.

(v) As Γ_0 and Γ/Γ_0 are both cyclic, we can write

$$(1.8) \qquad \chi = \operatorname{ind}_\Sigma^\Gamma \beta, \qquad \Sigma \supset \Gamma_0, \qquad \beta \text{ Abelian}.$$

Define in the usual manner an action of Γ on the Abelian characters ϕ of Γ_0, by conjugation, i.e., by the formula $(\phi\gamma)(\delta) = \phi(\gamma\delta\gamma^{-1})$. Then the Σ in (1.8) is the stabilizer in Γ of $\operatorname{res}_{\Gamma_0}^\Sigma \beta = \eta$. In this situation we have

1.2. Lemma. *There is a homomorphism*

$$g: \Gamma \to \operatorname{Gal}(\mathbb{Q}(p)(\eta)/\mathbb{Q}(p))$$

with $\eta\gamma = \eta^{g(\gamma)}$. *Also,* $\operatorname{Ker} g = \Sigma$ *and* $\mathbb{Q}(p)(\operatorname{res}_{\Gamma_0}^\Gamma \chi) = \mathbb{Q}(p)(\eta)^{\operatorname{Im} g}$.

The only part of the Lemma which is perhaps not obvious is the last equation. This follows from that fact that

$$\operatorname{res}_{\Gamma_0}^\Gamma \chi = \sum_{\gamma \bmod \Sigma} \eta\gamma = \sum_{h \in \operatorname{Im} g} \eta^h. \qquad \square$$

We return to the proof of (v) and we now use the inductive property. Replacing β by $\beta - \varepsilon_\Sigma$, a character of degree zero, we get the formula

$$(1.9) \qquad \tau^*(\chi)\tau^*(\beta)^{-1} = \tau^*(\operatorname{ind}_\Sigma^\Gamma \varepsilon_\Sigma)\tau^*(\varepsilon_\Sigma)^{-1},$$

which holds for any subgroup Σ of Γ. As, however, $\Sigma \supset \Gamma_0$, we conclude that $\operatorname{ind}_\Sigma^\Gamma \varepsilon_\Sigma = \sum \alpha_i$ is the sum of non-ramified Abelian characters α_i. We verify easily that $\tau^*(\alpha_i) = -1$. Thus

$$(1.9\text{a}) \qquad \tau^*(\chi)\tau^*(\beta)^{-1} = (-1)^{[\Gamma:\Sigma]-1}.$$

Therefore

$$(1.10) \qquad \mathbb{Q}(\tau^*(\chi)) = \mathbb{Q}(\tau^*(\beta)).$$

If β is non-ramified, then $\tau^*(\beta) = -1$, $\beta = \chi$, and (v) is trivially true. So now assume β to be ramified. Let for the moment $H = L^{\Sigma}$. Let A be the Artin map $H^* \to \Sigma^{ab}$. For $u \in \mathfrak{o}_H^*$, Au lies in the image of $\Gamma_0 \to \Sigma^{ab}$, and $\theta_\beta(u) = \beta(Au) = \eta(Au)$. Also now $y(\beta) = 1$ and so

$$\tau^*(\beta) = \sum_u \theta_\beta(u)\psi_H(uc^{-1}). \tag{1.11}$$

Thus $\tau^*(\beta) \in \mathbb{Q}(p)(\eta)$. Moreover, if $\gamma \in \Gamma$, then in the notation of Lemma 1.2,

$$\tau^*(\beta)^{g(\gamma)} = \sum \theta_\beta(u)^{g(\gamma)}\psi_H(uc^{-1}) = \sum \theta_\beta(u^{\gamma^{-1}})\psi_H(uc^{-1}),$$

as $\theta_{\beta\gamma}(u) = \theta_\beta(u^{\gamma^{-1}})$. But $\psi_H(u^\gamma c^{-1}) = \psi_H((uc^{-1})^\gamma) = \psi_H(uc^{-1})$, and hence $\tau^*(\beta)^{g(\gamma)} = \tau^*(\beta)$. Therefore $\tau^*(\beta) \in \mathbb{Q}(p)(\eta)^{\mathrm{Im}\, g}$. By Lemma 1.2 and (1.10) we now get the assertion (v).

(vi) Let α be a non-ramified Abelian character of Γ. Consider all pairs (Δ, χ), with $\Delta \subset \Gamma$, $\chi \in R_\Delta$. Define

$$\tau'(\Delta, \chi) = \tau^*((\mathrm{res}_\Delta^\Gamma \alpha)\chi).$$

This is additive, inductive in degree zero. An easy calculation shows that for χ Abelian it coincides with $\tau^*(\chi)$. Therefore $\tau' = \tau^*$, as we had to show. □

We shall now state the two basic theorems on τ^* and then show that they imply Theorem 11. These theorems are variants of Theorem 2 in [Ty7]. Their proofs will occupy the subsequent sections.

In both theorems we shall consider a tame Galois extension L/F of local fields with Galois group Γ. (Thus now $F' = F$.) For any rational prime l, we shall write τ_l^* for the composite map

$$R_\Gamma \to \mathbb{Q}^{c*} \to (\mathbb{Q}^c)_l^*,$$

the semi-local component of τ^* at l.

Theorem 30. *Suppose that $l \neq p$. Then*

$$\tau_l^* \in \mathrm{Det}(\mathfrak{o}_{\mathbb{Q}(p),l}\Gamma^*).$$ □

Remark. The reader who follows our argument carefully will observe that in fact we shall prove the stronger result

$$\tau_l^* \in \mathrm{ind}_{\Gamma_0}^\Gamma(\mathrm{Det}(\mathfrak{o}_{\mathbb{Q}(p),l}\Gamma_0^*)).$$

Although Theorem 30 would remain valid for τy^{-1} in place of τ^*, this is not so for this stronger version.

While Theorem 30 deals with the case where localization of values of τ^* takes place with respect to a prime l different from the residue class characteristic p of the given field F, the next theorem covers the case $l = p$. We now let E' be a Galois extension of finite degree of $\mathbb{Q}_p = \mathbb{Q}_l$, containing L and the p-th roots of unity, and which further has the property that all representations of Γ and of its subgroups over $\mathbb{Q}_p^c$ are equivalent to representations over E'. We shall write $\mathfrak{o}_{E',1}$ for the ring of integers in the maximal tame (over $\mathbb{Q}_p$) subfield of E'.

Theorem 31. *Let $j: \mathbb{Q}^c \to \mathbb{Q}_p^c$ be an embedding. Let a be a free generator of $\mathfrak{o}_L$ over $\mathfrak{o}_F\Gamma$. Then, for some $u = u(j) \in \mathfrak{o}_{E',1}\Gamma^*$,*

$$(\tau^*(\chi)^j)^{-1}(\mathcal{N}_{F/\mathbb{Q}_p}(a \mid \chi^j)) = \mathrm{Det}_{\chi^j}(u),$$

for all $\chi \in R_\Gamma$. □

Deduction of Theorem 11 from Theorems 30 and 31. Let N/K be a tame Galois extension of number fields with Galois group Γ. Let $\mathcal{T}_{\mathbb{Q}}$ be a finite set of (finite) rational primes, including all those which are ramified in $N/\mathbb{Q}$, and, for convenience only, also those dividing order (Γ). Denote by $\mathcal{T}_K$ the set of prime divisors $\mathfrak{p}$ of K lying above those in $\mathcal{T}_{\mathbb{Q}}$. For every $\mathfrak{p} \in \mathcal{T}_K$ and all $\chi \in R_\Gamma$, write

(1.12) $$y^*(\chi_\mathfrak{p}) = y^*(K_\mathfrak{p}, \chi_\mathfrak{p}) = y(K_\mathfrak{p}, \chi_\mathfrak{p}) z(K_\mathfrak{p}, \chi_\mathfrak{p})^{-1},$$

with y and z defined as in (1.1), and in (1.6), respectively, for $F = K_\mathfrak{p}$. Here, as usual, $\chi_\mathfrak{p}$ is the local component of the Galois character χ. In other words if $\mathfrak{q}$ is some prime divisor of N above $\mathfrak{p}$, then $\chi_\mathfrak{p}$ is the restriction of χ to $\Gamma_\mathfrak{p} = \mathrm{Gal}(N_\mathfrak{q}/K_\mathfrak{p})$. The map $\chi \mapsto y^*(K_\mathfrak{p}, \chi_\mathfrak{p})$ clearly lies in $\mathrm{Hom}^+_{\Omega_{\mathbb{Q}}}(R_\Gamma, \mu(\mathbb{Q}^c))$. Now define, for all $\chi \in R_\Gamma$,

(1.13) $$y^*(\chi) = y^*(K, \chi) = \prod_{\mathfrak{p} \in \mathcal{T}_K} y^*(\chi_\mathfrak{p}).$$

Then $y^* \in \mathrm{Hom}^+_{\Omega_{\mathbb{Q}}}(R_\Gamma, \mu(\mathbb{Q}^c))$. Assuming Theorems 30 and 31, we shall establish Theorem 11 with this choice of y^*. There is to be a slight change of notation compared to Theorem 11; the field F in that theorem will now turn up under the symbol E. We take E then as a number field containing N and $\mathbb{Q}(p)$ for all p dividing order (Γ), and so that all representations of Γ and of its subgroups over $\mathbb{Q}^c$ are realizable, to within equivalence, over E. Moreover E is to be Galois over $\mathbb{Q}$.

The proof is analogous to that of Theorem 19, proceeding by "double localization", i.e., via localizing global valued functions of global characters both in their domain and in their range, possibly with respect to quite distinct prime divisors. We do however at this stage no longer need to be as pedantic as for Theorem 19.

From now on we fix a prime divisor l of order (Γ). The completion of E under any embedding $j: \mathbb{Q}^c \to \mathbb{Q}_l^c$ will be a subfield E' of $\mathbb{Q}_l^c$, the same for all j. We then fix a set J of embeddings j, in biunique correspondence with the prime divisors $\mathfrak{P}_j$ of E lying above l. Each such j gives rise to a surjection

$$j \otimes 1: E \otimes_{\mathbb{Q}} \mathbb{Q}_l \to E_{\mathfrak{P}_j} = E'.$$

These give rise to an isomorphism

$$E \otimes_{\mathbb{Q}} \mathbb{Q}_l \cong \prod_J E_{\mathfrak{P}_j}$$

of $\mathbb{Q}_l$-algebras, and so to an isomorphism

$$(1.14) \qquad (E \otimes_{\mathbb{Q}} \mathbb{Q}_l)^* \cong \prod_J E^*_{\mathfrak{P}_j}$$

of multiplicative groups. Without danger of confusion we shall identify these two groups. The $j \otimes 1$ are then to be viewed as the projections onto factors, and the j as the composites of those with the embedding $E^* \to (E \otimes_{\mathbb{Q}} \mathbb{Q}_l)^*$. Moreover the product (1.14) is naturally a subgroup of $\mathfrak{J}(E)$. For $y \in (E \otimes_{\mathbb{Q}} \mathbb{Q}_l)^*$, the image $y^{j\otimes 1}$ is just its $\mathfrak{P}_j$-component $y_{\mathfrak{P}_j}$, as idele, and for $\mathfrak{P} \neq \mathfrak{P}_j$ all j, $y_{\mathfrak{P}} = 1$. If $x \in E^*$ we shall write x_l for its image in $(E \otimes_{\mathbb{Q}} \mathbb{Q}_l)^*$. Thus $x^j = x_l^{(j\otimes 1)} = x_{\mathfrak{P}_j}$. The norm resolvent

$$\mathcal{N}_{K/\mathbb{Q}} g(\chi) = \prod_\sigma g(\chi^{\sigma^{-1}})^\sigma$$

is in the sequel formally defined in terms of some right transversal $\{\sigma\}$ of Ω_K in $\Omega_{\mathbb{Q}}$, for any homomorphism $g: R_\Gamma \to (E \otimes_{\mathbb{Q}} \mathbb{Q}_l)^*$.

Now let $\mathfrak{p}$ be a prime divisor of K above l, and $\mathfrak{q}$ a prime divisor of N above $\mathfrak{p}$. Write $\Gamma_\mathfrak{p} = \mathrm{Gal}(N_\mathfrak{q}/K_\mathfrak{p})$. Let $a_\mathfrak{p}$ be a free generator of $\mathfrak{o}_{N,\mathfrak{q}}$ over $\mathfrak{o}_{K,\mathfrak{p}}\Gamma_\mathfrak{p}$. With a suitable choice of a free generator $b_\mathfrak{p}$ of $\mathfrak{o}_{N,\mathfrak{p}}$ over $\mathfrak{o}_{K,\mathfrak{p}}\Gamma$ we have, by Theorem 19, a relation

$$(1.15) \qquad \mathcal{N}_{K/\mathbb{Q}}(b_\mathfrak{p} \mid \chi)_{\mathfrak{P}_j} = \mathcal{N}_{K_\mathfrak{p}/\mathbb{Q}_p}(a_\mathfrak{p} \mid \chi_\mathfrak{p}^j)(\det_\chi(\gamma(j, \mathfrak{p})))_{\mathfrak{P}_j},$$

holding for all $\chi \in R_\Gamma$, with $\gamma(j, \mathfrak{p}) \in \Gamma$ independent of χ.

We now apply Theorem 31 with $K_\mathfrak{p} = F$ and Γ replaced by $\Gamma_\mathfrak{p}$. In conjunction with (1.15), this yields equations

$$\mathcal{N}_{K/\mathbb{Q}}(b_\mathfrak{p} \mid \chi)_{\mathfrak{P}_j} = \tau^*(\chi_\mathfrak{p})_{\mathfrak{P}_j} \det_\chi(\gamma(j, \mathfrak{p}))_{\mathfrak{P}_j} \mathrm{Det}_{\chi_\mathfrak{p}^j}(u(j, \mathfrak{p})).$$

Here $u(j, \mathfrak{p}) \in \mathfrak{o}_{E',1}\Gamma_\mathfrak{p}^* \subset \mathfrak{o}_{E',1}\Gamma^*$. Thus also

$$\mathcal{N}_{K/\mathbb{Q}}(b_\mathfrak{p} \mid \chi)_{\mathfrak{P}_j} = \tau^*(\chi_\mathfrak{p})_{\mathfrak{P}_j} \det_\chi(\gamma(j, \mathfrak{p}))_{\mathfrak{P}_j} \mathrm{Det}_{\chi^j}(u(j, \mathfrak{p})).$$

Now define $u(\mathfrak{p}) \in \prod_j E_{\mathfrak{P}_j}\Gamma^*$ by

$$u(\mathfrak{p})_{\mathfrak{P}_j} = \gamma(j, \mathfrak{p}) u(j, \mathfrak{p}).$$

Then $\mathrm{Det}_\chi(u(\mathfrak{p}))_{\mathfrak{P}_j} = \mathrm{Det}_{\chi^j}(u(\mathfrak{p})_{\mathfrak{P}_j})$. Thus our equations now read

$$\mathcal{N}_{K/\mathbb{Q}}(b_\mathfrak{p} \mid \chi)_{\mathfrak{P}_j} = (\tau^*(\chi_\mathfrak{p}) \mathrm{Det}_\chi(u(\mathfrak{p})))_{\mathfrak{P}_j}.$$

This holds for all j. Therefore

$$(1.16) \qquad \mathcal{N}_{K/\mathbb{Q}}(b_\mathfrak{p} \mid \chi) = \tau^*(\chi_\mathfrak{p})_l \mathrm{Det}_\chi(u(\mathfrak{p})).$$

Now we vary $\mathfrak{p}$ over all prime divisors of K above l. Let, for each $\mathfrak{p}$, $b_l \in \mathfrak{o}_{N,l}$ have components $b_\mathfrak{p}$ in $\mathfrak{o}_{N,\mathfrak{p}}$, as above. Then b_l is a free generator of $\mathfrak{o}_{N,l}$ over $\mathfrak{o}_{K,l}\Gamma$, and in $(E \otimes_{\mathbb{Q}} \mathbb{Q}_l)^*$ we have the equations

$$\mathcal{N}_{K/\mathbb{Q}}(b_l \mid \chi) = \prod_{\mathfrak{p} \mid l} \mathcal{N}_{K/\mathbb{Q}}(b_\mathfrak{p} \mid \chi).$$

Write $u(l) = \prod_{\mathfrak{p}\mid l} u(\mathfrak{p})$. From (1.16) we now obtain

$$(1.17)\qquad \mathcal{N}_{K/\mathbb{Q}}(b_l \mid \chi) = \mathrm{Det}_\chi(u(l)) \prod_{\mathfrak{p}\mid l} \tau^*(\chi_\mathfrak{p})_l.$$

Going back to the definition of the $u(\mathfrak{p})$ we also see that

$$(1.18)\qquad u(l) \in \mathfrak{O}_{E,l}\Gamma^*,$$

where $\mathfrak{O}_{E,l} = \prod_J \mathfrak{o}_{E,\mathfrak{P}_j,1}$, as defined for Theorem 11 (with E in place of F).

Note in passing that as usual one shows that a formula (1.17) for one free generator b_l of $\mathfrak{o}_{N,l}$ over $\mathfrak{o}_{K,l}\Gamma$ implies a similar formula for any free generator.

Next let $\mathfrak{p} \in \mathcal{T}_K$, but $\mathfrak{p} \nmid l$. Now we apply Theorem 30, with the appropriate changes in notation. If in particular $\Gamma_\mathfrak{p}$ is the local Galois group, we deduce the existence of an element $w(\mathfrak{p}) \in \mathfrak{o}_{\mathbb{Q}(p),l}\Gamma_\mathfrak{p}^* \subset \mathfrak{o}_{\mathbb{Q}(p),l}\Gamma^*$, so that

$$\tau^*(\chi_\mathfrak{p})_l = \mathrm{Det}_{\chi_\mathfrak{p}}(w(\mathfrak{p})).$$

Here we may again replace $\chi_\mathfrak{p}$ by χ on the right. Moreover as $\mathbb{Q}(p)/\mathbb{Q}$ is tame at l, we have $\mathfrak{o}_{\mathbb{Q}(p),l} \subset \mathfrak{O}_{E,l}$. Thus we have, for all $\chi \in R_\Gamma$,

$$(1.19)\qquad \tau^*(\chi_\mathfrak{p})_l = \mathrm{Det}_\chi(w(\mathfrak{p})), \qquad w(\mathfrak{p}) \in \mathfrak{O}_{E,l}\Gamma^*.$$

Such equations hold for each $\mathfrak{p} \in \mathcal{T}_K$, $\mathfrak{p} \nmid l$.

Finally observe that if $\mathfrak{p} \notin \mathcal{T}_K$, and $\chi \in R_\Gamma$ then $\tau(\chi_\mathfrak{p}) = 1$. Thus

$$\tau(\chi) = \prod_{\mathfrak{p}\in\mathcal{T}_K} \tau(\chi_\mathfrak{p}).$$

From this equation, from Eqs. (1.13), and (1.17)–(1.19), and from the definition of v_l:

$$v_l(\chi) = \tau(\chi)_l^{-1} \mathcal{N}_{K/\mathbb{Q}}(b_l \mid \chi),$$

we conclude that

$$\begin{aligned} v_l(\chi) y^*(\chi)_l &= \mathcal{N}_{K/\mathbb{Q}}(b_l \mid \chi)\tau(\chi)_l^{-1} y^*(\chi)_l \\ &= \mathcal{N}_{K/\mathbb{Q}}(b_l \mid \chi) \prod_{\mathfrak{p}\in\mathcal{T}_K} \tau^*(\chi_\mathfrak{p})_l^{-1} \\ &= \mathrm{Det}_\chi\Bigl(u(l) \prod_{\substack{\mathfrak{p}\in\mathcal{T}_K \\ \mathfrak{p}\nmid l}} w(\mathfrak{p})^{-1}\Bigr). \end{aligned}$$

This is the required result:

$$(v(\chi) y^*(\chi))_l \in \mathrm{Det}\, \mathfrak{O}_{E,l}\Gamma^*. \qquad \square$$

Remark. The last result remains true on replacing $y^*(\chi)$ by $y(\chi)^{-1}$, where

$$(1.20)\qquad y(\chi) = \prod_{\mathcal{T}_K} y(K_\mathfrak{p}, \chi_\mathfrak{p})$$

is the product of the non-ramified characteristics over $\mathscr{T}_K$. For,

$$y(\chi)y^*(\chi)^{-1} = z(\chi) = \prod_{\mathscr{T}_K} z(K_\mathfrak{p}, \chi_\mathfrak{p}),$$

and

$$z \in \mathrm{Hom}^+_{\Omega_\mathbb{Q}}(R_\Gamma, \mu(\mathbb{Q}^c)) \cap \mathrm{Det}\, \mathfrak{U}\mathbb{Z}\Gamma. \tag{1.21}$$

Thus z lies in both the groups, which appear in the denominator when $D(\mathbb{Z}\Gamma)$ is represented as a quotient of $\mathrm{Hom}^+_{\Omega_\mathbb{Q}}(R_\Gamma, \mathfrak{U}(\mathbb{Q}^c))$. It can thus be switched from one to the other and thus indeed

$$\chi \mapsto y(\chi)_l^{-1} v(\chi)_l \quad \text{lies in} \quad \mathrm{Det}\, \mathfrak{O}_{F,l}\Gamma^*. \tag{1.22}$$

To prove (1.21), we go back to the definition of $z(K_\mathfrak{p}, \cdot)$ (cf. (1.6)). Fix for each $\mathfrak{p} \in \mathscr{T}_K$, an element $c_\mathfrak{p}$ of $K_\mathfrak{p}^*$ so that $c_\mathfrak{p}\mathfrak{o}_{K,\mathfrak{p}} = \mathfrak{D}_{K,\mathfrak{p}}\mathfrak{p}$. Then $z(K_\mathfrak{p}, \chi_\mathfrak{p}) = \theta_{\det(\chi_\mathfrak{p})}(c_\mathfrak{p})$. For each $\mathfrak{p}$, let $\delta_\mathfrak{p} \in \Gamma_\mathfrak{p}$ be an element whose image in $\Gamma_\mathfrak{p}^{ab}$ is the image of $c_\mathfrak{p}$ under the local Artin map. Then $z(K_\mathfrak{p}, \chi_\mathfrak{p}) = \det_{\chi_\mathfrak{p}}(\delta_\mathfrak{p}) = \det_\chi(\delta_\mathfrak{p})$. Let δ be the product of the $\delta_\mathfrak{p}$, in some order; then

$$z(\chi) = \det_\chi(\delta).$$

Thus $z \in \mathrm{Det}(\Gamma)$, which implies (1.21).

As mentioned already, the choice of y^* in place of y is imposed on one by the requirement of the proofs of the local theorems.

§2. Proof of Theorem 31

We shall first get the proof of this theorem out of the way – the two basic tools already being readily available, namely the local main theorem, Theorem 23, and the theorem dealing with non-ramified basefield extensions, Theorem 25.

We shall have to consider here both the group R_Γ, and the group $R_{\Gamma,l}$, of virtual characters over $\mathbb{Q}^c$, and over $\mathbb{Q}_l^c$, respectively. Throughout this section L/F is a tame Galois extension of non-Archimedean local fields of residue class characteristic $p = l$, with Galois group Γ. Γ_0 is as always, the inertia subgroup of Γ. $B = L^{\Gamma_0}$ is the maximal non-ramified extension of F in L. We choose again a Galois extension of $\mathbb{Q}_l$ of finite degree, now to be denoted be E (as global fields no longer appear) which contains L as well as the p-th roots of unity, and so that all representations of Γ and of Γ_0 over $\mathbb{Q}_l^c$ are equivalent to representations over E. Finally M is the maximal tame extension of $\mathbb{Q}_l$ in E. For the benefit of the reader we set out the relevant groups in a diagram, incorporating already information contained in the subsequent paragraph.

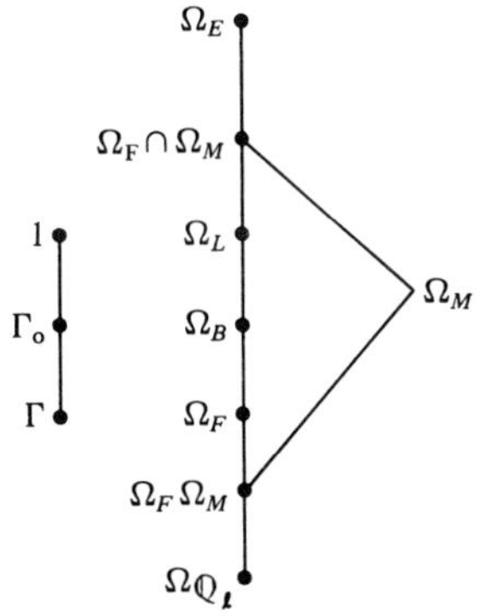

As always, one sees easily, that it suffices to prove the equation of Theorem 31 with a particular choice of transversal of Ω_F in $\Omega_{\mathbb{Q}_l}$, to be used to define $\mathcal{N}_{F/\mathbb{Q}_l}$. We first choose a right transversal $\{\pi_i\}$ of $\Omega_L \cap \Omega_M$ in Ω_M. As L/F is tame, and Ω_M is the wild (or first) ramification group, $\Omega_L \cap \Omega_M = \Omega_F \cap \Omega_M$. Hence $\{\pi_i\}$ is also a right transversal of Ω_F in $\Omega_F \Omega_M$; the latter is indeed a subgroup of $\Omega_{\mathbb{Q}_l}$, as Ω_M is normal. Now let $\{\omega_j\}$ be a right transversal of $\Omega_F \Omega_M$ in $\Omega_{\mathbb{Q}_l}$. Then $\{\pi_i \omega_j\}$ is a right transversal of Ω_F in $\Omega_{\mathbb{Q}_l}$. Let X and Y be $\Omega_{\mathbb{Q}_l}$-modules, Y written multiplicatively. For

(2.1) $$f \in \mathrm{Hom}_{\Omega_L}(X, Y)$$

define

(2.2) $$\mathcal{N} f = \prod_{i,j} f^{\pi_i \omega_j},$$

where, as usual, $f^\sigma(x) = f(x^{\sigma^{-1}})^\sigma$, $x \in X$, $\sigma \in \Omega_{\mathbb{Q}_l}$. Note that this is not the usual norm operator. For we take a (special!) right transversal of Ω_F in $\Omega_{\mathbb{Q}_l}$, but we start with maps f, which are not necessarily Ω_F-homomorphisms.

2.1. Lemma. *With f as in* (2.1), $\mathcal{N} f \in \mathrm{Hom}_{\Omega_M}(X, Y)$.

Proof. For each j, $f^{\omega_j} \in \mathrm{Hom}_{\omega_j^{-1}\Omega_L \omega_j}(X, Y)$, hence

$$\prod_i (f^{\omega_j})^{\omega_j^{-1}\pi_i \omega_j} \in \mathrm{Hom}_{\Omega_M}(X, Y),$$

as $\{\omega_j^{-1}\pi_i\omega_j\}$ is a right transversal of $\omega_j^{-1}\Omega_L\omega_j \cap \Omega_M$ in Ω_M (recall that Ω_M is normal in $\Omega_{\mathbb{Q}_l}$). Therefore

$$\prod_j \prod_i (f^{\omega_j})^{\omega_j^{-1}\pi_i\omega_j} = \prod_{i,j} f^{\pi_i\omega_j} \in \mathrm{Hom}_{\Omega_M}(X, Y)$$

as asserted. □

2.2. Lemma.

$$\mathrm{Det}(\mathfrak{o}_M \Gamma_0^*) = \mathrm{Hom}(R_{\Gamma_0,l}, \mathfrak{o}_M^*),$$

as subgroups of $\mathrm{Hom}(R_{\Gamma_0,l}, \mathbb{Q}_l^{c*})$.

Proof. This is trivial. One can either use general results, or direct verification. We shall indicate both methods of proof.

As the order of Γ_0 is not divisible by p, $\mathfrak{o}_M\Gamma_0$ is the maximal order of $M\Gamma_0$. Therefore the argument leading to Proposition I.2.2 now tells us that $\mathrm{Det}(\mathfrak{o}_M\Gamma_0^*) = \mathrm{Hom}_{\Omega_M}(R_{\Gamma_0,l}\mathfrak{U}(\mathbb{Q}_l^c))$. Let for the moment Φ be the set of Abelian characters of Γ_0 in $\mathbb{Q}_l^c$. By hypothesis, E contains their values; but as $p \nmid$ order (Γ_0), these values actually lie in M. As Γ_0 is Abelian, this implies that $R_{\Gamma_0,l}$ is fixed under Ω_M, and hence

$$\mathrm{Hom}_{\Omega_M}(R_{\Gamma_0,l}\mathfrak{U}(\mathbb{Q}_l^c)) = \mathrm{Hom}(R_{\Gamma_0,l}, \mathfrak{o}_M^*).$$

Alternatively, with Φ as above, we have by restricting from R_{Γ_0} to Φ, an isomorphism $\mathrm{Hom}(R_{\Gamma_0,l}, \mathbb{Q}_l^{c*}) \cong \mathrm{Map}(\Phi, \mathbb{Q}_l^{c*})$, and this isomorphism maps both $\mathrm{Det}(\mathfrak{o}_M\Gamma_0^*)$ and $\mathrm{Hom}(R_{\Gamma_0,l}, \mathfrak{o}_M^*)$ onto $\mathrm{Map}(\Phi, \mathfrak{o}_M^*)$. □

Now we come to the actual proof of Theorem 31. Let a be again a free generator of $\mathfrak{o}_L$ over $\mathfrak{o}_F\Gamma$, and let now b be a free generator of $\mathfrak{o}_L$ over $\mathfrak{o}_B\Gamma_0$. By Theorem 25, we have for all $\chi \in R_\Gamma$,

(2.3) $$(b \mid \mathrm{res}^{\Gamma}_{\Gamma_0}\chi^j) = (a \mid \chi^j)\,\mathrm{Det}_{\chi^j}(\lambda),$$

where $\lambda \in \mathfrak{o}_B\Gamma^*$. Lemma 2.1 applies to everyone of the three maps involved in Eq. (2.3). As on similar occasions, we write by abuse of notation,

$$\mathcal{N}(b \mid \mathrm{res}^{\Gamma}_{\Gamma_0}\chi^j), \qquad \mathcal{N}(a \mid \chi^j)$$

for the values $\mathcal{N}f(\chi^j)$, if f is the map $\chi \mapsto (b \mid \mathrm{res}^{\Gamma}_{\Gamma_0}\chi^j)$, or $\chi \mapsto (a \mid \chi^j)$, respectively, and $\mathcal{N}f$ is defined by (2.2). By Theorem 13, $\mathcal{N}\,\mathrm{Det}(\lambda) \in \mathrm{Det}(\mathfrak{o}_M\Gamma^*)$. This has two consequences:

I. *In order to prove that the map* $\chi^j \mapsto (\tau^*(\chi)^j)^{-1}\mathcal{N}(a \mid \chi^j)$ *lies in* $\mathrm{Det}(\mathfrak{o}_M\Gamma^*)$, *it suffices to prove the same for* $\chi^j \mapsto (\tau^*(\chi)^j)^{-1}\mathcal{N}(b \mid \mathrm{res}^{\Gamma}_{\Gamma_0}\chi^j)$.

For, by (2.3), the map $\chi^j \mapsto \mathcal{N}(b \mid \mathrm{res}^{\Gamma}_{\Gamma_0}\chi^j)\mathcal{N}(a \mid \chi^j)^{-1}$ certainly lies in $\mathrm{Det}(\mathfrak{o}_M\Gamma^*)$. Its values are units. Therefore by Theorem 23, we also have the fact:

II. *The values* $(\tau^*(\chi)^j)^{-1}\mathcal{N}(b \mid \mathrm{res}^{\Gamma}_{\Gamma_0}\chi^j)$ *are units, for all* $\chi \in R_\Gamma$.

Next we show

III. *For all* $\chi \in R_\Gamma$, $\tau^*(\chi)^j$ *and* $\mathcal{N}(b \mid \mathrm{res}^{\Gamma}_{\Gamma_0}\chi^j)$ *lie in* M^*.

Proof. As far as the $\tau^*(\chi)^j$ are concerned, we may assume χ to be irreducible. By Proposition 1.1 (v), $\tau^*(\chi)$ lies in $Q(p)(\mathrm{res}^{\Gamma}_{\Gamma_0}\chi)$. As the values of $\mathrm{res}^{\Gamma}_{\Gamma_0}\chi$ are sums of roots of unity of order prime to p, p is tamely ramified in $\mathbb{Q}(p)(\mathrm{res}^{\Gamma}_{\Gamma_0}\chi)$, hence the image of this field under j – which does lie in E, by hypothesis on E – actually lies in M.

We have just observed that the virtual $\mathbb{Q}_l^c$-characters η of Γ_0 have values in M. Applying Lemma 2.1 to $\eta \mapsto (b \mid \eta)$, we conclude that $\mathcal{N}(b \mid \eta^\omega) = (\mathcal{N}(b \mid \eta))^\omega$, for all $\omega \in \Omega_M$. Therefore finally $\mathcal{N}(b \mid \eta) \in M^*$, as we had to show. □

In the sequel let Φ be again the set of Abelian characters of Γ_0 over $\mathbb{Q}_l^c$, i.e., of homomorphisms $\Gamma_0 \to M^*$.

IV. *Let* $\phi \in \Phi$. *The values of* $\tau^*(\chi)$ *and similarly those of* $\mathcal{N}(b \mid \operatorname{res}_{\Gamma_0}^{\Gamma} \chi^j)$ *remain the same for all irreducible characters* χ^j *which occur in* $\operatorname{ind}_{\Gamma_0}^{\Gamma} \phi$.

Proof. If χ^j is such an irreducible character, then any other irreducible character occurring in $\operatorname{ind}_{\Gamma_0}^{\Gamma} \phi$ is of form $\chi^j \alpha^j$, where α is an Abelian character of Γ/Γ_0 inflated to Γ, i.e., is non-ramified. By Proposition 1.1 (vi), $\tau^*(\chi) = \tau^*(\chi\alpha)$. Trivially $\operatorname{res}_{\Gamma_0}^{\Gamma} \chi^{j\omega^{-1}} = \operatorname{res}_{\Gamma_0}^{\Gamma} (\chi\alpha)^{j\omega^{-1}}$, for all $\omega \in \Omega_{\mathbb{Q}_l}$. Thus $\mathcal{N}(b \mid \operatorname{res}_{\Gamma_0}^{\Gamma} \chi^j) = \mathcal{N}(b \mid \operatorname{res}_{\Gamma_0}^{\Gamma} (\chi\alpha)^j)$. □

As Φ freely generates $R_{\Gamma_0,l}$ as Abelian group we shall in the sequel identify

(2.4) $$\operatorname{Map}(\Phi, \cdot) = \operatorname{Hom}(R_{\Gamma_0,l}, \cdot).$$

Let $\phi \in \Phi$, and let χ^j be an irreducible character of Γ occurring in $\operatorname{ind}_{\Gamma_0}^{\Gamma} \phi$. By statement IV the element

(2.5) $$r(\phi) = (\tau^*(\chi)^j)^{-1} \mathcal{N}(b \mid \operatorname{res}_{\Gamma_0}^{\Gamma} \chi^j)$$

is uniquely defined. By statements II and III,

(2.6) $$r(\phi) \in \mathfrak{o}_M^*.$$

The group Γ acts on Φ, where we write

$$(\phi\gamma)(\delta) = \phi(\gamma\delta\gamma^{-1}), \qquad \text{for} \quad \phi \in \Phi, \quad \gamma \in \Gamma, \quad \delta \in \Gamma_0.$$

As $\operatorname{ind}_{\Gamma_0}^{\Gamma} \phi = \operatorname{ind}_{\Gamma_0}^{\Gamma} \phi\gamma$, and by (2.5) we have $r(\phi) = r(\phi\gamma)$. Let Φ' be a set of representatives of the orbits of Φ under the action of Γ. Now write

(2.7) $$s(\phi) = \begin{cases} r(\phi), & \text{if} \quad \phi \in \Phi', \\ 1, & \text{if} \quad \phi \in \Phi, \quad \phi \notin \Phi'. \end{cases}$$

By (2.6), and with identification (2.4), we see that $s \in \operatorname{Hom}(R_{\Gamma_0,l}, \mathfrak{o}_M^*)$. By Lemma 2.2,

(2.8) $$s \in \operatorname{Det}(\mathfrak{o}_M \Gamma_0^*).$$

Now let χ be an irreducible character of Γ. Then $\operatorname{res}_{\Gamma_0}^{\Gamma} \chi = \sum \phi\gamma$, where $\phi \in \Phi'$ and γ runs over Γ modulo the stabilizer of ϕ. Moreover χ occurs in $\operatorname{ind}_{\Gamma_0}^{\Gamma} \phi$. Thus by (2.6), (2.7),

$$\begin{aligned} (\tau^*(\chi)^j)^{-1} \mathcal{N}(b \mid \operatorname{res}_{\Gamma_0}^{\Gamma} \chi^j) &= r(\phi) \\ &= \prod s(\phi\gamma) \qquad \text{(product over } \gamma \text{ in } \Gamma \text{ - modulo the stabilizer of } \phi) \\ &= s(\operatorname{res}_{\Gamma_0}^{\Gamma} \chi) = (\operatorname{ind}_{\Gamma_0}^{\Gamma} s)(\chi) \qquad \text{(see (II §3))}. \end{aligned}$$

Thus the map

$$\chi \mapsto (\tau^*(\chi)^j)^{-1} \mathcal{N}(b \,|\, \mathrm{res}^{\Gamma}_{\Gamma_0} \chi^j)$$

is the same as $\mathrm{ind}^{\Gamma}_{\Gamma_0} s$. By (2.8) and Theorem 12 it lies in $\mathrm{Det}(\mathfrak{o}_M \Gamma^*)$, as we had to show. By statement I, this completes the proof of Theorem 31. □

§3. Reduction Steps for Theorem 30

The proof of Theorem 30 involves several relatively self-contained and clearly distinct intermediary results, with lengthy and sometimes quite technical proofs of their own, and based on different ideas. It is therefore useful to separate out formally the various aspects. The real arithmetic core will be the logarithmic evaluation and the congruences, to be derived in §5 and §6.

In the present section the proof of Theorem 30 will be reduced to that of a theorem (Theorem 32) which deals with a twisted and localized form of τ^* and a special type of Galois group. Essentially this is reduction to l-groups. It will be of interest to note how closely parallel most individual steps are to those in the proof of Theorem 10A in Chap. II §6.

Throughout we consider again tame Galois extensions, of non-Archimedean local fields, of residue class characteristic p. Otherwise the notation differs from that in §2. In particular l is now a prime different from p. The symbol M will from now on always stand for the field $\mathbb{Q}_l(p)$ of p-th roots of unity over $\mathbb{Q}_l$. Initially we have to consider both the rings of virtual characters over $\mathbb{Q}^c$, and over $\mathbb{Q}_l^c$, denoting them for the moment for a group Γ in the usual way, i.e., by R_Γ, and by $R_{\Gamma,l}$ respectively.

In using methods of going up or going down, we have always to remember that the definition of τ^* involves the choice of an element c (cf. (1.6), (1.7)) with certain properties. These properties are retained by c on transition to tame extensions only. It will thus always be assumed that all local fields of residue class characteristic p which we shall consider are tame over the field in which we have chosen c once and for all.

Let then L/F be a tame Galois extension of local fields of residue class characteristic p, with Galois group Γ. Using the localization lemma II.2.1, at the prime divisors above l, the proof of Theorem 30 amounts to showing that if $j\colon \mathbb{Q}^c \to \mathbb{Q}_l^c$ is an embedding, then the map $\chi \mapsto \tau^*(\chi)^j$ ($\chi \in R_\Gamma$) will lie in $\mathrm{Det}(\mathfrak{o}_M \Gamma^*)$, viewed as a subgroup of $\mathrm{Hom}_{\Omega_M}(R_\Gamma, \mathbb{Q}_l^{c*})$. Now we fix an embedding j once and for all. For the given Galois group Γ we write

(3.1) $$T_\Gamma(\chi^j) = \tau^*(\chi)^j.$$

As $j\colon R_\Gamma \to R_{\Gamma,l}$ is an isomorphism, this defines a unique map T_Γ on $R_{\Gamma,l}$. As $p \neq l$ and by (III.5.4), the $T_\Gamma(\chi)$ will lie in $\mathfrak{U}(\mathbb{Q}_l^c)$. Therefore, and by Proposition 1.1,

(3.2) $$T_\Gamma \in \mathrm{Hom}_{\Omega_M}(R_{\Gamma,l}, \mathfrak{U}(\mathbb{Q}_l^c)).$$

We can now restate Theorem 30, in the new notation:

Theorem 30A.

$$T_\Gamma \in \mathrm{Det}(\mathfrak{o}_M \Gamma^*).$$

(This group now viewed as a subgroup of the group in (3.2).) □

Having established the definition of the T_Γ in terms of τ^*, we shall henceforth almost exclusively by concerned with characters over $\mathbb{Q}_l^c$. Thus from now on we introduce

Change of notation. R_Γ is the ring of virtual characters over $\mathbb{Q}_l^c$, i.e., now $R_\Gamma = R_\Gamma(\mathbb{Q}_l^c)$. This will be kept until further notice.

Note that if $\alpha\colon \Gamma \to \mathbb{Q}_l^{c*}$ is a tame Abelian character, then

$$T_\Gamma(\alpha) = \sum_u \theta_\alpha(u)(\psi_F(uc^{-1})^j), \qquad \theta_\alpha : F^* \to \mathbb{Q}_l^{c*}, \tag{3.3}$$

u running over $\mathfrak{o}_F^*$ mod $\mathfrak{p}_F$. If α is ramified, this is immediate from the definition. If α is non-ramified one shows that each side of (3.3) has value -1; for the left hand side the argument was given in §1. (See the proof of (1.4)). Thus T_Γ really is an l-adic Gauss sum on a $\mathfrak{p}$-adic field (with $p \neq l$).

Suppose for the moment that in particular

$$\Gamma = \Pi \times \Sigma, \tag{3.4}$$

Π an l-group, Σ a cyclic group of order d prime to l. For every pair (χ, ϕ) with $\chi \in R_\Pi$ and $\phi\colon \Sigma \to \mathbb{Q}_l^{c*}$ an Abelian character, we define as in Chap. II §6 (leading up to (6.19)) a virtual character $\chi \times \phi \in R_\Gamma$, with values

$$(\chi \times \phi)(\pi\sigma) = \chi(\pi)\phi(\sigma), \qquad \pi \in \Pi, \qquad \sigma \in \Sigma. \tag{3.5}$$

Identifying

$$\Pi = \Gamma/\Sigma, \tag{3.6}$$

we view R_Π as group of virtual characters of Γ/Σ, and then of course

$$\chi \times \varepsilon_\Sigma = \mathrm{inf}_\Pi^\Gamma \chi. \tag{3.7}$$

With these notations, we now define

$$T_{\Gamma,\phi}(\chi) = T_\Gamma(\chi \times \phi), \tag{3.8}$$

viewing $T_{\Gamma,\phi}$ as a map on R_Π. Let in the sequel

$$\Phi = \Phi_d \tag{3.9}$$

be the set of faithful Abelian characters $\phi: \Sigma \to \mathbb{Q}_l^{c*}$, i.e., of Abelian characters of order d. Write

$$N = M(d) \tag{3.9a}$$

for the field of d-th roots of unity over M. Then we have

Theorem 32. *If Γ is of form $\Pi \times \Sigma$ and $\phi \in \Phi_d$, with $d =$ order (Σ) then*

$$T_{\Gamma,\phi} \in \mathrm{Det}(\mathfrak{o}_N \Pi^*).$$ □

The proof of this theorem will occupy §4–6. The remainder of the present section will be taken up with the derivation of Theorem 30 from Theorem 32. From now on the hypothesis (3.4) on Γ will again be dropped; Γ is just be Galois group of a tame extension L/F, as before. We shall establish four intermediate results. Recall that a group Δ is $\mathbb{Q}$-q-elementary, q a prime number, if $\Delta = \Sigma \rtimes \Pi$ (semidirect product) with Π a q-group and Σ a cyclic group of order prime to q. Δ is $\mathbb{Q}$-elementary if it is $\mathbb{Q}$-q-elementary for some q.

A. If Γ is $\mathbb{Q}$-q-elementary, with $q \neq l$, then for some $h \geqslant 0$

$$T_\Gamma^{l^h} \in \mathrm{Det}(\mathfrak{o}_M \Gamma^*). \tag{3.10}$$

B. If Δ is a $\mathbb{Q}$-l-elementary group, and Theorem 32 *holds for any group of type* (3.4), *then*

$$T_\Delta \in \mathrm{Det}(\mathfrak{o}_M \Delta^*).$$

C. If Γ is $\mathbb{Q}$-q-elementary, with $q \neq l$, and $T_\Delta \in \mathrm{Det}(\mathfrak{o}_M \Delta^*)$, for every $\mathbb{Q}$-l-elementary subgroup Δ of Γ then for some s prime to l

$$T_\Gamma^s \in \mathrm{Det}(\mathfrak{o}_M \Gamma^*). \tag{3.11}$$

D. If $T_\Delta \in \mathrm{Det}(\mathfrak{o}_M \Delta^*)$ for every $\mathbb{Q}$-elementary subgroup Δ of Γ, then $T_\Gamma \in \mathrm{Det}(\mathfrak{o}_M \Gamma^*)$. □

We first show that these assertions, together with Theorem 32 imply Theorem 30A. Firstly, by Theorem 32 and *B*, we have (replacing symbols) $T_\Delta \in \mathrm{Det}(\mathfrak{o}_M \Delta^*)$ for all $\mathbb{Q}$-l-elementary subgroups Δ of Γ. Therefore by *C*, we get (3.11), with Γ replaced by any $\mathbb{Q}$-q-elementary subgroup; by *A*, we get relation (3.10) as well, hence indeed – changing symbols again – $T_\Delta \in \mathrm{Det}(\mathfrak{o}_M \Delta^*)$ for all $\mathbb{Q}$-q-elementary subgroups Δ with $q \neq l$. If Theorem 32 holds, the same applies if Δ is $\mathbb{Q}$-l-elementary. Finally by *D*, $T_\Gamma \in \mathrm{Det}(\mathfrak{o}_M \Gamma^*)$. This is Theorem 30A. □

Proof of A. This is, as will be seen, analogous to the proof of assertion II in Chap. II §6.

We have

$$\Gamma = \Sigma \rtimes \Pi,$$

Π a q-group, Σ cyclic. Let Σ_l be the l-Sylow group of Σ, hence of Γ. It is normal in Γ. Put $\Gamma' = \Gamma/\Sigma_l$. The order of Γ' is prime to l, whence $\mathfrak{o}_M\Gamma'$ is a maximal order in $M\Gamma'$. By the same argument as that used in II §1 for Proposition I.2.2, or directly by Lemma II.1.5, we conclude that

$$\mathrm{Det}(\mathfrak{o}_M\Gamma') = \mathrm{Hom}_{\Omega_M}(R_{\Gamma'}, \mathfrak{U}(\mathbb{Q}_l^c)).$$

By (3.2), it follows now that

$$T_{\Gamma'} \in \mathrm{Det}(\mathfrak{o}_M\Gamma'^*).$$

The coinflation map $\mathrm{coinf} = \mathrm{coinf}_{\Gamma'}^{\Gamma}$, defined in (II.3.1) maps $\mathrm{Det}(\mathfrak{o}_M\Gamma^*) \to \mathrm{Det}(\mathfrak{o}_M\Gamma'^*)$ surjectively (see (II.6.6)). Therefore

$$T_{\Gamma'} = \mathrm{coinf}\,\mathrm{Det}(z), \qquad z \in \mathfrak{o}_M\Gamma^*. \tag{3.12}$$

But τ^*, and hence T, is clearly inflation invariant. Thus also

$$T_{\Gamma'} = \mathrm{coinf}\, T_\Gamma.$$

In conjunction with (3.12), this implies that

$$(T_\Gamma \cdot \mathrm{Det}(z)^{-1})(\chi) = 1 \qquad \text{for} \qquad \chi \in \inf_{\Gamma'}^{\Gamma} R_{\Gamma'}. \tag{3.13}$$

On the other hand, by Theorem 15A and Proposition 1.1 (iii),

$$(T_\Gamma \cdot \mathrm{Det}(z)^{-1})(\chi) \equiv 1 \pmod{\mathfrak{L}}, \tag{3.14}$$

if $\chi \in \mathrm{Ker}\, d_l$, where $\mathfrak{L}$ is the maximal ideal of $\mathbb{Z}_l^c$. But (cf. (II.6.7)) this, together with (3.13), implies that (3.14) in fact holds for $\chi \in R_\Gamma$. In other words $T_\Gamma \cdot \mathrm{Det}(z)^{-1} \in \mathrm{Hom}_{\Omega_M}(R_\Gamma, 1 + \mathfrak{L})$, i.e., $T_\Gamma \in \mathrm{Det}(\mathfrak{o}_M\Gamma^*)\mathrm{Hom}_{\Omega_M}(R_\Gamma, 1 + \mathfrak{L})$. By (3.2) T_Γ is of finite order mod $\mathrm{Det}(\mathfrak{o}_M\Gamma^*)$. But $\mathrm{Hom}_{\Omega_M}(R_\Gamma, 1 + \mathfrak{L})$ is a pro-l-group, so this order is a power of l. □

For the proof in C and D we shall need two Lemmas. The second of these is reminiscent of the argument in Chap. II, leading from (II.6.10 and 6.11) to (II.6.12), and was used in two places, quite analogous to C and D.

3.1. Lemma. *Let Δ be a subgroup of Γ. Then for some $v = v(\Gamma, \Delta) \in \mathfrak{o}_M^*$,*

$$\mathrm{res}_\Delta^\Gamma T_\Gamma = T_\Delta \cdot \mathrm{Det}(v).$$

Proof. From (1.9) we get the induction formula

$$\begin{aligned} \mathrm{res}_\Delta^\Gamma\, T_\Gamma(\beta) &= T_\Gamma(\mathrm{ind}_\Delta^\Gamma\, \beta) \\ &= T_\Delta(\beta)[T_\Delta(\varepsilon_\Delta)^{-1} \cdot T_\Gamma(\mathrm{ind}_\Delta^\Gamma\, \varepsilon_\Delta)]^{\deg(\beta)}, \end{aligned}$$

for $\beta \in R_\Delta$. Here ε_Δ is, as usual, the identity character of Δ. We know that $T_\Delta(\varepsilon_\Delta) = -1$. By Proposition 1.1 (ii), $T_\Gamma(\operatorname{ind}_\Delta^\Gamma \varepsilon_\Delta)^\omega = T_\Gamma((\operatorname{ind}_\Delta^\Gamma \varepsilon_\Delta)^\omega)$; for $\omega \in \Omega_M$. $\operatorname{ind}_\Delta^\Gamma \varepsilon_\Delta$ has values in $\mathbb{Q}_l^*$, i.e., is certainly fixed under Ω_M. Thus $T_\Gamma(\operatorname{ind}_\Delta^\Gamma \varepsilon_\Delta) \in M^*$; and it is a unit. Therefore

$$T_\Delta(\varepsilon_\Delta)^{-1} T_\Gamma(\operatorname{ind}_\Delta^\Gamma \varepsilon_\Delta) = v \in \mathfrak{o}_M^*,$$

and of course, for all β,

$$\operatorname{Det}_\beta(v) = v^{\deg(\beta)}. \qquad \square$$

Addendum to Lemma 3.1. *If $\Delta \supset \Gamma_0$ then $v = \pm 1$.*

Proof. By Proposition 1.1 (vi). $\square$

3.2. Lemma. *Suppose that there is a family $\mathscr{S}$ of subgroups Δ_i of Γ, such that*

$$T_{\Delta_i} \in \operatorname{Det}(\mathfrak{o}_M \Delta_i^*), \qquad \textit{for all} \qquad \Delta_i \in \mathscr{S},$$

and furthermore an integer $s \neq 0$ and virtual characters $\eta_i \in R_{\Delta_i}(\mathbb{Q}_l)$, for all $\Delta_i \in \mathscr{S}$, such that

$$s\varepsilon_\Gamma = \sum_{\mathscr{S}} \operatorname{ind}_{\Delta_i}^\Gamma \eta_i.$$

Then

$$T_\Gamma^s \in \operatorname{Det}(\mathfrak{o}_M \Gamma^*).$$

Proof. Write $v_i = v(\Gamma, \Delta_i)$ for the constants in Lemma 3.1 and let $T_{\Delta_i} = \operatorname{Det}(u_i)$, with $u_i \in \operatorname{Det}(\mathfrak{o}_M \Delta_i^*)$. For every $\chi \in R_\Gamma$ we have

$$s\chi = \sum_i \operatorname{ind}_{\Delta_i}^\Gamma (\eta_i \cdot \operatorname{res}_{\Delta_i}^\Gamma \chi).$$

Therefore

$$\begin{aligned} T_\Gamma^s(\chi) &= \prod_i T_\Gamma(\operatorname{ind}_{\Delta_i}^\Gamma(\eta_i \cdot \operatorname{res}_{\Delta_i}^\Gamma \chi)) \\ &= \prod_i [T_{\Delta_i}(\eta_i \cdot \operatorname{res}_{\Delta_i}^\Gamma \chi) \cdot (\operatorname{Det}(v_i)(\eta_i \cdot \operatorname{res}_{\Delta_i}^\Gamma \chi))] \qquad \text{(by Lemma 3.1)} \\ &= \prod_i (\operatorname{Det}(u_i v_i)(\eta_i \cdot \operatorname{res}_{\Delta_i}^\Gamma \chi)) \\ &= \prod_i (\bar{\eta}_i \operatorname{Det}(u_i v_i)(\operatorname{res}_{\Delta_i}^\Gamma \chi)). \end{aligned}$$

Here we have used the character action defined in (II.3.6). By Theorem 14,

$$\bar{\eta}_i \operatorname{Det}(u_i v_i) = \operatorname{Det}(w_i), \qquad w_i \in \mathfrak{o}_M \Delta_i^*.$$

Moreover, by Theorem 12, $\exists w_i' \in \mathfrak{o}_M\Gamma^*$, so that

$$\begin{aligned}\mathrm{Det}(w_i)(\mathrm{res}^{\Gamma}_{\Delta_i}\chi) &= (\mathrm{ind}^{\Gamma}_{\Delta_i}\mathrm{Det}(w_i))(\chi)\\ &= \mathrm{Det}(w_i')(\chi), \qquad \text{for all } \chi.\end{aligned}$$

Thus finally $T^s_\Gamma = \prod_i \mathrm{Det}(w_i') \in \mathrm{Det}(\mathfrak{o}_M\Gamma^*)$. □

Proof of C. Here we take $\mathscr{S}$ to be the set of $\mathbb{Q}$-l-elementary subgroups of Γ. Then indeed the hypotheses of the last Lemma will hold, for some s prime to l (cf. [Se2], Theorem 28). Thus we obtain C. □

Remark. We used the same fact at the corresponding place in II §6 (for assertion IV).

Proof of D. We now take $\mathscr{S}$ as the set of all $\mathbb{Q}$-elementary subgroups of Γ. We can then satisfy the hypothesis of Lemma 3.2 with $s = 1$ (cf. [Se2] Theorem 27). This gives us D. □

Remark. This is analogous to the proof in Chap. II §6 of "(6.1) for any Γ".

We finally come to the proof of B. This is similar to the proof of assertion III in Chap. II, §6, using the same tools and ideas, and proceeding more or less by the same individual steps. Accordingly we shall at this stage constantly refer back to Chap. II, §6 for notation and results. In particular

$$\Gamma = \Sigma \rtimes \Pi,$$

Π an l-group, Σ cyclic of order prime to l.

Let $m \mid \mathrm{order}(\Sigma)$. Restricting to $R^{(m)}_\Gamma \subset R_\Gamma$ (cf. (II.6.16)) we get an element

$$T_{\Gamma \mid R^{(m)}_\Gamma} \in \mathrm{Hom}_{\Omega_M}(R^{(m)}_\Gamma, \mathfrak{U}(\mathbb{Q}^c_l)).$$

Theorem 30A is then equivalent with

$$T_{\Gamma \mid R^{(m)}_\Gamma} \in \mathrm{Det}(\mathfrak{o}_M\Gamma^*)_{\mid R^{(m)}_\Gamma}, \qquad \text{for all } m \mid \mathrm{order}(\Sigma). \tag{3.15}$$

We now fix our attention on one divisor d of order (Σ). We do not yet at this stage assume – as we did for the statement of Theorem 32, see also (3.9) – that d is actually the order of Σ. Let $\Phi = \Phi_d$ be the set of Abelian characters of Σ in $\mathbb{Q}^{c*}_l$ of precise order d. Π acts on Φ by conjugation. Denote by Π_1 the elementwise stabilizer of Φ under this action. Then

$$\Delta = \Sigma\Pi_1 = \Sigma \rtimes \Pi_1$$

is a subgroup of Γ, and this meaning of the symbol Δ will be kept for the remainder of the proof of B. We also put

$$\Pi/\Pi_1 = \Xi.$$

Then (cf. (II.6.23), (II.6.24) and Lemma II.6.5) the restriction map $\mathrm{res}_\Delta^\Gamma$ yields a commutative diagram with bijective rows and injective columns

$$\begin{array}{ccc} (\mathrm{Det}(\mathfrak{o}_M\Gamma^*))_{|R_\Gamma^{(d)}} & \to & ((\mathrm{Det}(\mathfrak{o}_M\Delta^*))_{|R_\Delta^{(d)}})^{\Xi} \\ \downarrow & & \downarrow \\ \mathrm{Hom}_{\Omega_M}(R_\Gamma^{(d)}, \mathbb{Q}_l^{c*}) & \to & (\mathrm{Hom}_{\Omega_M}(R_\Delta^{(d)}, \mathbb{Q}_l^{c*}))^{\Xi}. \end{array}$$

The Ξ-action is the obvious one, carefully described in Chap. II §6. It follows, that in order to prove (3.15) for $d = m$, it suffices to show that

$$(\mathrm{res}_\Delta^\Gamma T_\Gamma)_{|R_\Delta^{(d)}} \in \mathrm{Det}(\mathfrak{o}_M\Delta^*)_{|R_\Delta^{(d)}}.$$

By Lemma 3.1 this simplifies to

$$T_{\Delta|R_\Delta^{(d)}} \in \mathrm{Det}(\mathfrak{o}_M\Delta^*)_{|R_\Delta^{(d)}}. \tag{3.16}$$

Next observe that, as $(d, l) = 1$, the surjection $\mathfrak{o}_M\Delta \to \mathfrak{o}_M\bar{\Delta}$ ($\bar{\Delta} = \Delta/\Sigma^d$) will split. Inflation $R_{\bar{\Delta}} \to R_\Delta$ restricts to an isomorphism $R_{\bar{\Delta}}^{(d)} \to R_\Delta^{(d)}$. We thus get a commutative diagram with bijective rows and injective columns

$$\begin{array}{ccc} (\mathrm{Det}(\mathfrak{o}_M\Delta^*))_{|R_\Delta^{(d)}} & \to & (\mathrm{Det}(\mathfrak{o}_M\bar{\Delta}^*))_{|R_{\bar{\Delta}}^{(d)}} \\ \downarrow & & \downarrow \\ \mathrm{Hom}_{\Omega_M}(R_\Delta^{(d)}, \mathbb{Q}_l^{c*}) & \to & \mathrm{Hom}_{\Omega_M}(R_{\bar{\Delta}}^{(d)}, \mathbb{Q}_l^{c*}). \end{array}$$

We therefore may, and will assume from now on that $\Delta = \bar{\Delta}$, i.e., that Σ is of order d and Φ_d is the set of faithful Abelian characters of Σ. Moreover this implies that

$$\Delta = \Pi_1 \times \Sigma.$$

Thus, apart from the change in notation (see (3.4)) we are now in the situation in which Theorem 32 applies.

We shall next use the isomorphism (II.6.2) and apply Lemma II.6.4. There are however some minor changes in notation. Firstly the field L of II. §6 is now denoted by N, as defined in (3.9a). Secondly we do not need here the prefix Π_1 in front of Det. Moreover we replace the left hand side in Lemma II.6.4 by $\mathrm{Det}(\mathfrak{o}_M\Delta^*)_{|R_\Delta^{(d)}}$, as we may do by (II.6.17). We then end up with a commutative diagram with bijective rows and injective columns.

$$\begin{array}{ccc} (\mathrm{Det}(\mathfrak{o}_M\Delta^*))_{|R_\Delta^{(d)}} & \to & \mathrm{Map}_{\Omega_M}(\Phi, \mathrm{Det}(\mathfrak{o}_N\Pi_1^*)) \\ \downarrow & & \downarrow \\ \mathrm{Hom}(R_\Delta^{(d)}, \mathbb{Q}_l^{c*}) & \to & \mathrm{Map}(\Phi, \mathrm{Hom}(R_{\Pi_1}, \mathbb{Q}_l^{c*})) \end{array} \tag{3.17}$$

By definition, the image of $T_{\Delta|R_\Delta^{(d)}}$ under the bottom row is

$$T' \colon \phi \mapsto T_{\Delta,\phi},$$

with $T_{\Delta,\phi}$ as defined in (3.8). As T_Δ is fixed under Ω_M, so is T'. Hence, by Theorem 32,

$$T' \in \mathrm{Map}_{\Omega_M}(\Phi, \mathrm{Det}(\mathfrak{o}_N\Pi_1^*)).$$

Using diagram (3.17), we now get (3.16). This concludes the proof of B. □

§4. Strategy for Theorem 32

We have to repeat at the new level what was said in the introduction to the preceding §3. We shall yet again separate out distinct aspects of the proof. It will be reduced to three other theorems, whose proofs – as far as they will be given – will follow in §5 and §6. These theorems form the real arithmetic core of this part of the theory. In the present section we shall, after going over the notation, first of all describe briefly the strategy of the proof of Theorem 32 following [Ty7]. After this we shall go over the individual steps and complete the proof, modulo three specific results, which will be stated here and to which we shall return in the subsequent sections.

Notations. R_Γ, R_Π etc. are again the rings of virtual characters over $\mathbb{Q}_l^c$ (not over $\mathbb{Q}^c$). L/F is a tame Galois extension of non-Archimedean local fields with Galois group Γ, of residue class characteristic $p \neq l$. Γ is of the special form

$$\Gamma = \Pi \times \Sigma$$

as in (3.4), i.e., with Π an l-group and Σ cyclic of order prime to l. We use again the notations (3.1) and (3.5)–(3.9a). In particular ϕ is now a faithful Abelian character of Σ with values in $\mathbb{Q}_l^{c*}$. Also $M = \mathbb{Q}_l(p)$ and N is the extension of $\mathbb{Q}_l$ as in (3.9a).

The first part of the proof of Theorem 32 relies heavily on the integral logarithm for local group rings, introduced in Chap. II, §5, and in particular on Theorem 17. The required detailed result has already been formulated as Corollary 2 to that theorem. We shall use the notation of that section.

We are studying the two maps in the following diagram

$$\begin{array}{ccc} \mathrm{Hom}_{\Omega_N}(R_\Pi, \mathfrak{U}(\mathbb{Q}_l^c)) & \xrightarrow{\text{coinf}} & \mathrm{Hom}_{\Omega_N}(R_{\Pi^{ab}}, \mathfrak{U}(\mathbb{Q}_l^c)) \\ {\scriptstyle \Lambda}\downarrow & & \\ N\mathfrak{C}_\Pi & & \end{array} \tag{4.1}$$

Here coinf is the map coming from the surjection $\Pi \to \Pi^{ab}$, and Λ is the logarithmic map into the N-space on the set $\mathfrak{C}_\Pi$ of conjugacy classes of Π. The element $T_{\Gamma,\phi}$ lies in the domain of both these maps.

We shall almost explicitly produce an element t of $\mathfrak{o}_N\Pi_0^*$, where Π_0 is the inertia subgroup of $\Pi = \Gamma/\Sigma$, such that

$$\operatorname{coinf} T_{\Gamma,\phi} = \operatorname{coinf} \mathrm{Det}(t),$$

with t viewed as element of $\mathfrak{o}_N\Pi^*$. Thus

$$T_{\Gamma,\phi} \cdot \mathrm{Det}(t)^{-1} \in \operatorname{Ker} \operatorname{coinf}. \tag{4.2}$$

We shall then show that t can be further adjusted, so that, while preserving (4.2), we also get

$$T_{\Gamma,\phi} \cdot \mathrm{Det}(t)^{-1} \in \operatorname{Ker} \Lambda. \tag{4.3}$$

This is the step where the logarithm of Gauss sums enters.

By Theorem 17 we get

(4.4a) $$T_{\Gamma,\phi} = \mathrm{Det}(t)u,$$

where u is a torsion element of $\mathrm{Hom}_{\Omega_N}(R_\Pi, \mathfrak{U}(\mathbb{Q}_l^c))$, i.e.,

(4.4b) $$u \in \mathrm{Hom}_{\Omega_N}(R_\Pi, \mu(\mathbb{Q}_l^c)), \qquad \mu = \text{roots of unity}.$$

Moreover, by (4.2),

(4.4c) $$u \in \mathrm{Ker\,Coinf}.$$

The second part of the proof consists in showing that $u = 1$. It is here that the congruences for Galois Gauss sums and for group determinants come in. We shall state certain congruences for the values of $T_{\Gamma,\phi}$, and analogous congruences for the values of elements in $\mathrm{ind}_{\Pi_0}^{\Pi}(\mathrm{Det}\,\mathfrak{o}_N\Pi_0^*)$. By (4.4a), u can be written as the product of an element in the latter group, and of $T_{\Gamma,\phi}$, and one uses this fact to derive congruences for u. These together with (4.4b), (4.4c) then imply that $u = 1$. The three results referred to in the introduction, which will form the contents of §5 and §6 are the logarithmic evaluation of $T_{\Gamma,\phi}$ and the two congruence theorems mentioned last.

Remark. Ideally, one would want to say, that the congruences on $T_{\Gamma,\phi}$ and on group determinants force these to lie in a subgroup $C(\Pi)$ of $\mathrm{Hom}_{\Omega_N}(R_\Pi, \mathfrak{U}(\mathbb{Q}_l^c))$, whose intersection with the two subgroups occurring in (4.4b) and (4.4c) is $= 1$. Then of course we would have $u \in C(\Pi)$, and thus $u = 1$. This is "nearly" true, but unfortunately one of the sets of congruences, namely that for semidihedral characters, can at present not be made to fit into this neat multiplicative pattern. We shall also see that semidihedral characters present a special difficulty in the context of §7.

We shall now turn to details. As always, the subscript $_0$ denotes inertia groups.

A. $\exists t_1 \in \mathfrak{o}_N\Pi_0^*$, *so that*

$$T_{\Gamma,\phi}(\alpha) = \mathrm{Det}(t_1)(\alpha)$$

for all Abelian characters α *of* Π.

Proof. Let A be the Artin map $F^* \to \Gamma^{ab}$. Write

(4.5) $$t_1'' = \sum_y A(y) \cdot \psi_F(c^{-1}y)^j \in \mathfrak{o}_M\Gamma^{ab} \qquad (\text{sum over } y \in \mathfrak{o}_F^* \bmod \mathfrak{p}_F),$$

where j is the embedding already used in §3, see in particular (3.1). Here ψ_F is the canonical additive character of F, and c has the usual meaning, and has been chosen for the definition of $z(F, \cdot)$ and of τ^* (cf. (1.6)). For a ramified Abelian character β of Γ, one immediately verifies that $\mathrm{Det}(t_1'')(\beta) = T_{\Gamma^{ab}}(\beta) = T_\Gamma(\beta)$. If, on the other hand, β is non-ramified Abelian, then $\mathrm{Det}(t_1'')(\beta) = \sum_y \psi_F(c^{-1}y)^j = -\psi_F(0)^j = -1$, and also $T_\Gamma(\beta) = -1$. Thus in all cases

(4.6) $$T_{\Gamma^{ab}}(\beta) = T_\Gamma(\beta) = \mathrm{Det}(t_1'')(\beta).$$

By Proposition 1.1 (i), t_1'' is a unit of $\mathfrak{o}_M(\Gamma^{ab})$. The kernel of the map $\mathfrak{o}_M(\Gamma) \to \mathfrak{o}_M\Gamma^{ab}$ is contained in the radical. Therefore, choosing $t_1' \in \mathfrak{o}_M(\Gamma)$ with image t_1'' in $\mathfrak{o}_M\Gamma^{ab}$, we know that $t_1' \in \mathfrak{o}_M\Gamma^*$. Moreover the $A(y)$ in (4.5) all lie in $(\Gamma^{ab})_0$. As $\Gamma_0 \to (\Gamma^{ab})_0$ is surjective, we may take the element t_1' above to lie in $\mathfrak{o}_M\Gamma_0$. Thus we now have, by (4.6),

$$\text{(4.7)} \qquad \begin{cases} T_\Gamma(\beta) = \mathrm{Det}(t_1')(\beta), & \text{for all Abelian characters } \beta \text{ of } \Gamma, \\ \text{with} \quad t_1' \in \mathfrak{o}_M\Gamma_0^*. \end{cases}$$

Now define

$$G = G_\phi\colon \mathrm{Hom}(R_\Gamma, \mathfrak{U}(\mathbb{Q}_l^c)) \to \mathrm{Hom}(R_\Pi, \mathfrak{U}(\mathbb{Q}_l^c))$$

by

$$G(f)(\chi) = f(\chi \times \phi)$$

(cf. (3.5)).

By (3.8),

$$\text{(4.8)} \qquad G_\phi(T_\Gamma) = T_{\Gamma,\phi}.$$

We are now again working in the framework of II §6 (care with changes in notation!) and of the proof of B in §3. (In fact G_ϕ is the composite

$$\begin{array}{ll} \mathrm{Hom}(R_\Gamma, \mathfrak{U}(\mathbb{Q}_l^c)) & \\ \quad\downarrow & \text{(see (II.6.16))} \\ \mathrm{Hom}(R_\Gamma^{(d)}, \mathfrak{U}(\mathbb{Q}_l^c)) & \\ \quad\downarrow & \text{(see (II.6.22))} \\ \mathrm{Map}(\Phi, \mathrm{Hom}(R_\Pi, \mathfrak{U}(\mathbb{Q}_l^c)) & \\ \quad\downarrow & \text{evaluation at } \phi \\ \mathrm{Hom}(R_\Pi, \mathfrak{U}(\mathbb{Q}_l^c)).) & \end{array}$$

On the other hand, define a homomorphism

$$\tilde{G}_\phi\colon \mathfrak{o}_M(\Gamma) \to \mathfrak{o}_N(\Pi)$$

of $\mathfrak{o}_M$-algebras, by

$$\text{(4.9)} \qquad \tilde{G}_\phi\left(\sum a_{\sigma,\pi}\sigma\pi\right) = \sum a_{\sigma,\pi}\phi(\sigma)\pi$$

(sum over all $\pi \in \Pi$, all $\sigma \in \Sigma$). Then

$$\text{(4.10)} \qquad \mathrm{Det}(\tilde{G}_\phi x) = G_\phi \,\mathrm{Det}(x), \qquad \text{for } x \in \mathfrak{o}_M\Gamma^*.$$

This is easily verified directly. It also follows from the results of II §6 (cf. Lemmas II.6.1 and II.6.2). Indeed $\tilde{G}_\phi$ has a decomposition, similar to the one given above for G_ϕ, now in terms of the map g_ϕ of Lemma II.6.1.

Now put

$$\text{(4.11)} \qquad t_1 = \tilde{G}_\phi t_1'.$$

As $\tilde{G}_\phi$ clearly maps $\mathfrak{o}_M\Gamma_0^*$ into $\mathfrak{o}_N\Pi_0^*$, we get $t_1 \in \mathfrak{o}_N\Pi_0^*$. If now α is an Abelian character of Π, put $\beta = \alpha \times \phi$. Applying (4.7), (4.8), (4.10), (4.11), we now obtain A. □

If Θ is any subgroup of Γ containing Σ, then it is of form $\Pi' \times \Sigma$, with Π' a subgroup of Π. Associated with any $g \in \mathrm{Hom}(R_\Theta, \cdot)$ we get $G_\phi g \in \mathrm{Hom}(R_{\Pi'}, \cdot)$, given by $G_\phi g(\chi) = g(\chi \times \phi)$, for $\chi \in R_{\Pi'}$, just as in the original case $\Theta = \Gamma$. Here again $(\chi \times \phi)(\pi\sigma) = \chi(\pi)\phi(\sigma)$ for $\pi \in \Pi'$, $\sigma \in \Sigma$. It follows immediately that for $\chi \in R_\Pi$,

(4.12a) $$\mathrm{res}_\Theta^\Gamma(\chi \times \phi) = (\mathrm{res}_{\Pi'}^\Pi(\chi)) \times \phi,$$

whence

(4.12b) $$G_\phi(\mathrm{ind}_\Theta^\Gamma g) = \mathrm{ind}_{\Pi'}^\Pi(G_\phi g).$$

In particular we put $G_\phi T_\Theta = T_{\Theta,\phi}$.

From now on we let Θ be the maximal Abelian subgroup of Γ containing $\Pi_0 \times \Sigma$. Let $\Theta \cap \Pi = \Pi'$. Then we get

B. $\Lambda_\Pi \mathrm{ind}_{\Pi'}^\Pi(T_{\Theta,\phi}) = [\Gamma : \Theta]\Lambda_\Pi(T_{\Gamma,\phi})$.

Proof. We shall work with congruences modulo roots of unity. By Theorem 17, B is equivalent with the congruence

(4.13) $$T_{\Theta,\phi}(\mathrm{res}_{\Pi'}^\Pi \chi) \equiv T_{\Gamma,\phi}(\chi)^{[\Gamma:\Theta]}, \qquad \text{for all} \qquad \chi \in R_\Pi.$$

By (4.12) this congruence is implied by

(4.14) $$T_\Theta(\mathrm{res}_\Theta^\Gamma \chi) \equiv T_\Gamma(\chi)^{[\Gamma:\Theta]}, \qquad \text{for all} \qquad \chi \in R_\Gamma,$$

and it is (4.14) which we shall prove.

By the addendum to Lemma 3.1,

$$T_\Theta(\beta) \equiv T_\Gamma(\mathrm{ind}_\Theta^\Gamma \beta), \qquad \text{for all} \qquad \beta \in R_\Theta,$$

whence

(4.15) $$T_\Theta(\mathrm{res}_\Theta^\Gamma \chi) \equiv T_\Gamma(\mathrm{ind}_\Theta^\Gamma \mathrm{res}_\Theta^\Gamma \chi).$$

By Frobenius reciprocity,

$$\mathrm{ind}_\Theta^\Gamma \mathrm{res}_\Theta^\Gamma \chi = \chi \cdot \mathrm{ind}_\Theta^\Gamma \varepsilon, \qquad \varepsilon \text{ the identity character of } \Theta.$$

As $\Theta \supset \Gamma_0$, we have

$$\mathrm{ind}_\Theta^\Gamma \varepsilon = \sum \alpha_i' \qquad ([\Gamma : \Theta] \text{ terms}),$$

where the α_i' are Abelian, non-ramified characters. By Proposition 1.1 (vi), $T_\Gamma(\chi \cdot \alpha_i') = T_\Gamma(\chi)$ and therfore

$$T_\Gamma(\mathrm{ind}_\Theta^\Gamma \mathrm{res}_\Theta^\Gamma \chi) = T_\Gamma(\chi)^{[\Gamma:\Theta]},$$

which in conjunction with (4.15) gives (4.14). □

For the next assertion we denote by $\mathfrak{C}_{\Pi,\Pi_0}$ the set of Π-conjugacy classes of elements in Π_0. We know from Theorem 17 that $\Lambda_\Pi \operatorname{ind}_{\Pi'}^{\Pi}(T_{\Theta,\phi}) \in N\mathfrak{C}_\Pi$. In fact we have

C. $\Lambda_\Pi \operatorname{ind}_{\Pi'}^{\Pi}(T_{\Theta,\phi}) \in l[\Gamma:\Theta]\mathfrak{o}_N\mathfrak{C}_{\Pi,\Pi_0}$.

This result will be deduced via the evaluation of $\Lambda_\Theta T_{\Theta,\phi}$, i.e., the explicit evaluation of the logarithm of the Gauss sum, in §5. From *B* and *C* we now conclude that

D. $\Lambda_\Pi(T_{\Gamma,\phi}) \in l\mathfrak{o}_N\mathfrak{C}_{\Pi,\Pi_0}$. □

A and *D* are the crucial results. Let $\mathfrak{a}$ be the commutator ideal of $\mathfrak{o}_N\Pi$. Let t_1 be the element occurring in *A*. Then $T_{\Gamma,\phi}\cdot \operatorname{Det}(t_1)^{-1}$, as an element of $\operatorname{Hom}_{\Omega_N}(R_\Pi, \mathfrak{U}(\mathbb{Q}_l^c))$ is of finite order modulo $\operatorname{Det}(\mathfrak{o}_N\Pi^*)$, and by *A* even modulo

$$\operatorname{Ker}[\operatorname{Det}(\mathfrak{o}_N\Pi^*) \to \operatorname{Det}((\mathfrak{o}_N\Pi^{ab})^*)] = \operatorname{Det}(1+\mathfrak{a}).$$

By Theorem 17 therefore

$$\Lambda_\Pi(T_{\Gamma,\phi}\cdot \operatorname{Det}(t_1^{-1})) \in c(\mathfrak{a}) \otimes_{\mathfrak{o}_N} N.$$

The final result of this stage of the proof is

E. $\exists t' \in (1+\mathfrak{a}) \cap \mathfrak{o}_N\Pi_0$ *with the property that for* $t = t_1 t'$,

$$\Lambda_\Pi(T_{\Gamma,\phi}\operatorname{Det}(t)^{-1}) = 0.$$

The proof of *E* depends essentially on *D*. It goes by induction on order (Π), the case of Abelian Π clearly being trivial. It consists of purely technical computations of logarithmic values and of various submodules of $N\Pi$, of a type which have been given in II §5. The details will be omitted here. See [Ty7], p. 57–58. □

For the next part of the proof we recall some basic facts on representations of a finite group G, with normal cyclic subgroup G_0 and cyclic quotient group G/G_0. If J is a normal subgroup of G, β a character of J, and g an element of G, we define βg by $\beta g(x) = \beta(gxg^{-1})$. This is a character of J. It only depends on the coset $\tilde{g}$ of g mod J and we denote it by $\beta\tilde{g}$.

4.1. Lemma. *For every irreducible character χ of G there is a subgroup J of G, containing G_0, and an Abelian character β of J with $\operatorname{ind}_J^G\beta = \chi$. This determines J uniquely, and determines β to within the set $\{\beta\tilde{g}\}$. The characters $\operatorname{res}_{G_0}^J(\beta\tilde{g})$ are distinct, and $\operatorname{res}_J^G\chi = \sum\beta\tilde{g}$.*

Indeed $\operatorname{res}_{G_0}^G\chi$ is a sum of Abelian characters. Let α be one of these. Then let J be the stabilizer of α under the action by G. Then α extends to an Abelian character β of J, contained in $\operatorname{res}_J^G\chi$. As G/J acts on the conjugacy class of β without fixed points, $\operatorname{ind}_J^G\beta$ is irreducible, hence $=\chi$. □

Let now in particular G be a 2-group, and consider irreducible characters χ of degree 2. Thus $[G:J] = 2$; we let now $\tilde{g}$ be the coset different from J. Write

$$2^r = \operatorname{order}(\operatorname{res}^J_{G_0} \beta).$$

By the Lemma, $r \geqslant 2$. We distinguish between three types of characters χ, corresponding to the three automorphisms of order 2 of the cyclic group of order 2^r (when $r \geqslant 3$). We shall abbreviate $\operatorname{res}^J_{G_0} = \operatorname{res}$.

(i) χ is of *inversion type*: $\operatorname{res}(\beta\tilde{g}) = \operatorname{res}(\beta)^{-1}$.
(ii) χ is of *semi-Abelian type*: $\operatorname{res}(\beta\tilde{g}) = \operatorname{res}(\beta)^{1+2^{r-1}}$, and $r \geqslant 3$.
(iii) χ is of *semi-dihedral type*: $\operatorname{res}(\beta\tilde{g}) = \operatorname{res}(\beta)^{-1+2^{r-1}}$, and $r \geqslant 3$.

We shall apply all this to our Galois group $\Pi = \operatorname{Gal}(L^\Sigma/F)$ of local fields, a group of l-power order. As the extension is tame, $\Pi = G$ satisfies the hypothesis of the Lemma, with $\Pi_0 = G_0$ the inertia group. Given a non-Abelian irreducible character χ of Π, we shall write Ψ for the group J of the Lemma, and put throughout

$$\chi = \operatorname{ind}^\Pi_\Psi \alpha, \qquad \alpha \text{ Abelian } (\Psi \supset \Pi_0). \tag{4.16}$$

Whenever Ψ' is a subgroup of Π containing Π_0 we shall always just write

$$\operatorname{res}^{\Psi'}_{\Pi_0}(\beta) = \operatorname{res}(\beta). \tag{4.17}$$

All other restriction operators will be explicitly indicated in the usual manner. If now, in (4.16), χ is of degree 2, then the non-trivial coset of Ψ in Π contains the Frobenius acting by $\operatorname{res}(\alpha) \mapsto \operatorname{res}(\alpha^{\mathbf{N}_F \mathfrak{p}_F})$. Thus we have:

$$\left\{\begin{array}{c} \chi \text{ is of inversion type} \\ \chi \text{ is of semi-Abelian type} \\ \chi \text{ is of semi-dihedral type} \end{array}\right\} \Leftrightarrow \mathbf{N}_F \mathfrak{p}_F \equiv \left\{\begin{array}{c} -1 \\ 1 + 2^{r-1}\ (r \geqslant 3) \\ -1 + 2^{r-1}\ (r \geqslant 3) \end{array}\right\} \pmod{2^r} \tag{4.18}$$

We have four cases to consider, and in some of these we have to establish some auxiliary results. χ is always irreducible, given as in (4.16), and non-Abelian.

Case I. $\deg(\chi) > 2$. (This is the only case which can occur when l is odd.)

4.2. Lemma for Case I. *Let $\Psi^{(1)}$ be the unique subgroup of Π containing Ψ, with $[\Psi^{(1)}:\Psi] = l$. Then, if $\mathrm{v}^{\Psi^{(1)}}_\Psi = \mathrm{v}$ is the cotransfer, we have $\mathrm{v}\alpha(\pi) = \alpha(\pi)^l$ for $\pi \in \Pi_0$, and*

$$1 < \operatorname{order}(\operatorname{res}(\mathrm{v}\alpha)) < \operatorname{order}(\operatorname{res}(\alpha)).$$

(α defined in (4.16)).

Proof. Let l^r be the order of $\operatorname{res}(\alpha)$. Let $\delta \in \Psi^{(1)} \backslash \Psi$ and write $\operatorname{res}(\alpha)\delta = \operatorname{res}(\alpha)^x$ with $x \in \mathbb{Z}_l^*$. Then $x \not\equiv 1 \pmod{l^r}$, but $x^l \equiv 1 \pmod{l^r}$. Hence $x \equiv 1 \pmod{l}$ and thus $r \geqslant 2$. If $l = 2$ then $2 \mid [\Pi : \Psi^{(1)}]$, whence x is a square, i.e., $x \equiv 1 \pmod 8$ and $r \geqslant 4$. One now computes that $\operatorname{res}(\mathrm{v}\alpha) = \operatorname{res}(\alpha)^{x^l - 1/x - 1} = \operatorname{res}(\alpha)^l$. As $x \equiv 1 + al^{r-1} \pmod{l^r}$, with $(a, l) = 1$, the result follows. □

In the sequel it will often be useful to assume:

(4.19) The sequence $1 \to \Pi_0 \to \Pi \to \Pi/\Pi_0 \to 1$ splits.

This can be done without loss of generality. We only need to replace L, if necessary, by a non-ramified, l-power degree, extension of L to ensure that (4.19) holds.

Case II. $\deg(\chi) = 2$ *and* χ *is of semi-Abelian type.*

4.3. Lemma for Case II. *There is a subgroup* $\Psi^{(2)}$ *of* Π *with* $[\Pi : \Psi^{(2)}] = [\Pi_0 : \Pi_0 \cap \Psi^{(2)}] = 2$, *and an Abelian character* β *of* $\Psi^{(2)}$, *with*

$$\operatorname{ind}_{\Psi^{(2)}}^{\Pi} \beta = \chi, \qquad \operatorname{res}_{\Psi^{(3)}}^{\Psi^{(2)}} \beta = \operatorname{res}_{\Psi^{(3)}}^{\Psi} \alpha,$$

where we write $\Psi^{(3)} = \Psi^{(2)} \cap \Psi$. *Also*

$$\mathrm{v}\beta(\pi) = \beta(\pi^2) \qquad \textit{for all} \qquad \pi \in \Pi_0,$$

where $\mathrm{v} = \mathrm{v}_{\Psi^{(2)}}^{\Pi}$.

Proof. The existence of $\Psi^{(2)}$ is a consequence of (4.19). Π/Ψ induces an automorphism of order 2 on the multiplicative group of characters, generated by α, which gives rise to the automorphism $\operatorname{res}(\alpha) \mapsto \operatorname{res}(\alpha)^{1+2^{r-1}}$. This implies that order $(\alpha) = 2^r$ as well, and α goes into $\alpha^{1+2^{r-1}}$. Therefore $\operatorname{res}_{\Psi^{(3)}}^{\Psi} \alpha$ is fixed under the action of Π. Thus $\operatorname{res}_{\Psi^{(3)}}^{\Psi} \alpha = \operatorname{res}_{\Psi^{(3)}}^{\Psi^{(2)}} \beta$ for some Abelian character of $\Psi^{(2)}$. By Frobenius reciprocity and the character restriction formula, we now get

$$\langle \operatorname{ind}_{\Psi^{(2)}}^{\Pi} \beta, \operatorname{ind}_{\Psi}^{\Pi} \alpha \rangle = \langle \operatorname{res}_{\Psi^{(3)}}^{\Psi^{(2)}} \beta, \operatorname{res}_{\Psi^{(3)}}^{\Psi} \alpha \rangle = 1.$$

Hence indeed $\operatorname{ind}_{\Psi^{(2)}}^{\Pi} \beta = \chi$. The formula for $\mathrm{v}\beta$ is immediate. □

Case III. $\deg(\chi) = 2$ *and* χ *is of inversion type.*

Let order $(\operatorname{res}(\alpha)) = 2^r$. Assume that $\mathbf{N}_F \mathfrak{p}_F \equiv -1 + 2^s \pmod{2^{s+1}}$. Then $r \leqslant s$. Let F' be the fixed field of Ψ in L^Σ. Then $\mathbf{N}_{F'} \mathfrak{p}_{F'} \equiv 1 \pmod{2^{s+1}}$. This implies that there is a tame Abelian character $\tilde{\alpha}$ of F' with $\tilde{\alpha}^2 = \alpha$. By extending L, if necessary, we may suppose that $\tilde{\alpha} \in R_\Psi$, and that (4.19) still holds. Now we get as an obvious consequence:

4.4. Lemma for Case III. *With* $\tilde{\alpha}^2 = \alpha$, *let* $\tilde{\chi} = \operatorname{ind}_{\Psi}^{\Pi} \tilde{\alpha}$. *Then* $\tilde{\chi}$ *is irreducible and either of inversion or of semi-dihedral type.*

We also note that $N(\operatorname{res}(\tilde{\alpha}))$ is a quadratic extension of $N(\operatorname{res}(\alpha))$. □

Case IV. $\deg(\chi) = 2$ *and* χ *is of semi-dihedral type.*

Here we have to introduce some notation. We write

$$N' = N(\operatorname{res}(\alpha)).$$

As before order $(\mathrm{res}(\alpha)) = 2^r$. f is the automorphism which acts trivially on all 2-power roots of unity and is the Frobenius on $N/\mathbb{Q}_l$. Finally $\sum^*$ is always the sum over Abelian characters η of Π_0 with $\operatorname{Ker}\eta \supset \operatorname{Ker}(\mathrm{res}(\alpha^2))$, taken as extended in some way to Ψ.

Now we can state the two parallel congruence theorems, each dealing with each of the Cases I–IV and using the notations introduced above for each case. Furthermore, notations in Theorem 33 will also be used in Theorem 34.

Theorem 33. *Case I: Let* $\chi' = \mathrm{ind}_{\Psi(1)}^{\Pi}(\mathrm{v}\alpha)$. *Then*

$$T_{\Gamma,\phi}(\chi)T_{\Gamma,\phi}(\chi')^{-f} \equiv 1 \bmod \begin{cases} (l), & \text{if } l \neq 2, \\ (2\mathfrak{L}_2), & \text{if } l = 2, \end{cases}$$

with the usual meaning of $\mathfrak{L}_2$.

Case II. Let $\chi' = \mathrm{ind}_{\Psi(2)}^{\Pi}(\beta^2)$. *Then*

$$T_{\Gamma,\phi}(\chi) \cdot T_{\Gamma,\phi}(\mathrm{v}\beta)^{-f} = T_{\Gamma,\phi}(\chi')T_{\Gamma,\phi}(\mathrm{v}\beta^2)^{-f}.$$

Case III. If $\omega \in \Omega_N$ *fixes* $N(\mathrm{res}(\alpha))$, *but not* $N(\mathrm{res}(\tilde{\alpha}))$ *then*

$$T_{\Gamma,\phi}(\tilde{\chi} + \tilde{\chi}^{\omega})T_{\Gamma,\phi}(\chi)^{-f} \equiv 1 \pmod 4.$$

Case IV.

$$2^{1-r}t_{N'/N}(T_{\Gamma,\phi}(\chi)) - T_{\Gamma,\phi}(\det_\chi)^f \equiv \sum{}^* T_{\Gamma,\phi}(\mathrm{ind}_\Psi^\Pi \eta) \cdot 2^{1-r} - T_{\Gamma,\phi}(\det_{\mathrm{ind}(\varepsilon)}) \pmod 4,$$

where $\mathrm{ind}(\varepsilon) = \mathrm{ind}_\Psi^\Pi \varepsilon_\Psi$. *Also each side of this congruence is* $\equiv 0 \pmod 2$. □

Theorem 34. *Let throughout* $x \in \mathfrak{o}_N \Pi_0^*$, *viewed as an element of* $\mathfrak{o}_N \Pi^*$.

Case I.

$$\mathrm{Det}(x)(\chi) \cdot \mathrm{Det}(x)(\chi')^{-f} \equiv 1 \bmod \begin{cases} (l), & \text{if } l \neq 2, \\ (2\mathfrak{L}_2), & \text{if } l = 2. \end{cases}$$

Case II.

$$\mathrm{Det}(x)(\chi) \cdot \mathrm{Det}(x)(\mathrm{v}\beta)^{-f} \equiv \mathrm{Det}(x)(\chi') \cdot \mathrm{Det}(x)(\mathrm{v}\beta^2)^{-f} \pmod{2\mathfrak{L}_2}.$$

Case III.

$$\mathrm{Det}(x)(\tilde{\chi} + \tilde{\chi}^{\omega}) \cdot \mathrm{Det}(x)(\chi)^{-f} \equiv 1 \pmod{2\mathfrak{L}_2}.$$

Case IV.

$$\begin{aligned} &2^{1-r}t_{N'/N}(\mathrm{Det}(x)(\chi)) - \mathrm{Det}(x)(\det_\chi)^f \\ &\equiv \sum{}^* \mathrm{Det}(x)(\mathrm{ind}_\Psi^\Pi \eta)2^{1-r} - \mathrm{Det}(x)(\det_{\mathrm{ind}(\varepsilon)}) \pmod 4. \end{aligned}$$ □

We shall return to the last two theorems in §6. Here we shall use them to prove that the element

$$u = T_{\Gamma,\phi} \cdot \mathrm{Det}(t)^{-1}$$

(see E, and (4.4a)) is 1. We know that u takes root of unity values, by E (cf. (4.4b)). We shall prove that $u(\chi) = 1$, for all χ. It suffices to take χ irreducible. We then proceed by induction on $[\Pi_0 : \mathrm{Ker}(\mathrm{res}\,\chi)]$, where $\mathrm{Ker}(\mathrm{res}\,\chi)$ is the kernel of the restriction of the associated representation of Π_0. We know (cf. (4.4c) that $u(\chi) = 1$ if χ is Abelian. This starts the induction. We shall apply the preceding theorems, using their notation.

First suppose $\deg(\chi) > 2$. By Theorems 33, 34,

$$u(\chi)u(\chi')^{-f} \equiv 1 \bmod \begin{cases} (l), & \text{if} \quad l \neq 2, \\ (2\mathfrak{L}_2), & \text{if} \quad l = 2. \end{cases}$$

But $\mathrm{Ker}(\mathrm{res}\,\chi') \supsetneqq \mathrm{Ker}(\mathrm{res}\,\chi)$, by Lemma 4.2. So $u(\chi') = 1$. Hence

$$u(\chi) \equiv 1,$$

modulo an ideal which distinguishes roots of unity. Therefore $u(\chi) = 1$.

Next let $\deg(\chi) = 2$, χ of semi-Abelian type. By the preceding theorems,

$$u(\chi)u(\mathrm{v}\beta)^{-f} \equiv u(\chi')u(\mathrm{v}\beta^2)^{-f} \pmod{2\mathfrak{L}_2}.$$

But $u(\mathrm{v}\beta) = u(\mathrm{v}\beta^2) = 1$, only Abelian characters being involved, and $u(\chi') = 1$ by induction hypothesis. Thus $u(\chi) \equiv 1 \pmod{2\mathfrak{L}_2}$, whence $u(\chi) = 1$.

Next we show that, whenever χ is of inversion or semi-dihedral type, then

(4.20) $$u(\chi) = \pm 1.$$

Certainly $\chi - \mathrm{ind}_{\Psi}^{\Pi}\,\varepsilon_{\Psi} \in \mathrm{Ker}\,d_2$, and similarly $(\chi - \mathrm{ind}_{\Psi}^{\Pi}\,\varepsilon_{\Psi}) \times \phi \in \mathrm{Ker}\,d_2$. By Theorem 15.A and Proposition 1.1,

$$u(\chi - \mathrm{ind}_{\Psi}^{\Pi}\,\varepsilon_{\Psi}) \equiv 1 \pmod{\mathfrak{L}_2}.$$

But $\mathrm{ind}_{\Psi}^{\Pi}\,\varepsilon_{\Psi}$ is sum of Abelian characters, so

$$u(\chi) \equiv 1 \pmod{\mathfrak{L}_2}.$$

Thus $u(\chi)$ is of 2-power order. But we also know that $T_{\Gamma,\phi}(\chi)$ (by Proposition 1.1), and $\mathrm{Det}(x)(\chi)$ lie in $N(\mathrm{res}(\chi))$. The only 2-power roots of unity in this field are ± 1. This then gives (4.20).

Now assume in particular that χ is of inversion type. By the theorem,

$$u(\tilde{\chi} + \tilde{\chi}^{\omega}) \equiv u(\chi)^f \pmod{2\mathfrak{L}_2}.$$

Also $u(\tilde{\chi}^{\omega}) = u(\tilde{\chi})^{\omega}$ (ω lies in Ω_N), and by (4.20), $u(\tilde{\chi})^{\omega} = u(\tilde{\chi}) = \pm 1$. Thus, $u(\tilde{\chi} + \tilde{\chi}^{\omega}) = 1$, and hence $u(\chi) = 1$.

Finally suppose χ is of semi-dihedral type. By case IV in Theorem 34, and substituting $\mathrm{Det}(t) = u^{-1}T_{\Gamma,\phi}$, we get

(4.21) $$\begin{cases} 2^{1-r}t_{N'/N}(u^{-1}T_{\Gamma,\phi}(\chi)) - u^{-1}T_{\Gamma,\phi}(\det_{\chi})^f \\ \equiv \sum{}^* u^{-1}T_{\Gamma,\phi}(\mathrm{ind}_{\Psi}^{\Pi}\,\eta)2^{1-r} - u^{-1}T_{\Gamma,\phi}(\det_{\mathrm{ind}(\varepsilon)})^f \pmod{4}. \end{cases}$$

But $u(\det_\chi) = u(\det_{\mathrm{ind}(\varepsilon)}) = 1$, as the characters are Abelian, and $u(\mathrm{ind}_\Psi^\Pi \eta) = 1$ for all η, by induction hypothesis. Substituting this into (4.21), and then subtracting from the congruence in Theorem 33 Case IV, we get

$$2^{1-r} t_{N'/N}((u(\chi)^{-1} - 1) T_{\Gamma,\phi}(\chi)) \equiv 0 \pmod 4.$$

We know, however, by (4.20), that $u(\chi)^{-1} - 1 \in \mathbb{Q}_l$, so

$$(u(\chi)^{-1} - 1) \cdot 2^{1-r} t_{N'/N}(T_{\Gamma,\phi}(\chi)) \equiv 0 \pmod 4.$$

On the other hand, because the left-hand side of the congruence in Theorem 33 Case IV is known to be divisible by 2, while $T_{\Gamma,\phi}(\det_\chi)$ is a unit, it follows that $2^{1-r} t_{N'/N}(T_{\Gamma,\phi}(\chi))$ is a unit. Therefore finally $u(\chi) = 1$. □

§5. Gauss Sum Logarithm

Here the logarithm of l-adic Abelian Gauss sums will be computed, and the result then applied in the proof of §4, C. The notation is the same as in §4. In particular Θ is again the maximal Abelian subgroup containing $\Pi_0 \times \Sigma$, hence containing Γ_0. Put $\Theta = \Pi' \times \Sigma$ and let $H = L^\Theta$ be the fixed field of Θ. As Π' is Abelian, $\mathfrak{C}_{\Pi'} = \Pi'$. We write $\Lambda_{\Pi'} T_{\Theta,\phi}$ as an element of $N\Pi'$ in the form

(5.1) $$\Lambda_{\Pi'} T_{\Theta,\phi} = \sum_{\pi \in \Pi'} s_\pi \pi, \qquad s_\pi \in N.$$

Theorem 35. (a) *If* $\pi \notin \Pi_0$, *then* $s_\pi = 0$.
(b) *If* $\Pi_0 \neq 1$, *then* $s_1 = [(l-1)/2] \log \mathbf{N}_H \mathfrak{p}_H$.
(c) *If* $\pi \in \Pi_0, \pi \neq 1$, π *not a generator of* Π_0, *then* $s_\pi = 0$.
(d) *If* π *generates* Π_0, *then* $s_\pi \in l\mathfrak{o}_N$.
(e) *If* $\pi \in \Pi'$, $\gamma \in \Pi$, *then* $s_{\gamma^{-1}\pi\gamma} = s_\pi$. □

We shall first show that the Theorem implies §4, C. By the Corollary to Proposition II.5.5 (prior to which the relevant symbols were defined),

$$\Lambda_\Pi \mathrm{ind}_{\Pi'}^\Pi T_{\Theta,\phi} = P_\Pi^{\Pi'} \Lambda_{\Pi'} T_{\Theta,\phi},$$

i.e.,

(5.2) $$\begin{cases} \Lambda_\Pi \mathrm{ind}_{\Pi'}^\Pi T_{\Theta,\phi} = \sum_{c \in \mathfrak{C}_\Pi} s'_c c, \\ s'_c = \sum_{\pi \in \Pi', \pi \mapsto c} s_\pi. \end{cases}$$

By the theorem we only have to consider the identity class c_1 in the case $\Pi_0 \neq 1$, and the classes c^* of generators π of Π_0. Now Π/Π' acts faithfully by conjugation on Π_0, and so c^* has $[\Pi : \Pi'] = [\Gamma : \Theta]$ elements. Hence, by Theorem 35 (d), (e), $s_{c^*} \in l[\Gamma : \Theta]\mathfrak{o}_N$. As for c_1, we use (b). As H/F is non-ramified, $\mathbf{N}_H \mathfrak{p}_H = \mathbf{N}_F \mathfrak{p}_F^{[\Gamma:\Theta]}$. The result now follows by observing, that always $\log \mathbf{N}_F \mathfrak{p}_F \equiv 0 \pmod{l}$, and for $l = 2$ even $\log \mathbf{N}_F \mathfrak{p}_F \equiv 0 \pmod{4}$. □

Proof of Theorem 35. For any Abelian character α of Π', we use the same symbol α for its extension to a homomorphism $N\Pi' \to \mathbb{Q}_l^c$ of N-algebras. We then have the inversion formula (with α running over the Abelian characters of Π'),

$$\operatorname{order}(\Pi')s_\pi = \sum_\alpha \alpha(\Lambda_{\Pi'} T_{\Theta,\phi}) \cdot \alpha(\pi^{-1}).$$

Now apply Proposition II.5.5 to the right-hand side, and we get

$$\operatorname{order}(\Pi')s_\pi = \sum_\alpha [\log(T_{\Theta,\phi}(l\alpha) \cdot T_{\Theta,\phi}(-\alpha^l)^f]\alpha(\pi^{-1}). \tag{5.3}$$

To prove (a), suppose $\pi \notin \Pi_0$. Choose a non-ramified Abelian character β of Π' with $\beta(\pi) \neq 1$. By Proposition 1.1,

$$T_{\Theta,\phi}(l\alpha\beta)T_{\Theta,\phi}(-(\alpha\beta)^l)^f = T_{\Theta,\phi}(l\alpha)T_{\Theta,\phi}(-\alpha^l)^f.$$

Hence we get from (5.3)

$$\operatorname{order}(\Pi')s_\pi = \operatorname{order}(\Pi')s_\pi\beta(\pi).$$

Thus $s_\pi = 0$.

Next assume $\Pi_0 \neq 1$. Let β_i run through the Abelian characters of Π_0 of order dividing l, and all extended to Π'. Let $\pi \in \Pi_0$, π not a generator of Π_0. Then $\beta_i(\pi) = 1$. We thus get from (5.3)

$$l \operatorname{order}(\Pi')s_\pi = \sum_\alpha \left(\sum_i \log(T_{\Theta,\phi}(l\alpha\beta_i)T_{\Theta,\phi}(-\alpha^l)^f) \right) \alpha(\pi^{-1}). \tag{5.4}$$

Here we have also used the fact that, as β_i^l is non-ramified, $T_{\Theta,\phi}(-\alpha^l\beta_i^l) = T_{\Theta,\phi}(-\alpha^l)$. We shall now use again congruences modulo roots of unity, recalling that $\log x$ is constant within each congruence class.

Firstly we go back to the definition of $T_{\Theta,\phi}$ in terms of T_Θ (cf. (3.8)). The Frobenius element f fixes α and maps ϕ into ϕ^l. From the Galois action formula for Galois Gauss sums (cf. Theorem 20B) we get

$$T_{\Theta,\phi}(-\alpha^l)^f \equiv T_\Theta(-\alpha^l \times \phi^l).$$

Therefore the inner sum $\sum_i$ in (5.4) is now

$$= \log\left[\prod_i T_\Theta(\beta_i\alpha \times \phi)^l \cdot T_\Theta((\alpha \times \phi)^l)^{-l}\right].$$

By a formula of Davenport-Hasse (cf. [DH] (0.9.)),

$$\left(\prod_i T_\Theta(\beta_i\alpha \times \phi)\right) T_\Theta((\alpha \times \phi)^l)^{-1} \equiv \prod_i T_\Theta(\beta_i \times \varepsilon_\Sigma).$$

On the other hand, using the formula for the absolute value of Gauss sums (cf. (I.5.7a)), we obtain

$$T_\Theta(\beta_i \times \varepsilon_\Sigma)T_\Theta(\beta_i^{-1} \times \varepsilon_\Sigma) \equiv \mathbf{N}_H\mathfrak{p}_H, \qquad \text{for } \beta_i \text{ non-trivial on } \Pi_0.$$

Thus, pairing β_i with β_i^{-1}, we get

$$\prod_i T_\Theta(\beta_i \times \varepsilon_\Sigma) \equiv \mathbf{N}_H\mathfrak{p}_H^{l-1/2}.$$

Putting everything together, we now have

$$\text{(5.5)} \qquad \sum_i \log(T_{\Theta,\phi}(l\alpha\beta_i)T_{\Theta,\phi}(-\alpha^l)^f) = (l(l-1)/2)\log \mathbf{N}_H\mathfrak{p}_H.$$

Substitute this into (5.4). If first $\pi \neq 1$, the expression vanishes, i.e., we get $s_\pi = 0$, as stated in (c). Finally if $\pi = 1$ we get (b).

Next (d) follows from Theorem 16 and Corollary 4 to Theorem 17, writing N in place of M.

Finally an easy calculation shows that if $\gamma \in \Pi, \pi \in \Pi'$

$$\text{order}(\Pi')s_{\gamma^{-1}\pi\gamma} = \sum_\alpha [\log(T_{\Theta,\phi}(l(\alpha\gamma))T_{\Theta,\phi}(-(\alpha\gamma)^l)^f)]\alpha(\pi^{-1}).$$

Hence also

$$\text{(5.6)} \qquad \text{order}(\Pi')s_{\gamma^{-1}\pi\gamma} = \sum_\alpha [\log(T_\Theta((l\alpha \times \phi)\gamma)T_\Theta(-(\alpha^l \times \phi)\gamma)^f)]\alpha(\pi^{-1}),$$

where we have used for the last equation the fact that for all $\eta \in R_{\Pi'}$, all $\gamma \in \Pi \subset \Gamma$ one has $(\eta \times \phi)\gamma = \eta\gamma \times \phi$. But by the inductivity of T in degree zero (i.e., by Proposition 1.1), $T_\Theta((\eta \times \phi)\gamma - \eta \times \phi) = T_\Gamma(0) = 1$. We can thus omit all the γ's in (5.6), i.e., indeed $s_{\gamma^{-1}\pi\gamma} = s_\pi$, as we had to show. □

§6. The Congruence Theorems

This section contains the complete proofs of Theorems 33 and 34, for cases I and II only. This should suffice to give the essential flavour. The proofs, as will be seen, are in fact highly computational. For details of the other two cases see Taylor's original paper [Ty7], §9 and §10. Except where otherwise mentioned – as in Lemmas 6.3 and 6.4 below – the general notation is the same as in §4, and we shall also use in both cases I, and II, the special notations introduced for these in the appropriate places of §4.

Recall that for any subgroup Δ of Γ which contains Σ i.e., is of form $\Psi \times \Sigma$ with $\Psi \subset \Pi$, and for any l-adic Abelian character ϕ of Σ, we have defined a Gauss sum homomorphism $T_{\Delta,\phi}$, by $T_{\Delta,\phi}(\chi) = T_\Delta(\chi \times \phi)$, for $\chi \in R_\Psi$.

6.1. Lemma. *Let $\Psi, \Psi^{(1)}$ be normal subgroups of Π containing Π_0, with $\Psi \subset \Psi^{(1)}$, $[\Psi^{(1)} : \Psi] = l$; if $l = 2$ assume also that $\Pi \neq \Psi^{(1)}$. Let α be a (genuinely) ramified*

Abelian character of Ψ *with the property that* $\mathrm{v}_{\Psi}^{\Psi^{(1)}}\alpha = \mathrm{v}\alpha$ *is also ramified. Write* $\Delta = \Psi \times \Sigma$, $\Delta^{(1)} = \Psi^{(1)} \times \Sigma$. *Then for every Abelian character* ϕ *of* Σ,

$$T_{\Delta,\phi}(\alpha) \equiv T_{\Delta^{(1)},\phi}(\mathrm{v}\alpha)^f \bmod \begin{cases} (l), & \text{if } l \neq 2, \\ (2\mathfrak{L}_2), & \text{if } l = 2. \end{cases}$$

Proof. The first part of the proof, up to and including (6.7), does not use the assumption that $\Pi_0 \subset \Psi$, nor in the case $l = 2$ the assumption that $\Pi \neq \Psi^{(1)}$. This should be remembered for a subsequent application.

We go over to the multiplicative characters θ_α etc. of the underlying local fields, as defined in Chap. I, §5, but now with values in $\mathbb{Q}_l^{c*}$, not in $\mathbb{Q}^{c*}$ (see also (3.3)). Here some care is needed. Let $H = L^\Delta$, $H^{(1)} = L^{\Delta^{(1)}}$ be the fixed fields; these are cyclic extensions of F of l-power degree. We write the character $\alpha \times \phi$ of Δ as a product of two other characters of Δ, namely

(6.1) $$\alpha \times \phi = (\alpha \times \varepsilon_\Sigma)(\varepsilon_\Psi \times \phi).$$

Here, moreover, $\varepsilon_\Psi \times \phi$ is the restriction to Δ of the character $\varepsilon_\Pi \times \phi$ of Γ. Write, by abuse of notation, $\theta_{\varepsilon \times \phi}$ for the multiplicative character of F, associated with $\varepsilon_\Pi \times \phi$. In other words

$$\theta_{\varepsilon \times \phi}\colon F^* \to \mathbb{Q}_l^{c*}$$

is the homomorphism with

(6.2) $$(\varepsilon_\Pi \times \phi)(Ax) = \theta_{\varepsilon \times \phi}(x), \qquad A \text{ the Artin map } F^* \to \Gamma^{ab}.$$

Then the multiplicative character of H associated with

(6.3) $$\varepsilon_\Psi \times \phi = \mathrm{res}_\Delta^\Gamma(\varepsilon_\Pi \times \phi)$$

is $\theta_{\varepsilon \times \phi} \circ N_{H/F}\colon H^* \to \mathbb{Q}_l^{c*}$. Let moreover $\theta_{\alpha \times \varepsilon}$ be the multiplicative character of H associated with $\alpha \times \varepsilon_\Sigma$. Then

(6.4) $$T_{\Delta,\phi}(\alpha) = T_\Delta(\alpha \times \phi) = \sum_u \theta_{\alpha \times \varepsilon}(u)\theta_{\varepsilon \times \phi}(N_{H/F}u)\psi_H(uc^{-1})^j,$$

where we sum over a set $\{u\}$ of repesentatives of $(\mathfrak{o}_H/\mathfrak{p}_H)^*$. Here as always $c \in F^*$, with $c\mathfrak{o}_F = \mathfrak{p}_F\mathfrak{D}_F$, and j is the given embedding $\mathbb{Q}^c \to \mathbb{Q}_l^c$.

Next observe that

$$(\mathrm{v}_\Psi^{\Psi^{(1)}}\alpha) \times \varepsilon_\Sigma = \mathrm{v}_\Delta^{\Delta^{(1)}}(\alpha \times \varepsilon_\Sigma),$$

and the multiplicative character associated with this is $\mathrm{res}_{H^{(1)}}^H \theta_{\alpha \times \varepsilon}$, the restriction to $H^{(1)*}$. Accordingly

(6.5) $$T_{\Delta^{(1)},\phi}(\mathrm{v}\alpha) = \sum_{u^{(1)}} \theta_{\alpha \times \varepsilon}(u^{(1)})\theta_{\varepsilon \times \phi}(N_{H^{(1)}/F}u^{(1)})\psi_{H^{(1)}}(u^{(1)}c^{-1})^j,$$

where $\{u^{(1)}\}$ is a set of representatives of $(\mathfrak{o}_{H^{(1)}}/\mathfrak{p}_{H^{(1)}})^*$.

Having accomplished the translation into Gauss sums for l-adic multiplicative characters, we can now set to work. We can certainly choose the set $\{u^{(1)}\}$ in (6.5) as a subset of the set $\{u\}$ in (6.4). Note that then $N_{H/F}u^{(1)} = (N_{H^{(1)}/F}u^{(1)})^l$, and $\psi_H(u^{(1)}c^{-1}) = \psi_{H^{(1)}}(lu^{(1)}c^{-1})$. Accordingly we can write

$$(6.6)\quad \begin{cases} T_{\Delta,\phi}(\alpha) = \sum_{u^{(1)}} \theta_{\alpha\times\varepsilon}(u^{(1)})\theta_{\varepsilon\times\phi}(N_{H^{(1)}/F}u^{(1)})^l\psi_{H^{(1)}}(lu^{(1)}c^{-1})^j + T' \\ T' = \sum_u{}' \theta_{\alpha\times\varepsilon}(u)\theta_{\varepsilon\times\phi}(N_{H/F}u)\psi_H(uc^{-1})^j, \end{cases}$$

with $\sum'$ extending over $\{u\}\backslash\{u^{(1)}\}$. Now $\psi_{H^{(1)}}(lu^{(1)}c^{-1}) = \psi_{H^{(1)}}(u^{(1)}c^{-1})^l$. Thus by comparison with (6.5) we now have

$$(6.7)\qquad T_{\Delta,\phi}(\alpha) = T_{\Delta^{(1)},\phi}(\mathrm{v}\alpha)^f + T'.$$

We now turn our attention to T'. From now on we use the full hypotheses of the Lemma, in particular that $\Psi \supset \Pi_0$. The aim is to show that

$$T' \equiv 0 \mod \begin{cases} (l), & \text{if } l \neq 2, \\ (2\mathfrak{L}_2), & \text{if } l = 2. \end{cases}$$

As H/F is non-ramified, Γ/Δ is also the Galois group of the residue class field extension. Moreover $\Delta^{(1)}/\Delta$ is its unique minimal subgroup $\neq 1$, and it thus acts on $\{u\}\backslash\{u^{(1)}\}$ without fixed points. From here on we give the detailed argument only for odd l; for $l = 2$ it is similar, but more complicated. We cut the sum for T' up into sums over $\Delta^{(1)}/\Delta$-orbits, which thus are of the form

$$\sum_{\delta\in\Delta^{(1)}/\Delta} \theta_{\alpha\times\varepsilon}(u^\delta)\cdot\theta_{\varepsilon\times\phi}(N_{H/F}u^\delta)\psi_H(u^\delta c^{-1})^j.$$

But

$$N_{H/F}(u^\delta) = N_{H/F}(u), \qquad \psi_H(u^\delta c^{-1}) = \psi_H((uc^{-1})^\delta) = \psi_H(uc^{-1}).$$

Therefore T' is a sum of terms

$$(6.8)\qquad \left(\sum_{\Delta^{(1)}/\Delta} \theta_{\alpha\times\varepsilon}(u^\delta)\right)(\theta_{\varepsilon\times\phi}(N_{H/F}u)\psi_H(uc^{-1})^j),$$

and what we have to show is that

$$(6.9)\qquad \sum_{\Delta^{(1)}/\Delta} \theta_{\alpha\times\varepsilon}(u^\delta) \equiv 0 \pmod{l}.$$

As $\Delta^{(1)}/\Delta$ is cyclic of prime order l, either the $\theta_{\alpha\times\varepsilon}(u^\delta)$ are all equal, in which case (6.9) is clearly satisfied, or else the left-hand side of (6.9) is $\theta_{\alpha\times\varepsilon}(u)\times$ sum of all l-th roots of unity, and then it is of course just zero.

In the case $l = 2$ one needs a stronger result, and starts with a stronger hypothesis. Moreover there are more cases to consider, as the group of order 2 can act in several ways on 2^n-th roots of unity. □

6.2. Lemma. *Let Ψ be a normal subgroup of Π of index l, with $\Psi\Pi_0 = \Pi$, and let $\Delta = \Psi \times \Sigma$. Let α be a (genuinely) ramified Abelian character of Ψ. Then, with* $\mathrm{v}_\Psi^\Pi\alpha = \mathrm{v}\alpha$,

$$T_{\Delta,\phi}(\alpha) = T_{\Gamma,\phi}(\mathrm{v}\alpha)^f.$$

Proof. In the first part of the proof of Lemma 6.1 put $\Delta^{(1)} = \Gamma$, $\Psi^{(1)} = \Pi$. As now $H/H^{(1)}$ $(= L^{\Delta}/F)$ is totally ramified, and so $\mathfrak{o}_{H^{(1)}}/\mathfrak{p}_{H^{(1)}} \cong \mathfrak{o}_H/\mathfrak{p}_H$, the restriction of a tame, but ramified multiplicative character θ of H to $H^{(1)}$ is still ramified. Thus $\mathrm{v}\alpha$ is ramified. In the same way as for Lemma 6.1, we get (6.7). But now $T' = 0$, since the set $\{u\}\backslash\{u'\}$ in (6.6) is empty. □

Interlude. We shall momentarily revert to characters with global values and our original Galois Gauss sums. In this situation we take moreover $\phi = \varepsilon_\Sigma$ the identity character, and now write α instead of $\alpha \times \varepsilon$. We get, as in (6.6),

$$(6.10) \qquad \begin{cases} \tau(\alpha) = \sum_{u^{(1)}} \theta_\alpha(u^{(1)}c^{-1})\psi_{H^{(1)}}(lu^{(1)}c^{-1}) + \tau', \\ \tau' = \sum_u{}' \theta_\alpha(uc^{-1})\psi_H(uc^{-1}). \end{cases}$$

As before, $\tau' = 0$ in the situation of Lemma 6.2, and $\tau' \equiv 0 \pmod{l}$ (or mod $(2\mathfrak{L}_2)$) in the situation of Lemma 6.1. Moreover as l is a unit in $\mathbb{Z}_p$, we can transform the first sum in (6.10) to give

$$\begin{aligned} \sum_{u^{(1)}} \theta_\alpha(u^{(1)}c^{-1})\psi_{H^{(1)}}(lu^{(1)}c^{-1}) &= \theta_\alpha(l)^{-1} \sum_{u^{(1)}} \theta_\alpha(u^{(1)}c^{-1})\psi_{H^{(1)}}(u^{(1)}c^{-1}) \\ &= \theta_\alpha(l)^{-1}\tau(\mathrm{v}\alpha). \end{aligned}$$

We sum up, the notation being the same as before, except that now we consider characters with values in $\mathbb{Q}^c$.

6.3. Lemma. *Let Δ, $\Delta^{(1)}$ be subgroups of Γ containing Γ_0, with $\Delta \subset \Delta^{(1)}$, $[\Delta^{(1)}:\Delta] = l$, and if $l = 2$ assume also that $\Gamma \neq \Delta^{(1)}$. Let α be a ramified Abelian character of Δ, with the property that $\mathrm{v}_\Delta^{\Delta^{(1)}}\alpha = \mathrm{v}\alpha$ is also ramified. Then*

$$\tau(\alpha) \equiv \theta_\alpha(l)^{-1}\tau(\mathrm{v}\alpha) \bmod \begin{cases} (l), & \text{if } l \neq 2, \\ (2\mathfrak{L}_2), & \text{if } l = 2. \end{cases}$$ □

6.4. Lemma. *Let Δ be a normal subgroup of Γ of index l, with $\Delta\Gamma_0 = \Gamma$. Let α be a ramified Abelian character of Δ. Then*

$$\tau(\alpha) = \theta_\alpha(l)^{-1}\tau(\mathrm{v}_\Delta^\Gamma\alpha).$$ □

From now on we return to l-adic characters and revert to the notation used before the interlude.

Proof of Theorem 33, Case I. In the notation used for this case in §4, we apply Lemma 6.1, whose hypotheses are satisfied by Lemma 4.2. In addition we shall show that

$$(6.11) \qquad T_{\Gamma,\phi}(\chi)T_{\Delta,\phi}(\alpha)^{-1} = (T_{\Gamma,\phi}(\chi')T_{\Delta^{(1)},\phi}(\mathrm{v}\alpha)^{-1})^f.$$

In conjunction with Lemma 6.1, this yields the required equation. (6.11) itself is a consequence of

6.5. Lemma. *Let Π' be a subgroup of Π containing Π_0. Let $\Gamma' = \Pi' \times \Sigma$ and let β be an Abelian character of Π', $\operatorname{ind}_{\Pi'}^{\Pi} \beta = \eta$. Then*

$$T_{\Gamma,\phi}(\eta) T_{\Gamma',\phi}(\beta)^{-1} = (-1)^{[\Gamma:\Gamma']-1}.$$

Proof. We have

$$\begin{aligned}\eta \times \phi &= [\operatorname{ind}_{\Gamma'}^{\Gamma}(\beta \times \varepsilon_\Sigma)] \cdot (\varepsilon_\Pi \times \phi)\\ &= \operatorname{ind}_{\Gamma'}^{\Gamma}[(\beta \times \varepsilon_\Sigma) \cdot \operatorname{res}_{\Gamma'}^{\Gamma}(\varepsilon_\Pi \times \phi)] \qquad \text{(Frobenius reciprocity)}\\ &= \operatorname{ind}_{\Gamma'}^{\Gamma}(\beta \times \phi).\end{aligned}$$

Thus

$$\begin{aligned}T_{\Gamma,\phi}(\eta) T_{\Gamma',\phi}(\beta)^{-1} &= T_\Gamma(\operatorname{ind}_{\Gamma'}^{\Gamma}(\beta \times \phi)) T_{\Gamma'}(\beta \times \phi)^{-1}\\ &= T_\Gamma(\operatorname{ind}_{\Gamma'}^{\Gamma} \varepsilon_{\Gamma'}) T_{\Gamma'}(\varepsilon_{\Gamma'})^{-1} \qquad \text{(induction formula).}\end{aligned}$$

But as $\Gamma' \supset \Gamma_0$, i.e., $L^{\Gamma'}/F$ is non-ramified, $\operatorname{ind}_{\Gamma'}^{\Gamma} \varepsilon_{\Gamma'}$ is the sum of $[\Gamma : \Gamma']$ non-ramified Abelian characters, each having T-value $= T(\varepsilon_{\Gamma'}) = -1$. □

To get (6.11) observe that

$$(-1)^{[\Gamma:\Delta]} = (-1)^{[\Gamma:\Delta^{(1)}]}. \qquad \square$$

Proof of Theorem 33, Case II. We first use Lemma 6.2, which tells us that (in the case II notation, and putting $\Delta^{(2)} = \Psi^{(2)} \times \Sigma$)

$$\text{(6.12)} \qquad \begin{cases} T_{\Gamma,\phi}(\mathrm{v}\beta)^f = T_{\Delta^{(2)},\phi}(\beta),\\ T_{\Gamma,\phi}(\mathrm{v}\beta^2)^f = T_{\Delta^{(2)},\phi}(\beta^2).\end{cases}$$

As in the proof of Lemma 6.5, we note that

$$(\operatorname{ind}_{\Psi^{(2)}}^{\Pi} \beta) \times \phi = \operatorname{ind}_{\Delta^{(2)}}^{\Gamma}(\beta \times \phi),$$

and analogously for β^2 in place of β. Therefore, and by (6.12),

$$\begin{aligned}T_{\Gamma,\phi}(\chi) \cdot T_{\Gamma,\phi}(\mathrm{v}\beta)^{-f} &= T_\Gamma(\operatorname{ind}_{\Delta^{(2)}}^{\Gamma}(\beta \times \phi)) \cdot T_{\Delta^{(2)}}(\beta \times \phi)^{-1},\\ T_{\Gamma,\phi}(\chi') \cdot T_{\Gamma,\phi}(\mathrm{v}\beta^2)^{-f} &= T_\Gamma(\operatorname{ind}_{\Delta^{(2)}}^{\Gamma}(\beta^2 \times \phi)) \cdot T_{\Delta^{(2)}}(\beta^2 \times \phi)^{-1}.\end{aligned}$$

But by Lemma 3.1, both right hand sides are equal to the induction factor $v = v^{\deg(\beta)} = v^{\deg(\beta^2)}$.

We need yet another Lemma.

6.6. Lemma. *Let $\{X_i\}$ be indeterminates (i running through a finite index set), and let μ_i be l-th roots of unity. Then*

$$\prod_{k=1}^{l} \left(\sum_i X_i \mu_i^k \right) \equiv \sum_i X_i^l \pmod{l}$$

in $\mathbb{Z}[X_1, \ldots, X_i, \ldots]$.

Proof. The congruence holds mod $\mathfrak{L}_l$. But both polynomials have coefficients in $\mathbb{Z}$, hence the congruence holds mod l. □

Proof of Theorem 34, Case I. We have $\mathrm{res}^{\Pi}_{\Pi_0}\chi = \sum \mathrm{res}(\alpha)\gamma$, sum over Π mod Ψ, and analogously for $\mathrm{v}\alpha$. Thus we have to prove that, for $x \in \mathfrak{o}_N \Pi_0^*$,

$$(6.13)\qquad \begin{cases} \displaystyle\prod_{\gamma \in \Pi/\Psi^{(1)}} \left[\left(\prod_{\delta \in \Psi^{(1)}/\Psi} (\mathrm{res}(\alpha)\delta\gamma)(x)\right)(\mathrm{res}(\mathrm{v}\alpha)\gamma)(x)^{-f}\right] \\ \equiv 1 \bmod \begin{cases} (l), & \text{if } l \neq 2, \\ (2\mathfrak{L}_2), & \text{if } l = 2. \end{cases} \end{cases}$$

Fix a generator δ of $\Psi^{(1)}$ mod Ψ. In the notation of the proof of Lemma 4.2, we have $\mathrm{res}(\alpha)\delta = \mathrm{res}(\alpha)^{1+al^{r-1}}$, where $l^r = \mathrm{order}(\mathrm{res}(\alpha))$. Hence $\mathrm{res}(\alpha)\delta^k = \mathrm{res}(\alpha)^{1+akl^{r-1}}$. Let σ be a generator of Π_0, and write x in the form $\sum x_i\sigma^i$, $x_i \in \mathfrak{o}_N$. Put in the above Lemma $X_i = x_i\alpha(\sigma^i)$, $\mu_i = \alpha(\sigma^{-i})\alpha\delta(\sigma^i)$. Then one verifies that $\mu_i^k = \alpha(\sigma^{-i})\alpha\delta^k(\sigma^i)$. Thus reverting to the notation used in (6.13), and applying Lemma 6.6, we have

$$\prod_{\delta \in \Psi^{(1)}/\Psi} (\mathrm{res}(\alpha)\delta)(x) \equiv \sum_i x_i^l \alpha^l(\sigma^i) \pmod{l},$$

or by Lemma 4.2,

$$(6.14)\qquad \prod_{\delta \in \Psi^{(1)}/\Psi} (\mathrm{res}(\alpha)\delta)(x) \equiv \mathrm{res}(\mathrm{v}\alpha)(x)^f \pmod{l}.$$

Next observe that Π/Ψ acts faithfully on the orbit of $\mathrm{res}(\alpha)$. In particular there is a map $g\colon \Pi/\Psi \to \mathrm{Gal}(N(\mathrm{res}\,\alpha)/N)$, such that

$$\mathrm{res}(\alpha)\gamma = \mathrm{res}(\alpha)^{g(\gamma)}.$$

As $\mathrm{v}\alpha(\sigma) = \alpha(\sigma^l)$ for all $\sigma \in \Pi_0$, also

$$\mathrm{res}(\mathrm{v}\alpha)\gamma = \mathrm{res}(\mathrm{v}\alpha)^{g(\gamma)}.$$

From (6.14) it follows that the left hand side of (6.13) is of the form $\prod_{\gamma \in \Pi/\Psi^{(1)}} (1 + ly)^{g(\gamma)}$, y an integer in $N(\mathrm{res}\,\alpha)$. For l odd this gives the result. For $l = 2$, $1 + 2y^{g(\gamma)} \equiv 1 + 2y \pmod{2\mathfrak{L}_2}$. Also the number of factors is $[\Pi : \Psi^{(1)}]$ i.e. is even. Thus the left hand side of (6.13) is of the form $(1 + 2y)^{2m} \bmod 2\mathfrak{L}_2$. □

Proof of Theorem 34, Case II. We let η be any Abelian character of $\Psi^{(2)}$. One verifies that always

$$(6.15)\qquad \mathrm{v}\eta(\sigma) = \eta(\sigma^2) \qquad \text{for} \qquad \sigma \in \Pi_0 \qquad (\mathrm{v} = \mathrm{v}^{\Pi}_{\Psi^{(2)}}).$$

Next observe that

$$\mathrm{res}^{\Pi}_{\Pi_0} \mathrm{ind}^{\Pi}_{\Psi^{(2)}} \eta = \eta' + \eta'\mu,$$

where μ is a quadratic character of Π_0, and $\mathrm{res}^{\Pi_0}_{\Pi_0{}^2}\eta' = \mathrm{res}^{\Psi^{(2)}}_{\Pi_0{}^2}\eta$. Let then $x = \sum x_i\sigma^i \in \mathfrak{o}_N\Pi_0^*(x_i \in \mathfrak{o}_N)$. Then

$$\mathrm{Det}(x)(\mathrm{ind}^{\Pi}_{\Psi^{(2)}}\eta) = (\sum x_i\eta'(\sigma^i))(\sum x_i\eta'(\sigma^i)(-1)^i).$$

Substituting $\mu_i = (-1)^i$ and $X_i = x_i\sigma^i$ in Lemma 6.6, we get

$$\mathrm{Det}(x)(\mathrm{ind}^{\Pi}_{\Psi(2)}\eta) = \sum x_i^2\,\eta'(\sigma^{2i}) + 2\eta' g(\ldots, x_i\sigma^i, \ldots)$$

where $g(\cdots X_i \cdots) \in \mathbb{Z}[\ldots, X_i, \ldots]$, i.e.

$$\mathrm{Det}(x)(\mathrm{ind}^{\Pi}_{\Psi(2)}\eta) = \sum x_i^f \eta'(\sigma^{2i}) + 2\eta'[g(\ ) + \sum \tfrac{1}{2}(x_i^2 - x_i^f)\sigma^{2i}]$$

where of course $\frac{1}{2}(x_i^2 - x_i^f) \in \mathfrak{o}_N$. By (6.15),

$$\mathrm{Det}(x)(\mathrm{v}\eta)^f = \sum x_i^f\,\eta'(\sigma^{2i}).$$

Therefore

$$\text{(6.16)}\qquad \begin{cases} \mathrm{Det}(x)(\mathrm{ind}^{\Pi}_{\Psi(2)}\eta)/\mathrm{Det}(x)(\mathrm{v}\eta)^f \\ = 1 + 2\eta'[g(\ ) + \sum \frac{1}{2}(x_i^2 - x_i^f)\sigma^{2i})/(\sum x_i^f\sigma^{2i})]. \end{cases}$$

Now $\sum x_i\sigma^i \mapsto \sum x_i^f\sigma^{2i}$ is a ring homomorphism, thus preserving units. Therefore the denominator on the right hand side of (6.16) is indeed a unit, i.e., $\exists a \in \mathfrak{o}_N\Pi_0$ with

$$\mathrm{Det}(x)(\mathrm{ind}^{\Pi}_{\Psi(2)}\eta)/\mathrm{Det}(x)(\mathrm{v}\eta)^f = 1 + 2\eta'(a)$$

for all η. The required congruence now follows by taking $\eta = \beta$ and $\eta = \beta^2$ and observing that $\beta'(a) \equiv \beta'^2(a) \pmod{\mathfrak{L}_2}$. □

§7. The Arithmetic Theory of Tame Local Galois Gauss Sums

This section contains a brief survey of the arithmetic theory of tame, local Galois Gauss sums, which was developed over the last ten years in conjunction with the theory of Galois module structure. Already, throughout the preceding chapters, and in particular in Chaps. III and IV a number of theorems were established, which could rightly be viewed, as falling under this heading. In the present section, however, we look at the whole theory systematically from a different point of view. Although the original motivation for studying this topic came largely from the study of Galois modules, the connection between the two aspects works both ways, and indeed has produced interesting arithmetic information on Galois Gauss sums – as has for instance been seen in the preceding §6.

Our first aim in this section is to collect systematically results from the preceding chapters and sections and to supplement these, so as to show that they form a coherent pattern of a theory. Secondly we shall derive an internal characterization of tame Galois Gauss sums, within the given local field, by their arithmetic properties. This is of some conceptual importance. For, if one takes as starting point the existence theorem – which should really be called the induction theorem (i.e., Theorem 18) – then one would have to know all tame Abelian characters of all tame extension fields of a given local field F (or at least all tame Abelian characters of all non-ramified extension fields of the given field F) in order to compute all tame Galois Gauss sums of F. In fact, as we shall show, the latter are uniquely determined as functions of tame Galois characters over F by the arithmetic properties within F, which we shall list. – Our third aim is to use our approach to give (or rather to outline) a new proof of Theorem 18, for tame characters and tame extensions only. We shall formally define the tame local Galois Gauss sums in F by an explicit canonical formula. Most of the arithmetic properties are then established on this basis, and these properties are then used to prove inductivity in degree zero.

Most of this section comes from [FT]. In fact the developments since the publication of this paper, and in particular those in [Ty7], described in the preceding sections, would enable one in principle to use a different approach. We shall discuss this briefly at the end.

Throughout we consider non-Archimedean local fields of fixed residue class characteristic p. Characters now take values in $\mathbb{Q}^c$, no longer in $\mathbb{Q}_l^c$ as in the preceding sections. We emphasize that we have now changed our point of view, compared to that taken in Chaps. III and IV. We do not at this stage assume the existence theorem 18. Indeed the contents of this section are in principle logically independent of all previous results, although we shall continuously refer back to previous statements and proofs, which in all cases can be seen to remain applicable here. We thus have first of all to produce a good definition of a tame Galois Gauss sum with which to begin working. Our starting point is the Gauss sum of a tame multiplicative character θ of a non-Archimedean local field F, as defined in (I.5.5), (I.5.6). Thus in the notation used there,

$$(7.1)\qquad \tau(\theta) = \tau(F,\theta) = \begin{cases} \theta(\mathfrak{D}_F)^{-1}, & \text{if } \theta \text{ is non-ramified,} \\ \sum\limits_{u \bmod \mathfrak{p}_F} \theta(uc^{-1})\psi_F(uc)^{-1}, & \text{if } \theta \text{ is ramified.} \end{cases}$$

Now let χ be an irreducible, tame Galois character of F. By Lemma 4.1, with G_0 always the inertia group, there exists a non-ramified extension E/F and an Abelian Galois character α of E, so that $\chi = \operatorname{ind}_E^F \alpha$, where we write here ind_E^F for induction from Ω_E to Ω_F. The character $\det(\operatorname{ind}_E^F \varepsilon_E) = \det(\operatorname{ind} \varepsilon)$ is non-ramified and we put

$$\theta_{\det(\operatorname{ind}\varepsilon)} = \rho_{E/F}.$$

Now define

$$(7.2)\qquad \tau(\chi) = \tau(F,\chi) = \tau(E,\theta_\alpha)\rho_{E/F}(\mathfrak{D}_F)^{-1}.$$

Uniqueness: By Lemma 4.1, E is unique, and α and so also θ_α is unique modulo the action of $\operatorname{Gal}(E/F)$. By inspection of (7.1) one verifies that this action does not affect $\tau(E,\theta_\alpha)$.

Thus (7.2) gives a unique Galois Gauss sum, for any pair (F,χ) as above. Now extend τ by additivity to a homomorphism on virtual characters, i.e.,

$$(7.3)\qquad \tau(F,\chi+\chi') = \tau(F,\chi)\tau(F,\chi').$$

Throughout this section this is the definition of τ with which we work. On this basis, the existence theorem has to be restated as

Theorem 18A. *If E/F is a tame extension, χ a tame Galois character of E, then*

$$\tau(F, \operatorname{ind}_E^F \chi) = \tau(E, \chi)\tau(F, \operatorname{ind}_E^F \varepsilon_E)^{\deg(\chi)}. \qquad \square$$

It is clear that Theorem 18A, together with definitions (7.1)–(7.3) imply all of Theorem 18.

We shall now give a list of properties of $\tau(\chi)$, for tame Galois characters χ, which one deduces from the definition given here. Of course, in all this, the definition of the tame conductor (cf. (I.5.26)) and of the norm resolvent will be assumed. For the reader's convenience, we shall in each relevant case give a reference to the place where the result occurs in some form in our previous account. The method of proof will be outlined later in this section.

A. The absolute value is given by

$$|\tau(F, \chi)| = \mathbf{N}_F \mathfrak{f}(\chi)^{1/2},$$

(see (III.2.4)). (Recall here that $|\ |$ is the absolute value in $\mathbb{C}$.) $\square$

B. The twisting rule: If β is Abelian and non-ramified, then

$$\tau(F, \beta\chi) = \tau(F, \chi)\tau(F, \beta)^{\deg(\chi)}\theta_\beta(\mathfrak{f}(\chi))^{-1}.$$

(See Lemma III.6.1; the present statement is however more precise.) $\square$

C. Galois action. $\tau(\chi) \in \mathbb{Q}^c$ (as a subfield of $\mathbb{C}$), and for all $\omega \in \Omega_\mathbb{Q}$

$$\tau(\chi^{\omega^{-1}})^\omega = \tau(\chi)\theta_{\det\chi}(u_p\omega),$$

where $u_p\colon \Omega_\mathbb{Q} \to \mathbb{Z}_p^*$ is the homomorphism given by the action of $\Omega_\mathbb{Q}$ on the p^n-th roots of unity for all n (cf. III §3) (see Theorem 20B(i)). $\square$

D. The resolvent relation. For every embedding $j\colon \mathbb{Q}^c \to \mathbb{Q}_p^c$,

$$(\tau(F, \chi)^j) = \mathscr{N}_{F/\mathbb{Q}_p} P(\chi^j),$$

and $\tau(F, \chi)$ is a unit outside p (see Theorem 23). $\square$

Next with definition (1.1) of the non-ramified characteristic

E. Let l be a prime number. Whenever $\chi \in \operatorname{Ker} d_l$, then

$$\tau(F, \chi) \equiv y(F, \chi) \pmod{\mathfrak{L}_l}$$

(see Theorem 29). $\square$

In addition we shall get a further basic property, which has not occurred before in this precise form; it is a variant of the congruences in Theorem 33. We shall state it as a separate theorem.

We call a character χ corresponding to a representation T of Ω_F an *l-character*, if $\Omega_F/\operatorname{Ker} T$ is an l-group.

Theorem 36A (cf. [FT]). *Let l be odd and χ an irreducible non-Abelian l-character. Then*

$$\tau(F, \chi) \equiv \tau(F, \det_\chi)\theta_{\det\chi}(\deg(\chi))^{-1} \pmod{l}. \qquad \square$$

Note that the hypothesis implies that $p \neq l$ (this is also true if $l = 2$), and hence $\deg(\chi)$, as an element of F^*, is a unit of $\mathfrak{o}_F$.

In the case $l = 2$ we need some definitions. First suppose $p \equiv -1 \pmod 4$; more precisely put $p \equiv -1 + 2^{N-1} \pmod{2^N}$ with $N \geqslant 3$. Let β be a character of $\mathbb{F}_{p^2}^*$ of order 2^N, and let $\mathbb{F}_{p^2} = \mathbb{F}_p(d)$, with $d^2 \in \mathbb{F}_p$. Then one has (cf. [FT] Proposition 1)

7.1. Proposition. *The class $\lambda_p \equiv [\sum_{x\in\mathbb{F}_p} \beta(1 + xd)] \bmod 2\mathfrak{L}_2$ is independent of the particular choice of β and of d as above.* $\square$

Next we define a character v_t on the rational 2-adic units x with $x \equiv \pm 1 \pmod{2^{t-1}}$ (for $t \geqslant 3$), by

(7.4)
$$v_t(x) = \begin{cases} 1 \\ -1 \end{cases} \quad \text{if} \quad x \equiv \begin{cases} \pm 1 \\ \pm 1 + 2^{t-1} \end{cases} \pmod{2^t}.$$

Recall also the definition of a character of inversion type, in §4 (after Lemma 4.1).

Theorem 36B (cf. [FT]). *Let χ be an irreducible, non-Abelian 2-character of F.*

(i) *If χ is of inversion type (hence $\deg(\chi) = 2$), and is a faithful character of a Galois group of order 2^t then $\mathbf{N}_F\mathfrak{p}_F \equiv -1 \pmod{2^{t-1}}$, and*

$$\tau(\chi) \equiv \tau(\det_\chi)v_t(\mathbf{N}_F\mathfrak{p}_F)\theta_{\det\chi}(\mathfrak{p}_F)^{-1} \pmod{2\mathfrak{L}_2}.$$

(ii) *If χ is not of inversion type and $\mathbf{N}_F\mathfrak{p}_F \equiv 1 \pmod 4$ then*

$$\tau(\chi) \equiv -\tau(\det_\chi)\left(\frac{2}{p}\right)_{\mathbb{Q}}^{f(F/\mathbb{Q}_p)} \theta_{\det\chi}(\deg(\chi))^{-1} \pmod{2\mathfrak{L}_2}.$$

(iii) *If χ is not of inversion type and $\mathbf{N}_F\mathfrak{p}_F \equiv -1 \pmod 4$, then*

$$\tau(\chi) \equiv -\tau(\det_\chi)\lambda_p^{f(F/\mathbb{Q}_p)}\theta_{\det\chi}(\deg(\chi))^{-1} \pmod{2\mathfrak{L}_2}. \qquad \square$$

Here $f(F/\mathbb{Q}_p)$ is the residue class degree.

Remark. If $\deg(\chi) = 2$ then cases (ii) and (iii) coincide with semi-Abelian, and semi-dihedral type, respectively, as defined in §4.

Let now $S_{(F)}$ be the additive group of tame virtual Galois characters χ of F with $\deg(\chi) = 0$, $\det_\chi = \varepsilon_F$. Then we get

Theorem 37 (cf. [FT]). *Let τ' be a homomorphism from the additive group of tame virtual Galois characters of F, into $\mathbb{Q}^{c*}$, with the following properties*:

(i) *If α is Abelian, then τ' is given as in* (7.1) *with $\theta = \theta_\alpha$ (with τ replaced by τ' of course).*

(ii) *With τ replaced by τ', A–E and the formulae in Theorem 36 are satisfied for all $\chi \in S_{(F)}$.*

Then

$$\tau' = \tau(F, \cdot). \qquad \square$$

Remark. Following [DH] one can also characterize the Gauss sums of Abelian characters intrinsically by arithmetic properties, rather than use the explicit formulae (7.1). The properties above by themselves will not suffice to determine Gauss sums of Abelian characters uniquely; the grip they give is much better on the essentially non-Abelian part $S_{(F)}$ of the ring of virtual characters.

We shall now discuss the derivation of the stated results. We shall first give the proof of two cases of Theorem 36, starting from the definitions in the present section.

Proof of Theorem 36A. This follows immediately by repeated application of Lemma 6.3 which is valid in our situation, and the observation that if β is an Abelian character of a group Δ, of odd index in a group Γ, then

$$\mathrm{v}_{\Delta/\Gamma}\beta = \det(\mathrm{ind}_\Delta^\Gamma \beta). \qquad \square$$

Proof of Theorem 36B(i). The congruence $\mathbf{N}_F\mathfrak{p}_F \equiv -1 \pmod{2^{t-1}}$ comes from (4.18). Indeed if E/F is quadratic and non-ramified, and $\chi = \mathrm{ind}_E^F \alpha$, α Abelian, then E is also a maximal cyclic extension of F in the fixed field of $\mathrm{Ker}\,\alpha$. Thus E is the maximal non-ramified extension, and α is totally ramified. Hence, in the notation of (4.18), $r = t - 1$.

Let now $\theta = \theta_\alpha$ and let $E = F(d)$, $d^2 \in F^*$. We may suppose that $d^2 \in \mathfrak{o}_F^*$. By definitions (7.1), (7.2) we have

$$\tau(F, \chi) = \tau(\theta)\rho_{E/F}(\mathfrak{D}_F)^{-1}. \tag{7.5}$$

The element c occurring in (7.1) may be chosen to lie in F. Since the restriction $\mathrm{res}_F^E\,\theta$ of θ to F^* is non-ramified, we get

$$\tau(\theta) = \mathrm{res}_F^E\,\theta(\mathfrak{D}_F\mathfrak{p}_F)^{-1}\left[\sum_{x,y}(\theta(x)\theta(1+yd)\psi_F(2xc^{-1})) + \sum_x \theta(xd)\right].$$

Here x runs through a set of representatives of $\mathfrak{o}_F^* \bmod \mathfrak{p}_F$ and y through one of $\mathfrak{o}_F \bmod \mathfrak{p}_F$. Note for the above, that as $t_{E/F}d = 0$, we have $\psi_E(xd) = 1$ and $\psi_E(x(1+yd)) = \psi_F(2x)$.

Now observe that $\theta(x) = 1$. Thus

$$\sum_x \theta(x)\psi_F(2xc^{-1}) = \sum_x \psi_F(xc^{-1}) = -1.$$

It follows that

$$(7.6)\qquad \tau(\theta) = \operatorname{res}_F^E \theta(\mathfrak{D}_F\mathfrak{p}_F)^{-1}\left[-\left(\sum_y \theta(1+yd)\right) + \theta(d)(\mathbf{N}_F\mathfrak{p}_F - 1)\right].$$

But, summing over $u \in \mathfrak{o}_E^*$ mod $\mathfrak{p}_E$ we get

$$0 = \sum \theta(u) = \sum_{x,y} (\theta(x)\theta(1+yd)) + \sum_x \theta(xd)$$

$$= (\mathbf{N}_F\mathfrak{p}_F - 1)\left[\left(\sum_y \theta(1+yd)\right) + \theta(d)\right].$$

Hence, by (7.6),

$$(7.7)\qquad \tau(\theta) = \operatorname{res}_F^E \theta(\mathfrak{D}_F\mathfrak{p}_F)^{-1}(\mathbf{N}_F\mathfrak{p}_F)\theta(d).$$

Since d is a $(\mathbf{N}_F\mathfrak{p}_F + 1)/2$-nd, but not a $\mathbf{N}_F\mathfrak{p}_F + 1$-st power residue mod $\mathfrak{p}_E$ we get

$$(7.8)\qquad \theta(d) = v_t(\mathbf{N}_F\mathfrak{p}_F).$$

As $\theta_{\det\chi} = \rho_{E/F} \cdot \operatorname{res}_F^E \theta$, and as $\rho_{E/F}(\mathfrak{p}_F) = -1$, we now get from (7.5), (7.7) and (7.8) that

$$(7.9)\qquad \tau(F,\chi) = -\mathbf{N}_F\mathfrak{p}_F \cdot \theta_{\det\chi}(\mathfrak{p}_F)^{-1} \cdot v_t(\mathbf{N}_F\mathfrak{p}_F)\tau(F,\det\chi).$$

Since $-\mathbf{N}_F\mathfrak{p}_F \equiv 1 \pmod 4$, we get the required result.

For the proofs of the remaining parts *B*(ii) and (iii) of Theorem 36 we refer the reader to the original paper (cf. [FT] §5). □

We shall now turn to the remainder of the proofs. We note first that, as E/F is non-ramified, the induced $\operatorname{ind}_E^F \varepsilon_E$ of the trivial character of Ω_E is the sum of the non-ramified Abelian characters α_i of $\operatorname{Gal}(E/F)$, and so, by (7.1), (7.3)

$$\tau(\operatorname{ind}_E^F \varepsilon_E) = \Pi\theta_i(\mathfrak{D}_F)^{-1}, \qquad \theta_i = \theta_{\alpha_i},$$

whence

$$\tau(\operatorname{ind}_E^F \varepsilon_E) = \theta_{\det\operatorname{ind}\varepsilon_E}(\mathfrak{D})^{-1} = \rho_{E/F}(\mathfrak{D})^{-1}.$$

Thus (7.2) can be rewritten as

$$(7.9a)\qquad \tau(F,\chi) = \tau(E,\theta_\alpha)\tau(\operatorname{ind}_E^F \varepsilon_E).$$

Statements *A* to *C* can now be established in exactly the same way as were the corresponding results earlier. See the proof of (III.2.4) in Chap. III, §2, the proof of Lemma III.6.1, and the proof of Theorem 20B(i). One only needs the induction formula (7.9a) for E/F non-ramified and χ irreducible, i.e., for the case when it is immediate from our definitions (7.1)–(7.3).

The fact that the $\tau(F,\chi)$ are units outside p follows from the fact that the Gauss sums for multiplicative characters θ are obviously algebraic integers, which by *A* divide p. For the resolvent formula D one needs of course the separate theory of resolvents. One can proceed in one of two ways, to reduce the proof to the Abelian case. Either one uses the induction property of norm resolvents, in the situation in which (7.9a) is already valid (see note [4] in Chap. III), or one proceeds via non-ramified basefield extension. Here one has to establish the rule given in Theorem 25, and this presents no difficulty. In any case one has to establish such a rule for the Gauss sums (7.2) of multiplicative characters and can then complete the proof along the lines given in III §7.

The next step is to establish the induction formula

$$\tau(F, \mathrm{ind}_E^F \eta) = \tau(E, \eta)\tau(F, \mathrm{ind}_E^F \varepsilon_E)^{\deg(\eta)} \tag{7.10a}$$

for any tame Galois character η under the hypothesis that

$$E/F \text{ is non-ramified.} \tag{7.10b}$$

If $\mathrm{ind}_E^F \eta$ is irreducible this is an almost immediate consequence of (7.2) and of the tower formula for the characters $\rho_{E/F}$. The other case needs a more careful analysis but is also routine (see [FT] §3 (b)).

On the basis of (7.10) one now deduces a weakened form of assertion E. For a given prime l we let H_l be the group of virtual characters generated by characters of form $\mathrm{ind}_{E'}^F(\beta - \beta')$ of the following form: (i) E' is any non-ramified extension of F, (ii) β is a tame Abelian character of $\Omega_{E'}$ and $\beta = \beta'\beta''$ where β' is of order prime to l and β'' of l-power order. Clearly $H_l \subset \mathrm{Ker}\, d_l$. For any l we then prove

$$\tau(\chi) \equiv y(\chi) \pmod{\mathfrak{L}_l}, \qquad \text{if} \quad \chi \in H_l. \tag{7.11}$$

Here $y(\chi)$ is the non-ramified characteristic (cf. (1.1)) and the proof follows exactly the lines of that of Theorem 29. The full statement E is deduced in a similar fashion *after* full inductivity in degree zero has been established.

Now we turn to the proof of Theorem 18A. We consider, for any tame extension E/F and any tame Galois character η of E the map

$$\eta \mapsto \tau(E, \eta)\tau(F, \mathrm{ind}_E^F \varepsilon_E)^{\deg(\eta)}\tau(F, \mathrm{ind}_E^F \eta)^{-1} = \tau(E/F, \eta). \tag{7.12}$$

(E no longer has the special connotation of the definition (7.2).) Our aim is to prove that $\tau(E/F, \eta) = 1$. By inductivity of norm resolvents (cf. III note [4]) and by D, $\tau(E/F, \eta)$ is certainly a unit.

By A and the inductive properties of conductors, $|\tau(E/F, \eta)| = 1$. By C one can then conclude that

$$\eta \mapsto \tau(E/F, \eta)$$

is a $\Omega_{\mathbb{Q}}$-homomorphism into roots of unity. Just as for the proof of Theorem 32, one then uses congruences to pin these roots of unity down to value 1. There are two types of congruences which come into play, those in E and those in Theorem 36. We shall concentrate here on this aspect of the proof.

It will be convenient to work again in the context of a given tame Galois extension L/F with Galois group Γ, and with E the fixed field of a subgroup Δ. In view of the hypersolubility of tame Galois groups we may suppose also that

(7.13a) $$[\Gamma : \Delta] = l, \qquad \text{a prime,}$$

and by (7.10) that E/F is totally ramified, i.e., that

(7.13b) $$\Delta\Gamma_0 = \Gamma.$$

This implies then that $l \neq p$. Using a result of Davenport-Hasse (cf. [DH] (0.9_1)), one can also prove that $\tau(E/F, \eta) = 1$ when Γ is Abelian (cf. [FT] §6 (b)). Thus we shall suppose, that

(7.13c) $$\Gamma \text{ is non-Abelian.}$$

We then proceed by induction on order (Γ), i.e., we assume the induction formula for Galois groups of smaller order.

F. To prove that always $\tau(E/F, \eta) = 1$, *we may assume* η *to be Abelian.*

Proof. We let Σ (respectively Θ) be the centralizer of Γ_0 (respectively of Δ_0) in Γ. (Note that we now have a new meaning for these symbols!) We thus have a tower of Galois groups

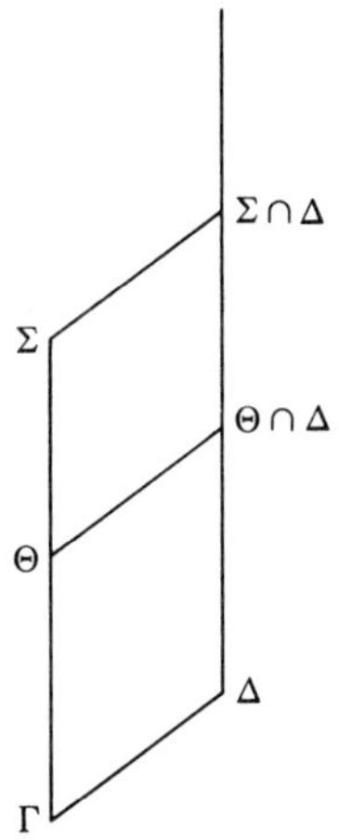

We certainly may assume η to be an irreducible character of Δ and by induction hypothesis, that it is faithful. Thus $\eta = \operatorname{ind}_{\Psi}^{\Delta} \beta$, where $\Psi \supset \Delta_0$ and β is Abelian. Moreover $\operatorname{res}_{\Delta_0}^{\Psi} \beta$ will now be faithful. As Ψ is the stabilizer of $\operatorname{res}_{\Delta_0}^{\Psi} \beta$, we have $\Psi = \Theta \cap \Delta$.

If β is non-ramified, then in any case $\Theta \cap \Delta = \Delta$ and $\eta = \beta$ is Abelian. So we suppose that β is ramified, i.e., $\Delta_0 \neq 1$. Put $\xi = \operatorname{ind}_{\Theta \cap \Delta}^{\Theta} \beta$. By non-ramified induction (cf. (7.10))

$$\tau(\mathrm{ind}_\Theta^\Gamma \xi) = \tau(\xi)\tau(\mathrm{ind}_\Theta^\Gamma \varepsilon_\Theta)^l,$$

where we omit the field symbol in τ. If $\Theta = \Gamma$, then again $\Theta \cap \Delta = \Delta$ and η is Abelian. Otherwise, by our induction hypothesis, applied to Θ,

$$\tau(\xi) = \tau(\beta)\tau(\mathrm{ind}_{\Theta\cap\Delta}^\Theta \varepsilon_{\Theta\cap\Delta}).$$

What we have to show, is that

$$\begin{aligned}\tau(\mathrm{ind}_\Delta^\Gamma \eta) &= \tau(\eta)\tau(\mathrm{ind}_\Delta^\Gamma \varepsilon_\Delta)^{\deg(\eta)}\\ &= \tau(\beta)\tau(\mathrm{ind}_{\Theta\cap\Delta}^\Delta \varepsilon_{\Theta\cap\Delta})\tau(\mathrm{ind}_\Delta^\Gamma \varepsilon_\Delta)^{\deg(\eta)}\end{aligned}$$

(the last equation following by our induction hypothesis). Since $\mathrm{ind}_\Theta^\Gamma \xi = \mathrm{ind}_\Delta^\Gamma \eta$, this reduces to showing that

$$\tau(\mathrm{ind}_{\Theta\cap\Delta}^\Theta \varepsilon_{\Theta\cap\Delta})\tau(\mathrm{ind}_\Theta^\Gamma \varepsilon_\Theta)^l = \tau(\mathrm{ind}_{\Theta\cap\Delta}^\Delta \varepsilon_{\Theta\cap\Delta})\tau(\mathrm{ind}_\Delta^\Gamma \varepsilon_\Delta)^{\deg(\eta)}.$$

All four characters factor through the proper quotient Γ/Δ_0 of Γ, and the above equation follows then on using the induction hypothesis and evaluating $\tau(\mathrm{ind}_{\Theta\cap\Delta}^\Gamma \varepsilon_{\Theta\cap\Delta})$ in two different ways. □

So henceforth we assume η to be an Abelian character of Δ. We can also assume $\mathrm{res}_{\Delta_0}^\Delta \eta$ to be faithful, otherwise the induction hypothesis comes into play once more. As $(\Delta, \Delta) \subset \mathrm{Ker}\,\eta \cap \Delta_0$, Δ will then be Abelian, and so $\Theta \cap \Delta = \Delta$. Moreover, by (7.13c), $\Sigma \neq \Gamma$, hence $\Sigma \cap \Delta \neq \Delta$.

In the sequel we shall denote an l-Sylow group of a finite group G by $G(l)$. (In our case this is always unique.) In the situation we have reached we now obtain the following alternatives, η being a fixed Abelian character of Δ, as above.

G. (i) *Suppose* $\Delta_0(l) = 1$, *i.e.*, $\Gamma_0 = \Delta_0 \times \Gamma_0(l)$. *Then* Γ/Σ *acts faithfully on* $\Gamma_0(l)$, *and so on the non-identity Abelian characters of* $\Gamma_0(l)$. As $[\Gamma_0 : \Delta_0] = l$, $\Gamma_0(l)$ *must be of order* l, *whence* $l \neq 2$. *Let* $\{\alpha_i\}$ $(i = 1, \ldots, g)$ *be a set of representatives of the orbits of such characters, under this action, and extended to* Σ, *in such a way that their restriction to* $\Sigma \cap \Delta$ *is the identity. Let* $\eta_i = \mathrm{ind}_\Sigma^\Gamma \alpha_i$. *Let moreover* ξ_0 *be the character of* Γ *extending* η *which is the identity on* $\Gamma_0(l)$. *Let* $\xi_i = \eta_i\xi_0$. *Then the characters* η_i $(1 \leqslant i \leqslant g)$ *and* ξ_i $(0 \leqslant i \leqslant g)$ *are irreducible and distinct. Also*

$$\text{(7.14)}\qquad \begin{cases} \mathrm{ind}_\Delta^\Gamma \varepsilon_\Delta = \varepsilon_\Gamma + \sum_{i=1}^{g} \eta_i \\ \mathrm{ind}_\Delta^\Gamma \eta = \xi_0 + \sum_{i=1}^{g} \xi_i. \end{cases}$$

(ii) *Suppose that* $\Delta_0(l) \neq 1$ *and* l *is odd. Then* Γ/Σ *is of order* l. *It acts trivially on the* $\Gamma_0(q)$, *for* $q \neq l$. *A generator of* Γ/Σ *acts on* $\Gamma_0(l)$ *by*

$$\sigma \mapsto \sigma^{1+al^t}, \qquad (a, l) = 1,$$

if $l^{t+1} =$ *order* $(\Gamma_0(l))$. *Hence* Γ/Δ_0 *is Abelian.*

Let α_i' *be the distinct, non-identity Abelian characters of* Γ/Δ, *inflated to* Γ. *Then*

$$\operatorname{ind}_\Delta^\Gamma \varepsilon_\Delta = \sum_{i=1}^{l-1} \alpha_i' + \varepsilon_\Gamma. \tag{7.15}$$

Further, if β *is an Abelian character of* Σ, *with* $\operatorname{res}^\Sigma_{\Delta\cap\Sigma} \beta = \operatorname{res}^\Delta_{\Delta\cap\Sigma} \eta$ *(such a* β *exists) then* $\operatorname{ind}^\Gamma_\Sigma \beta = \operatorname{ind}^\Gamma_\Delta \eta$, *and this character is irreducible.*

We omit the case $l = 2$ under (ii), which is similar, but slightly more complicated. The proof of G is an exercise in character theory. For details see [FT] (6(d) and 6(f)). □

We shall now prove that

$$\tau(E/F, \eta)(= \tau(\eta)\tau(\operatorname{ind}^\Gamma_\Delta \varepsilon_\Delta)\tau(\operatorname{ind}^\Gamma_\Delta \eta)^{-1}) = 1 \tag{7.16}$$

in all cases which can arise.

First assume that $\Delta_0(l) = 1$. Use (7.14) and cancel $\tau(\operatorname{ind}^\Gamma_\Delta \varepsilon_\Delta)^g$. Then (7.16) can, on using (7.10), be rewritten as

$$\tau(\eta) \prod_{i=1}^{g} \tau(\alpha_i) = \tau(\xi_0) \prod_{i=1}^{g} \tau(\alpha_i \operatorname{res}^\Gamma_\Sigma \xi_0). \tag{7.17}$$

If η is non-ramified, hence ξ_0 is non-ramified, one verifies (7.17) quite easily, using the original definition (7.1) together with the twisting rule B. So we now assume η to be ramified. Then some prime q other than l and p divides order $(\operatorname{res}^\Delta_{\Delta_0} \eta)$. The aim is to show that the two sides in (7.17) are congruent mod $\mathfrak{L}_q$ and mod $\mathfrak{L}_l$. We already know that they are congruent modulo roots of unity. As roots of unity are distinct modulo $\mathfrak{L}_l\mathfrak{L}_q$, Eq. (7.17) follows. To establish the required congruences we shall use the weaker form (7.11) of E, both for l, as well as for q in place of l.

Let then, in the notation of G (i), $\xi_0 = \xi_0'\xi_0''$, where ξ_0'' is of q-power order, ξ_0' of order prime to q. Put $\eta' = \operatorname{res}^\Gamma_\Delta \xi_0'$. Then by (7.11),

$$\begin{cases} \tau(\eta) \equiv \tau(\eta'), \qquad \tau(\xi_0) \equiv \tau(\xi_0'), \\ \tau(\alpha_i \operatorname{res}^\Gamma_\Sigma \xi_0) \equiv \tau(\alpha_i \operatorname{res}^\Gamma_\Sigma \xi_0') \quad (\operatorname{mod} \mathfrak{L}_q). \end{cases} \tag{7.18}$$

On the other hand if we replace, in (7.17), η by η', ξ_0 by ξ_0' then these and the α_i may be viewed as characters of $\Gamma/\Gamma_0(q)$, a group of smaller order than Γ. By induction hypothesis the equation corresponding to (7.17) but with η', ξ_0' in place of η, ξ_0, will hold. This together with (7.18) implies that the two sides in (7.17) – as it stands – are congruent mod $\mathfrak{L}_q$.

Observe next that $\theta_\eta = \theta_{\xi_0} \circ N_{E/F}$. Thus

$$\tau(\eta) = \sum \theta_\eta(uc^{-1})\psi_E(uc^{-1}) = \sum \theta_{\xi_0}(uc^{-1})^l \psi_F(uc^{-1})^l,$$

which is clearly congruent to $\tau(\xi_0)^l$ mod l. For, any set of representatives $\{u\}$ of $\mathfrak{o}_F^*$ mod $\mathfrak{p}_F$ is also a set of representatives of $\mathfrak{o}_E^*$ mod $\mathfrak{p}_E$. Also, as usual, we choose $c \in F$. Moreover then $\psi_E(uc^{-1}) = \psi_F(luc^{-1}) = \psi_F(uc^{-1})^l$.

Now we apply (7.11) to $\alpha_i - \varepsilon_\Sigma$, and to $\alpha_i \operatorname{res}_\Sigma^\Gamma \xi_0 - \operatorname{res}_\Sigma^\Gamma \xi_0$, and we obtain

$$\tau(\alpha_i) \equiv -1, \qquad \tau(\alpha_i \operatorname{res}_\Sigma^\Gamma \xi_0) \equiv \tau(\operatorname{res}_\Sigma^\Gamma \xi_0) \pmod{\mathfrak{L}_l}.$$

From this congruence, together with the congruence $\tau(\eta) \equiv \tau(\xi_0)^l \pmod{l}$ established above, and the equation $\tau(\operatorname{res}_\Sigma^\Gamma \xi_0) = (-1)^{1+[\Gamma:\Sigma]}\tau(\xi_0)^{[\Gamma:\Sigma]}$, which is a consequence of one of the Davenport-Hasse identities ([DH] (0.8)), it now follows that the two sides in (7.17) are congruent mod $\mathfrak{L}_l$. Thus we get (7.16) in the situation described in G(i).

Now we consider the case $\Delta_0(l) \neq 1$, l odd (see G(ii)). If for some prime $q \neq l$, $\Delta(q) \neq 1$ the principle of the proof is similar to that in the preceding case and we shall not give the details here. We shall instead assume Δ to be an l-group, whence Γ is an l-group. What will be used here is Theorem 36 and the fact that any root of unity $\equiv 1 \pmod{l}$ is $= 1$ (since $l \neq 2$).

We use the notation in G(ii). As l is odd we know that $\rho_{F'/F}$ is the identity character, where F' is the fixed field of Σ. Thus, by (7.2), and G(ii),

$$\tau(\beta) = \tau(\operatorname{ind}_\Sigma^\Gamma \beta) = \tau(\operatorname{ind}_\Delta^\Gamma \eta). \tag{7.19}$$

Hence what we have to prove is that

$$\tau(\eta) \prod_{i=1}^{l-1} \tau(\alpha_i') = \tau(\beta). \tag{7.20}$$

By (7.19) and Theorem 36, and as $\det(\operatorname{ind}_\Delta^\Gamma \eta) = \mathrm{v}_{\Delta/\Gamma}\eta$ (the cotransfer),

$$\tau(\beta) \equiv \tau(\mathrm{v}_{\Delta/\Gamma}\eta)\theta_\eta(l)^{-1} \pmod{l}. \tag{7.21}$$

By Lemma 6.4,

$$\tau(\eta) = \tau(\mathrm{v}_{\Delta/\Gamma}\eta)\theta_\eta(l)^{-1}. \tag{7.22}$$

By (7.11), as applied to $\alpha_i' - \varepsilon_\Gamma$, we get that $\tau(\alpha_i') \equiv -1 \pmod{\mathfrak{L}_l}$. The α_i' occur in complex conjugate pairs. By A and C, $\tau(\alpha_i')\tau(\bar{\alpha}_i') = \pm \mathbf{N}_F\mathfrak{f}(\alpha_i)$. This is a rational integer $\equiv 1 \pmod{\mathfrak{L}_l}$, thus $\equiv 1 \pmod{l}$. (By (7.21), (7.22), we conclude that the two sides of (7.20) are congruent mod l, hence are equal. □

We finally give an outline of the proof of Theorem 37. In the notation of that theorem, put

$$\mu(\chi) = \tau(\chi)/\tau'(\chi).$$

We have to show that $\mu(\chi) = 1$. This is trivially true for Abelian χ. As μ is additive it thus suffices to show that

$$\mu(\chi) = 1 \qquad \text{for non-Abelian, irreducible } \chi. \tag{7.23}$$

We first show

H. $\mu(\chi)$ *is a root of unity, and for all* $\omega \in \Omega_{\mathbb{Q}}$, $\mu(\chi^{\omega}) = \mu(\chi)^{\omega}$.

Proof. $S_{(F)}$, together with the tame Abelian characters, generates the additive group $R_{(F)}$ of tame Galois characters of F. It follows that A–E and Theorem 36 hold quite generally with τ' replacing τ – for all $\chi \in R_{(F)}$. (We shall use this observation again below.) By A and D, $\mu(\chi)$ is a unit of absolute value 1 (referring to the distinguished absolute value). By C, μ commutes with $\Omega_{\mathbb{Q}}$. Hence $\mu(\chi)$ is always a root of unity. □

The situation we are in is again similar to that facing us in the proofs of Theorem 32, and of Theorem 18A: we have to use congruences to prove that a root of unity is 1. We again proceed by induction on the order of Galois groups. We consider an irreducible, non-Abelian, faithful character χ of the Galois group Γ of a tame extension L/F. Let $\chi = \mathrm{ind}_{\Sigma}^{\Gamma} \alpha$, α Abelian and $\Sigma \supset \Gamma_0$. Then $\mathrm{res}_{\Gamma_0}^{\Sigma} \alpha$ is faithful, so Σ is Abelian.

I. If Γ *is an l-group, then* $\mu(\chi) = 1$.

We give the proof for odd l. By Theorem 36, $\mu(\chi - \det \chi) \equiv 1 \pmod{l}$. Hence $\mu(\chi - \det \chi) = 1$. But as $\det \chi$ is Abelian, $\mu(\det \chi) = 1$. Therefore $\mu(\chi) = 1$. □

II. *If order* (Σ) *has two prime divisors, say* l *and* q, *then* $\mu(\chi) = 1$.

Let $\alpha = \alpha^* \alpha_l \alpha_q$, where α_l, α_q are of l-power and q-power order, respectively, and α^* is of order prime to lq. Let

$$\chi^{(l)} = \mathrm{ind}_{\Sigma}^{\Gamma}(\alpha_q \alpha^*), \qquad \chi^{(q)} = \mathrm{ind}_{\Sigma}^{\Gamma}(\alpha_l \alpha^*), \qquad \chi^{(0)} = \mathrm{ind}_{\Sigma}^{\Gamma}(\alpha^*).$$

Then

$$\xi = \chi - \chi^{(l)} - \chi^{(q)} + \chi^{(0)} \in \mathrm{Ker}\, d_l \cap \mathrm{Ker}\, d_q \cap S_{(F)}.$$

Hence $\mu(\xi) \equiv 1 (\mathfrak{L}_l \mathfrak{L}_q)$ by E, i.e., $\mu(\xi) = 1$. But $\chi^{(l)}, \chi^{(q)}, \chi^{(0)}$ are faithful on proper quotients of Γ. By induction hypothesis, $\mu(\chi^{(l)}) = \mu(\chi^{(q)}) = \mu(\chi^{(0)}) = 1$, whence $\mu(\chi) = 1$. □

The remaining case, when Σ is an l-group, but Γ is not, is treated in a similar fashion.

Remark 1. The formal similarity between methods used in the present section, and those which are applied in the proof of Theorem 32, indicates a deeper connection between the two sets of ideas.

In the preceding sections we used congruences to establish determinantal behaviour of Galois Gauss sums. But in a sense, the congruences both of earlier sections, as well as those in E and in Theorem 36 can in turn be viewed as reflecting determinantal properties.

Remark 2. In two different contexts, namely here in assertion B, and in the preceding sections in the definition of $T_{\Gamma,\phi}$, the twisting of Galois Gauss sums by certain Abelian characters plays a central role. This indicates an analogy to the role of twisting which first appeared in Weil's work on modular forms and Dirichlet series, and which is of importance in "Langland's philosophy". Indeed the original formulae of Langlands on twisting of root numbers arose in precisely this context.

Notes to Chapter IV

[1] There are congruences for resolvents, analogous to those for determinants and for adjusted Galois Gauss sums (cf. [F13], [F17]). These are no longer needed at this stage, although they played a part in the earlier development, and they are of course of some interest in themselves – reflecting arithmetic properties of resolvents. We shall briefly state the result on $\operatorname{Ker} d_l$ and indicate how it is derived.

Let then L/M be a tame Galois extension of non-Archimedean local fields, of residue class characteristic p, with Galois group Γ. Let l be a rational prime. We wish to show that if a is a free generator of $\mathfrak{o}_L$ over $\mathfrak{o}_M\Gamma$ and $\chi \in \operatorname{Ker} d_l$, then

(a) $$(a \mid \chi) \equiv 1 \pmod{\mathfrak{L}_l}.$$

Let $\chi = \psi - \phi$, ψ, ϕ characters of representations. As in the proof of Theorem 15A one can certainly prove that

(b) $$(a \mid \psi) \equiv (a \mid \phi) \pmod{\mathfrak{L}_l}.$$

If first $l \neq p$ then $(a \mid \phi)$ is a unit at $\mathfrak{L}_l$. We thus can divide through to get (a). Now assume that $l = p$. Then we apply Theorem 25. We let B be the fixed field of the inertia subgroup Γ_0 of Γ, and b a free generator of $\mathfrak{o}_L$ over $\mathfrak{o}_B\Gamma_0$. Then

(c) $$(a \mid \chi)_{L/M} = (b \mid \operatorname{res}^{\Gamma}_{\Gamma_0} \chi)_{L/B} \operatorname{Det}_\chi(c),$$

$c \in \mathfrak{o}_B\Gamma^*$. By Theorem 15A

(d) $$\operatorname{Det}_\chi(c) \equiv 1 \pmod{\mathfrak{L}_l}.$$

Moreover $\operatorname{res}^{\Gamma}_{\Gamma_0} \chi \in \operatorname{Ker} d_{l,\Gamma_0}$. But $(l, \operatorname{order}(\Gamma_0)) = (p, \operatorname{order}(\Gamma_0)) = 1$. Thus $\operatorname{Ker} d_{l,\Gamma_0} = 0$, hence $\operatorname{res}^{\Gamma}_{\Gamma_0} \chi = 0$, i.e., $(b \mid \operatorname{res}^{\Gamma}_{\Gamma_0} \chi) = 1$. This, together with (c) and (d), implies (a).

For "higher" congruences see [F13].

V. Root Number Values

§1. The Arithmetic of Quaternion Characters

The purpose of this chapter is to derive explicit formulae for the root numbers of tame symplectic characters and to deduce from these formulae results on the possible distributions of root number values and on the densities of such distributions. The interest in this type of problem comes of course from the fact that the root numbers provide a full set of invariants for stable Galois module structure of rings of integers, and also from the implications on zeros of Dedekind zeta functions at $s = \frac{1}{2}$. For this see the remark in I §1.

By Serre's induction theorem (cf. I note [1]) every symplectic character is a $\mathbb{Z}$-linear combination of "trivial" symplectic characters of the form $\chi + \bar{\chi}$, which are uninteresting, and of characters induced from quaternion characters (cf. (I.1.14)). Quaternion characters can moreover be handled explicitly, and for these we shall get very precise results on value distribution. All this provides the motivation to study the arithmetic of quaternion characters and quaternion extensions. We shall follow essentially the class field theoretic methods of [F7] and [F26]. A complementary earlier approach, more in the spirit of Kummer theory, is given in [DM].

Let throughout this section K be a number field. We are interested in the quaternion characters ψ of its absolute Galois group Ω_K. It will be convenient to allow here also the "degenerate quaternion characters" ψ, which in the description of (I.1.14) correspond to the value $m = 1$, which was at that stage excluded. Such a character is in fact of the form $\mathrm{Tr}(\eta) = \eta + \eta^{-1}$ with η Abelian of order 4; it is therefore still symplectic. Given a continuous quaternion representation T of Ω_K with character ψ (possibly degenerate), denote for the moment by N_ψ the fixed field of $\operatorname{Ker} T$. Thus $\mathrm{Gal}(N_\psi/K) \cong H_{4m}$, and from the structure of the latter group it follows that N_ψ is cyclic of order $2m$ over some quadratic extension L of K, and unless $m = 2$, this L is unique. Anyway from now on we shall fix a quadratic extension L of K, and study the quaternion characters ψ with N_ψ/L cyclic. It is clear then that ψ is induced from an Abelian character ξ of Ω_L, and we shall say that "ψ is induced from L".

Let w be the generator of $\mathrm{Gal}(L/K)$. For ξ a character of Ω_L define ξw by $\xi w(\omega) = \xi(\delta\omega\delta^{-1})$ for all $\omega \in \Omega_L$, where $\delta \in \Omega_L \backslash \Omega_K$. In the obvious notation, always $\mathrm{res}_L^K \mathrm{ind}_L^K \xi = \xi + \xi w$. Thus if ξ is Abelian, $\mathrm{ind}_L^K \xi$ quaternion, then certainly $\xi w = \xi^{-1}$, i.e.,

$$\xi \cdot \xi w = \varepsilon. \tag{1.1}$$

(Here and in the sequel ε, possibly with some subscript, always denotes an identity character of some group.) Thus, in the language of IV §4, $\mathrm{ind}_L^K \xi$ is of inversion type. Conversely (1.1), for Abelian ξ, implies that $\mathrm{ind}_L^K \xi$ is either quaternion or dihedral (including degenerate cases). To distinguish the two alternatives let

$$\kappa = \kappa_{L/K}\colon \Omega_K \to \pm 1$$

be the Abelian character with kernel Ω_L. By (1.1), $\det(\mathrm{ind}_L^K \xi) = \kappa$ or $= \varepsilon$, according to whether we are in the dihedral or the quaternion case. But

$$\det(\mathrm{ind}_L^K \xi) = \kappa \cdot \mathrm{v}_{L/K}\xi,$$

$\mathrm{v}_{L/K}$ the cotransfer. Thus $\mathrm{ind}_L^K \xi$ is quaternion or dihedral, according to whether $\mathrm{v}_{L/K}\xi = \kappa$ or $= \varepsilon$, and either of these equations will imply (1.1). Now we shall go over to the idele class character θ_η which correspond to the Abelian Galois character η. Write in particular

(1.2) $$\phi_{L/K} = \theta_\kappa \qquad (\kappa = \kappa_{L/K}).$$

Thus $\phi_{L/K}$ gives rise to an isomorphism

$$\mathfrak{J}(K)/K^* N_{L/K}\mathfrak{J}(L) \cong \pm 1 \qquad (N_{L/K} \text{ the norm}).$$

We shall say that an idele class character θ of L is of *quaternion type* if its restriction to $\mathfrak{J}(K)$ satisfies

(1.3) $$\theta_{|\mathfrak{J}(K)} = \phi_{L/K}.$$

(It is always understood that this property is defined relative to the given quadratic extension L/K.) We thus have shown:

1.1. Proposition. *Let ξ be an Abelian character of Ω_L. $\mathrm{ind}_L^K \xi$ is a quaternion character, possibly degenerate, if, and only if, θ_ξ is of quaternion type.* □

Analogously we define the property of an idele class character θ of L to be of dihedral type if

(1.4) $$\theta_{|\mathfrak{J}(K)} = \varepsilon_K,$$

and one gets an obvious analogue to Proposition 1.1. The Abelian characters of quaternion or dihedral type form a multiplicative group, those of dihedral type a subgroup of index 2 or 1, according to whether there exist characters of quaternion type (then forming the non-trivial coset) or not.

1.2. Proposition (cf. [DM], [F7]). *L has idele class characters of quaternion type precisely if all real prime divisors of K split into real prime divisors of L.*

The necessity is obvious. Two sufficiency proofs are given in [F7], one of them a "2 line proof" using the cohomology of the connected component of 1 in $\mathfrak{J}(L)/L^*$. □

From now on we restrict ourselves to tame Galois characters and tame idele class characters. We shall assume that L/K is tame. Everything said so far remains true if "character" is always replaced by "tame character". In the sequel the symbol $\mathfrak{p}$ will stand for prime divisors of K, and $\mathfrak{P}$ for those of L; we write $\mathfrak{P} \mid \mathfrak{p}$, even in the infinite case, if $\mathfrak{P}$ lies above $\mathfrak{p}$. Local components of an idele character θ are denoted by $\theta_{\mathfrak{P}}$; semilocal components, i.e., restrictions to $L_{\mathfrak{p}}^* \subset \mathfrak{J}(L)$ by $\theta_{\mathfrak{p}}$. If $\mathfrak{P}$ is a finite prime divisor and θ of quaternion or dihedral type, then the order of $\theta_{\mathfrak{P}}$ restricted to $\mathfrak{o}_{L,\mathfrak{P}}^*$ (the ramification index) only depends on $\mathfrak{p}$; we denote it by $e(\theta_{\mathfrak{p}})$. In particular $e(\theta_{\mathfrak{p}}) = 1$, precisely when θ is non-ramified above $\mathfrak{p}$.

Further notation in this chapter: The quadratic residue symbol in a number field M is written as $(-)_M$.

1.3. Proposition (cf. [F7], [F26]). *Suppose θ is of dihedral or quaternion type.*

(i) *If the finite prime divisor $\mathfrak{p}$ splits in L into $\mathfrak{P}, \mathfrak{P}'$, then*

$$\theta_{\mathfrak{P}}(x) = \theta_{\mathfrak{P}'}(x)^{-1}, \qquad \textit{for} \qquad x \in \mathfrak{o}_{K,\mathfrak{p}}^*$$

and

$$\mathbf{N}_K\mathfrak{p} \equiv 1 \pmod{e(\theta_{\mathfrak{p}})}.$$

(ii) *If the finite prime divisor $\mathfrak{p}$ is inert in L then*

$$\mathbf{N}_K\mathfrak{p} \equiv -1 \pmod{e(\theta_{\mathfrak{p}})}.$$

(iii) *If the finite prime divisor $\mathfrak{p}$ is ramified in L, with $\mathfrak{P} \mid \mathfrak{p}$, then*

$$e(\theta_{\mathfrak{p}}) = 1 \qquad (\theta \textit{ of dihedral type}),$$

$$e(\theta_{\mathfrak{p}}) = 2, \quad \theta_{\mathfrak{P}}(x) = \left(\frac{x}{\mathfrak{P}}\right)_L \quad \textit{for} \quad x \in \mathfrak{o}_{L,\mathfrak{P}}^* \qquad (\theta \textit{ of quaternion type})$$

(iv) *If $\mathfrak{p}$ is real, then either*

$$\theta_{\mathfrak{p}} = \varepsilon_{\mathfrak{p}}$$

or

$$\theta_{\mathfrak{P}} = \mathrm{sign}_{\mathfrak{P}} \qquad \textit{for all } \mathfrak{P} \textit{ above } \mathfrak{p}.$$

Proof. As $\theta_{\mathfrak{P}}$ is tame for $\mathfrak{P}$ finite, we may always view its restriction to $\mathfrak{o}_{L,\mathfrak{P}}^*$ as a character of $(\mathfrak{o}_{L,\mathfrak{P}}/\mathfrak{P})^*$. In cases (i) and (ii) the restriction of $\theta_{\mathfrak{p}}$ to $\mathfrak{o}_{K,\mathfrak{p}}^*$ is trivial. If $\mathfrak{p}$ is inert, it follows that the character of $(\mathfrak{o}_{L,\mathfrak{P}}/\mathfrak{P})^*$ given by $\theta_{\mathfrak{P}}$ vanishes on the subgroup $(\mathfrak{o}_{K,\mathfrak{p}}/\mathfrak{p})^*$, which is of index $\mathbf{N}_L\mathfrak{P} - 1/\mathbf{N}_K\mathfrak{p} - 1 = \mathbf{N}_K\mathfrak{p} + 1$. Thus $e(\theta_{\mathfrak{p}}) \mid \mathbf{N}_K\mathfrak{p} + 1$. If, on the other hand, $\mathfrak{p} = \mathfrak{P}\mathfrak{P}'$ splits, then we must trivially have $\theta_{\mathfrak{P}}(x)\theta_{\mathfrak{P}'}(x) = 1$ for $x \in \mathfrak{o}_{K,\mathfrak{p}}^*$, and also $e(\theta_{\mathfrak{p}}) \mid \mathbf{N}_L\mathfrak{P} - 1 = \mathbf{N}_K\mathfrak{p} - 1$.

If $\mathfrak{p}$ is ramified, then $(\mathfrak{o}_{K,\mathfrak{p}}/\mathfrak{p})^* \cong (\mathfrak{o}_{L,\mathfrak{P}}/\mathfrak{P})^*$. Thus the restriction of $\theta_{\mathfrak{P}}$ to $\mathfrak{o}_{L,\mathfrak{P}}^*$ is given by its restriction to $\mathfrak{o}_{K,\mathfrak{p}}^*$, which immediately yields the assertion.

For (iv) observe that always $\theta_{\mathfrak{p}} = \theta_{\mathfrak{p}}^{-1}$, so $\theta_{\mathfrak{p}}$ is fixed under $\mathrm{Gal}(L/K)$. This leaves precisely the two stated possibilities. □

Remark. In the usual way (see e.g., I §5) one can "extend" $\phi_{L/K}$ to a multiplicative function of fractional ideals in K, which do not involve any of the prime ideals $\mathfrak{p}$ ramified in L/K. Indeed if the idele b of K has components $b_{\mathfrak{p}} = 1$ at all such prime divisors, and at all infinite prime divisors, then $\phi_{L/K}(b)$ only depends on the ideal $\mathfrak{b}$, the content of b; we accordingly write $\phi_{L/K}(\mathfrak{b})$. As a multiplicative function of such fractional ideals, $\phi_{L/K}$ is characterized by its values on prime ideals $\mathfrak{p}$, non-ramified in L. For these we simply have

$$\phi_{L/K}(\mathfrak{p}) = \begin{cases} 1, & \text{if } \mathfrak{p} \text{ splits in } L, \\ -1, & \text{if } \mathfrak{p} \text{ is inert in } L. \end{cases} \tag{1.5}$$

From the last Proposition we get then

$$\mathbf{N}_K\mathfrak{p} \equiv \phi_{L/K}(\mathfrak{p}) \pmod{e(\theta_{\mathfrak{p}})}. \tag{1.6}$$

1.4. Proposition. *Assume that there exists an idele class character of L of quaternion type. Then the possible orders of such characters are precisely the multiples of a certain power 2^{s-1} of 2, with $s \geqslant 2$.*

This was first proved in [DM], without the tame hypothesis, it was also shown there that always $s \leqslant 5$. Our proof follows that in [F26].

If first θ is of quaternion type of order $2^{t-1}h$, h odd, then θ^h is of quaternion type of order 2^{t-1}. On the other hand, given any positive integer g, there will exist – as we shall show in the next lemma – a (tame) character β of dihedral type of order $2^{t-1}hg$ with $e(\beta_{\mathfrak{p}}) = 2^{t-1}hg$, for some $\mathfrak{p}$ with $e(\theta_{\mathfrak{p}}) = 1$. It follows then that $\theta\beta$ is of quaternion type of order $2^{t-1}hg$. □

The next lemma goes further than is required for the above proof, and will also be used subsequently.

Let $\mathscr{S}_K$ be a finite set of finite prime divisors of K, $\mathscr{S}_L$ the set of prime divisors of L above K, chosen so that the ideal classes in K of the $\mathfrak{p} \in \mathscr{S}_K$ generate the ideal class group of K, and the ideal classes in L of the $\mathfrak{P} \in \mathscr{S}_L$ generate the ideal class group of L. Denote here by $\mathscr{E}$ the group of $\mathscr{S}_L$-units, i.e., of elements of L^* which are units outside $\mathscr{S}_L$. Write again $\mathbb{Q}(n)$ for the field of n-th roots of unity over $\mathbb{Q}$ and $\mathbb{Q}^+(n)$ for its maximal real subfield.

"Density" in the next lemma is Tchebotareff density. For $e = +1$ or $e = -1$, and for a positive integer n, let $\mathscr{T}_{n,e}$ be the set of finite prime divisors $\mathfrak{p}$ of K, outside $\mathscr{S}_K$, and non-ramified in L, with the following properties:

$$\phi_{L/K}(\mathfrak{p}) = e, \tag{1.7a}$$

$$\mathbf{N}_K\mathfrak{p} \equiv e \pmod{n}. \tag{1.7b}$$

(1.7c) For all $\mathfrak{P} \mid \mathfrak{p}$, the elements of $\mathscr{E}$ are n-th power residues mod $\mathfrak{P}$.

1.5. Lemma. (i) *$\mathscr{T}_{n,1}$ has positive density.*

(ii) *If n is odd, $L \not\subset K\mathbb{Q}^+(n)$, $\mathbb{Q}(n) \not\subset K\mathbb{Q}^+(n)$ then $\mathscr{T}_{n,-1}$ has positive density.*

(iii) *If $\mathfrak{p} \in \mathscr{T}_{n,e}$, for some e, then there is an idele class character β of L, of order n, of dihedral type, with $e(\beta_{\mathfrak{p}}) = n$, and $e(\beta_{\mathfrak{q}}) = 1$ for all $\mathfrak{q} \neq \mathfrak{p}$.*

Proof. Adjoin all $c^{1/n}$ to L, for all $c \in \mathscr{E}$. The resulting field M is Galois over K and of finite degree, as $\mathscr{E}$ is finitely generated and $\mathrm{Gal}(L/K)$-stable. $\mathscr{T}_{n,1}$ consists of the finite prime divisors of K splitting in M – excluding always those in $\mathscr{S}_K$. Thus $\mathscr{T}_{n,1}$ has positive density.

Next, under the hypothesis of (ii), there are prime divisors $\mathfrak{p}$ satisfying (1.7a), (1.7b) with $e = -1$ – namely all those with a certain given Frobenius element g in $\mathrm{Gal}(L\mathbb{Q}(n)/K)$, namely the unique element in $\mathrm{Gal}(L\mathbb{Q}(n)/K\mathbb{Q}^+(n))$ which is not in $\mathrm{Gal}(L\mathbb{Q}(n)/K\mathbb{Q}(n))$ nor in $\mathrm{Gal}(L\mathbb{Q}(n)/L\mathbb{Q}^+(n))$. Now $L\mathbb{Q}(n) \subset M$ and $[M : L\mathbb{Q}(n)]$ is odd. Therefore g lifts to elements g' of order 2 in $\mathrm{Gal}(M/K)$, and the prime divisors $\mathfrak{p}$ of K, whose Frobenius class in M/K consists of such elements g', will make up the set $\mathscr{T}_{n,-1}$, again excluding those in $\mathscr{S}_K$. Thus $\mathscr{T}_{n,-1}$ has positive density.

Now observe that

$$\mathfrak{J}(L) = L^* \mathfrak{U}(L) \prod_{\mathfrak{s} \in \mathscr{S}_K} L_{\mathfrak{s}}^*, \tag{1.8}$$

$\mathfrak{U}(L)$ the group of unit ideles. Also clearly

$$L^* \cap [\mathfrak{U}(L) \prod L_{\mathfrak{s}}^*] = \mathscr{E}. \tag{1.9}$$

Let then $\mathfrak{p}$ satisfy the hypotheses of (iii) in Lemma 1.5. If $\mathfrak{P} \mid \mathfrak{p}$ then $\mathbf{N}_L \mathfrak{P} \equiv 1 \pmod{n}$. Accordingly there exists a character η of $\mathfrak{o}_{L,\mathfrak{P}}^*$ of order n. In accordance with (1.8) every idele can be written as cua, $c \in L^*$, $u \in \mathfrak{U}(L)$, $a \in \prod L_{\mathfrak{s}}^*$. If we have $cua = c'u'a'$, with the obvious meaning of the symbols, then – as $a_{\mathfrak{P}} = a'_{\mathfrak{P}} = 1$ – we have $(u(u')^{-1})_{\mathfrak{P}} = (c'c^{-1})_{\mathfrak{P}}$. By (1.9), $c'c^{-1} \in \mathscr{E}$. By (1.7c), $\eta((c'c^{-1})_{\mathfrak{P}}) = 1$. Therefore $\rho(cua) = \eta(c_{\mathfrak{P}})$ only depends on the given idele. Thus ρ is an idele class character. If $\mathfrak{p}$ is inert, put $\beta = \rho$. If $\mathfrak{p}$ splits, put $\beta = \rho^{-1} \cdot \rho w$, w the generator of $\mathrm{Gal}(L/K)$. □

Remark. The proof of (i), and of (iii) for $e = 1$, is essentially that of a standard result on the embedding problem.

§2. Root Number Formulae

Throughout this section L/K is a tame quadratic extension of number fields. We shall establish two formulae for the root numbers $W(\theta)$ of an idele class character of quaternion type. Here $W(\theta)$ is the constant in Hecke's functional equation, quoted in I §5. As stated there, the constant $W(\chi)$ in the functional equation of the Artin L-function is fully inductive, and satisfies $W(\chi) = W(\theta_\chi)$ for Abelian χ (cf. (I.5.15)). Therefore if $\theta = \theta_\xi$, $\mathrm{ind}_L^K \xi = \psi$, as in §1, we have $W(\theta) = W(\psi)$. We thus get explicit formulae for the root numbers of quaternion characters of Ω_K.

We shall continue to use the notation of §1. In addition we shall write

$$L = K(d), \qquad d^2 \in K^*. \tag{2.1}$$

For the ideal (d) this implies that

$$(2.2) \qquad (d) = \mathfrak{D}_{L/K}\mathfrak{a}, \qquad \mathfrak{a} \text{ a fractional ideal of } \mathfrak{o}_K.$$

Here $\mathfrak{D}_{L/K}$ is the relative different. Given any finite set of idele class characters θ of L, we may moreover choose d in such a way, that

$$(2.3) \qquad \mathfrak{a}_\mathfrak{p} = 1, \qquad \text{if some prime divisor above } \mathfrak{p} \text{ is ramified at } \theta.$$

We shall denote the relative discriminant of L/K by $\mathfrak{d}$, and write

$$(2.4) \qquad \mathfrak{f}(\theta)^* = \prod_{\mathfrak{P} \nmid \mathfrak{D}_{L/K}} \mathfrak{f}(\theta_\mathfrak{P}).$$

If θ is of quaternion or dihedral type then $\mathfrak{f}(\theta)$ is $\mathrm{Gal}(L/K)$-stable, hence for $\mathfrak{p}$ non-ramified in L/K, $\mathfrak{f}(\theta)_\mathfrak{p} = \mathfrak{p}$ or $= 1$. Thus $\mathfrak{f}(\theta)^*$ is an ideal of $\mathfrak{o}_K$, and $\phi_{L/K}(\mathfrak{f}(\theta)^*)$ is defined, e.g., via (1.5) and multiplicativity. If θ is of quaternion type then moreover

$$(2.5) \qquad \mathfrak{f}(\theta) = \mathfrak{f}(\theta)^*\mathfrak{D}_{L/K}.$$

2.1. Proposition. *Let θ be an idele class character of L of quaternion type, and suppose that d satisfies* (2.1)–(2.3). *Then*

$$W(\theta) = \left(\frac{2}{\mathfrak{d}}\right)_K \phi_{L/K}(\mathfrak{a})\phi_{L/K}(\mathfrak{f}(\theta)^*) \prod_{\mathfrak{P} \mid \mathfrak{D}_{L/K}} \theta_\mathfrak{P}(d). \qquad \square$$

Before proceeding to the proof we shall state a second formula. We first need some more notations. For $t \geqslant 3$, we define, as in (IV.7.4), a restricted residue class character v_t on 2-adic rational units x, satisfying $x \equiv \pm 1 \pmod{2^{t-1}}$, by

$$v_t(x) = \begin{cases} 1, & \text{if} \quad x \equiv \pm 1 \pmod{2^t}, \\ -1, & \text{if} \quad x \equiv \pm 1 + 2^{t-1} \pmod{2^t}. \end{cases}$$

If θ is of quaternion type and of 2-power order, we also write

$$(2.6) \qquad e(\theta_\mathfrak{p}) = 2^{s_\mathfrak{p}-1}, \qquad s_\mathfrak{p} = s_\mathfrak{p}(\theta).$$

Recall then that, by Proposition 1.3, the symbol $v_{s_\mathfrak{p}}(\mathbf{N}_K\mathfrak{p})$ is well defined whenever $s_\mathfrak{p} \geqslant 3$.

2.2. Proposition. *With θ, d, $\mathfrak{a}$ as in Proposition 2.1, assume further that θ is of 2-power order. Denote by $r(\theta)$ the number of real prime divisors $\mathfrak{p}$ of K, for which $\theta_\mathfrak{p} \neq \varepsilon_\mathfrak{p}$. Then*

$$W(\theta) = (-1)^{r(\theta)}\left(\frac{2}{\mathfrak{d}}\right)_K \prod{}^* \left[v_{s_\mathfrak{p}}(\mathbf{N}_K\mathfrak{p})\phi_{L/K}(\mathfrak{p})\right] \prod{}^{**} \left(\frac{-1}{\mathbf{N}\mathfrak{p}_K}\right)_\mathbb{Q}.$$

Here $\prod^*$ and $\prod^{**}$ are products over finite prime divisors $\mathfrak{p}$ non-ramified in L, $\prod^*$ over those with $s_\mathfrak{p} \geqslant 3$, $\prod^{**}$ over those with $s_\mathfrak{p} = 2$.

Proof of Propositions 2.1 and 2.2. We define the root number of semilocal components by

$$W(\theta_{\mathfrak{p}}) = \prod_{\mathfrak{P} \mid \mathfrak{p}} W(\theta_{\mathfrak{P}}).$$

Then (cf. (I.5.13))

$$W(\theta) = \prod_{\mathfrak{p}} W(\theta_{\mathfrak{p}})$$

(product over all prime divisors $\mathfrak{p}$ of K), and similarly, abbreviating $\phi_{L/K} = \phi$,

$$W(\phi) = \prod_{\mathfrak{p}} W(\phi_{\mathfrak{p}}), \qquad \theta(d) = \prod_{\mathfrak{p}} \theta_{\mathfrak{p}}(d).$$

But, as d is a principal idele, $\theta(d) = 1$, and classically $W(\phi) = 1$ for quadratic ϕ. Therefore we can write

$$(2.7) \qquad W(\theta) = \begin{cases} \prod_{\mathfrak{p}} [W(\theta_{\mathfrak{p}}) W(\phi_{\mathfrak{p}}) \theta_{\mathfrak{p}}(d)^{-1}] & \text{(a)} \\ \prod_{\mathfrak{p}} [W(\theta_{\mathfrak{p}}) W(\phi_{\mathfrak{p}})] & \text{(b)} \end{cases}$$

We shall prove Propositions 2.1 and 2.2 by evaluating the semilocal factors in (2.7) (a), and in (2.7) (b), respectively.

A. $\mathfrak{p}$ real. Then $W(\phi_{\mathfrak{p}}) = 1$ and $W(\theta_{\mathfrak{p}}) = \theta_{\mathfrak{p}}(d) = +1$ or $= -1$, according to whether $\theta_{\mathfrak{p}} = \varepsilon_{\mathfrak{p}}$ or not.

Proof. By Proposition 1.2, $\phi_{\mathfrak{p}} = \varepsilon_{\mathfrak{p}}$, whence by (I.5.1), $W(\phi_{\mathfrak{p}}) = 1$. Similarly $W(\theta_{\mathfrak{p}}) = 1 = \theta_{\mathfrak{p}}(d)$ if $\theta_{\mathfrak{p}} = \varepsilon_{\mathfrak{p}}$. By Proposition 1.3 and (I.5.1), $W(\theta_{\mathfrak{p}}) = -1$ in the remaining case; if $\mathfrak{P}, \mathfrak{P}'$ lie above $\mathfrak{p}$, then in this case $\theta_{\mathfrak{P}} = \mathrm{sign}_{\mathfrak{P}}$, $\theta_{\mathfrak{P}'} = \mathrm{sign}_{\mathfrak{P}'}$. Moreover the conjugate of d over K is $-d$ and so

$$\mathrm{sign}_{\mathfrak{P}}(d)\,\mathrm{sign}_{\mathfrak{P}'}(d) = \mathrm{sign}_{\mathfrak{P}}(d \cdot (-d)) = \mathrm{sign}_{\mathfrak{p}}(-1) = -1. \qquad \square$$

Complex prime divisors make no contribution. From now on we consider finite prime divisors.

B. $\mathfrak{p}$ non-ramified at θ. Then $W(\theta_{\mathfrak{p}}) W(\phi_{\mathfrak{p}}) = 1$ and $\theta_{\mathfrak{p}}(d) = \phi(\mathfrak{a}_{\mathfrak{p}})$, $\mathfrak{a}$ as in (2.2) *and* (2.3).

Proof. By (I.5.6), $W(\theta_{\mathfrak{p}}) = \theta(\mathfrak{D}_{L,\mathfrak{p}})$. But $\mathfrak{D}_{L/K,\mathfrak{p}} = 1$, i.e., $\mathfrak{D}_{L,\mathfrak{p}} = \mathfrak{D}_{K,\mathfrak{p}}$, and since θ is of quaternion type, $\theta(\mathfrak{D}_{K,\mathfrak{p}}) = \phi(\mathfrak{D}_{K,\mathfrak{p}})$. Hence indeed $W(\theta_{\mathfrak{p}}) = W(\phi_{\mathfrak{p}}) = \pm 1$, i.e., $W(\theta_{\mathfrak{p}}) W(\phi_{\mathfrak{p}}) = 1$. Next $d_{\mathfrak{p}} = u_{\mathfrak{p}} a_{\mathfrak{p}}$, $u_{\mathfrak{p}}$ a unit, $a_{\mathfrak{p}} \in K_{\mathfrak{p}}^*$. Therefore

$$\theta_{\mathfrak{p}}(d) = \theta_{\mathfrak{p}}(a_{\mathfrak{p}}) \theta_{\mathfrak{p}}(u_{\mathfrak{p}}) = \theta(a_{\mathfrak{p}}) = \phi(a_{\mathfrak{p}}) = \phi(\mathfrak{a}_{\mathfrak{p}}) = \phi(\mathfrak{a}_{\mathfrak{p}})^{-1}. \qquad \square$$

C. $\mathfrak{p}$ ramified in L. Then

$$W(\theta_{\mathfrak{p}}) = \left(\frac{2}{\mathfrak{p}}\right)_K W(\phi_{\mathfrak{p}}).$$

Proof. By (I.5.8) it suffices to show that

$$\tau(\theta_{\mathfrak{P}}) = \left(\frac{2}{\mathfrak{p}}\right)_K \tau(\phi_{\mathfrak{p}}).$$

The argument is the same as that used earlier for Lemma IV.6.4. We get

$$\tau(\theta_{\mathfrak{P}}) = \phi_{\mathfrak{p}}(2)^{-1}\tau(\phi_{\mathfrak{p}}).$$

Now we observe that

$$\phi_{\mathfrak{p}}(2)^{-1} = \phi_{\mathfrak{p}}(2) = \left(\frac{2}{\mathfrak{p}}\right)_K. \qquad \square$$

D. $\mathfrak{p} \mid \mathfrak{f}(\theta)^*$, $\mathfrak{p}$ *splits in* L. *Then* $\phi(\mathfrak{p}) = W(\phi_{\mathfrak{p}}) = 1$, *and*

$$\pm 1 = \theta_{\mathfrak{p}}(d) = W(\theta_{\mathfrak{p}}) = \begin{cases} \left(\dfrac{-1}{\mathbf{N}_K\mathfrak{p}}\right)_{\mathbb{Q}}, & \text{if } s_{\mathfrak{p}} = 2, \\ v_{s_{\mathfrak{p}}}(\mathbf{N}_K\mathfrak{p}), & \text{if } s_{\mathfrak{p}} \geqslant 3. \end{cases}$$

Proof. As $\phi_{\mathfrak{p}} = \varepsilon_{\mathfrak{p}}$, $\phi(\mathfrak{p}) = W(\phi_{\mathfrak{p}}) = 1$. By Proposition 1.3, and by (I.5.7a), (I.5.8),

$$W(\theta_{\mathfrak{p}}) = W(\theta_{\mathfrak{P}})W(\theta_{\mathfrak{P}}^{-1}) = \theta_{\mathfrak{P}}(-1) = (-1)^{(\mathbf{N}_K\mathfrak{p}-1)/2^{s_{\mathfrak{p}}-1}},$$

which gives the required formulae. Moreover,

$$\theta_{\mathfrak{p}}(d) = \theta_{\mathfrak{P}}(d)\theta_{\mathfrak{P}}(-d)^{-1} = \theta_{\mathfrak{P}}(-1). \qquad \square$$

E. $\mathfrak{p} \mid \mathfrak{f}(\theta)^*$, $\mathfrak{p}$ *inert in* L. *Then*

$$W(\theta_{\mathfrak{p}}) = \theta_{\mathfrak{p}}(d)\phi(\mathfrak{p})W(\phi_{\mathfrak{p}}) = \pm 1,$$

and

$$\theta_{\mathfrak{p}}(d) = \begin{cases} \left(\dfrac{-1}{\mathbf{N}_K\mathfrak{p}}\right)_{\mathbb{Q}} \cdot \phi(\mathfrak{p}), & \text{if } s_{\mathfrak{p}} = 2, \\ v_{s_{\mathfrak{p}}}(\mathbf{N}_K\mathfrak{p}), & \text{if } s_{\mathfrak{p}} \geqslant 3. \end{cases}$$

Proof. The first equation is a consequence of (I.5.8) and of (IV.7.7). Also d is a $(\mathbf{N}_K\mathfrak{p}+1)/2$-st power residue mod $\mathfrak{P}$, but not a $\mathbf{N}_K\mathfrak{p}+1$-st power residue. Therefore (see IV.7.8)

$$\theta_{\mathfrak{P}}(d) = (-1)^{(\mathbf{N}_K\mathfrak{p}+1)/2^{s_{\mathfrak{p}}-1}},$$

and this gives the required values. $\square$

The two propositions now follow by collecting factors from *A–E*. We note that $\theta_{\mathfrak{p}}(d) = \theta_{\mathfrak{p}}(d)^{-1} = \pm 1$, for all $\mathfrak{p}$ non-ramified in L. $\square$

§3. Density Results

In the present section we shall apply the explicit formulae of §2 to obtain results on densities of Galois extensions of $\mathbb{Q}$ with 2-power quaternion Galois group, with prescribed symplectic root number and certain prescribed ramification pattern. The aim is firstly to provide an ample supply of explicit numerical examples and to give a concrete illustration of the method, as based on the preceding sections, and secondly to show that, over $\mathbb{Q}$, rootnumber values are indeed essentially random, within the obvious restrictions imposed by general theorems. We shall also use the results of the present section in the proof of the basic distribution theorem, coming in §4. Throughout "density" is Tchebotareff density.

The quaternion group H_{2^s} of order 2^s $(s \geqslant 3)$ has exactly one $\Omega_{\mathbb{Q}}$-orbit of irreducible symplectic characters ψ, namely those which come from a faithful representation of degree 2. Accordingly a tame Galois extension $N/\mathbb{Q}$ with group H_{2^s} has exactly one root number

(3.1) $$W(N) = W(N/\mathbb{Q}, \psi).$$

In addition we need a second invariant of such fields with possible values ± 1, defined by

(3.2) $$\operatorname{sign}(N) = \begin{cases} 1, & \text{if } N \text{ is real,} \\ -1, & \text{if } N \text{ is imaginary.} \end{cases}$$

(q/p) is the quadratic symbol.

3.1. Proposition (cf. [F26]). *Let $s \geqslant 3$ and let q be a prime number* $\equiv 1 \pmod 4$. *Then, for any odd prime p with* $(q/p) = 1$, *there exists at most one Galois extension N of $\mathbb{Q}$, cyclic over $\mathbb{Q}(\sqrt{q})$, with* $\operatorname{Gal}(N/\mathbb{Q}) \cong H_{2^s}$, *and such that only p and q are ramified in N. Denote this field – if it exists – by $N[p]$. For $a, b = \pm 1$, let $\mathcal{M}_{a,b}$ be the set of primes p as above, for which $N[p]$ exists, and for which furthermore*

$$W(N[p]) = a, \qquad \operatorname{sign}(N[p]) = b.$$

Then $\mathcal{M}_{a,b}$ has positive density. □

3.2. Proposition (cf. [F26]). *Let $s \geqslant 3$ and let q_1, q_2 be distinct prime numbers* $\equiv -1 \pmod 4$. *Then for any odd prime p, with* $(q_1 q_2/p) = 1$ *there exist exactly two Galois extensions N of $\mathbb{Q}$ or none, cyclic over $\mathbb{Q}(\sqrt{q_1 q_2})$, with* $\operatorname{Gal}(N/\mathbb{Q}) \cong H_{2^s}$, *and such that only p, q_1 and q_2 are ramified in N. If they exist, then one is real, the other imaginary; denote these by $N^+\{p\}$, and $N^-\{p\}$, respectively. Then*

$$W(N^+\{p\})/W(N^-\{p\}) = -1.$$

For $a = \pm 1$, let $\mathcal{L}_a$ be the set of all primes p as above, for which the $N\{p\}$ exist, and with $W(N^+\{p\}) = a$. Then $\mathcal{L}_a$ has positive density. □

Remark 1. The primes q, q_1, q_2 are fixed once and for all, and so do not appear in our notation for the fields N.

Remark 2. Similar results can be proved for other real quadratic fields $\mathbb{Q}(\sqrt{D})$. But in the nature of things these results will in general not be nearly as sharp. The real quadratic fields considered here are precisely the tame ones with odd classnumber, a fact which is implicitly used for two distinct aspects of the proof.

Remark 3. Anyone who reads the proofs can without difficulty write down the actual density of each of the sets that occur in the two propositions.

Proof of Propositions 3.1 and 3.2. We shall prove the two propositions together, referring where necessary to the situation of 3.1, and of 3.2 respectively, as case I and case II. We put $D = q$ (case I), $D = q_1 q_2$ (case II), and write $L = \mathbb{Q}(\sqrt{D})$. Prime ideals in L, dividing D, will be denoted by $\mathfrak{q}$. We then have to consider an odd prime p with

(3.3) $$\left(\frac{D}{p}\right)_{\mathbb{Q}} = 1.$$

It will split in L, and we write $\mathfrak{P}$, $\mathfrak{P}'$ for the prime divisors above it. By Proposition 1.1, we shall have to look for an idele class character θ of L of quaternion type, of order 2^{s-1}, with appropriate ramification properties. By Proposition 1.3, we would have

(3.4) $$\theta_{\mathfrak{q}}(x) = \left(\frac{x}{\mathfrak{q}}\right)_L \qquad \text{for} \qquad x \in \mathfrak{o}_{L,\mathfrak{q}}^*,$$

and

(3.5) $$\begin{cases} \text{either} & \theta_\infty = \varepsilon_\infty, \\ \text{or} & \theta_\infty = \operatorname{sign}_\infty, \end{cases}$$

where we write $\operatorname{sign}_\infty$ for the semi-local character at ∞, which is the signature at each of the two infinite prime divisors of L. Moreover we need

(3.6) $\quad e(\theta_{\mathfrak{r}}) = 1 \qquad$ for every finite prime divisor $\mathfrak{r}$, not dividing Dp.

Now the classnumber of L is odd. Hence $\theta^{2^{s-2}}$ is still genuinely ramified somewhere; as $s - 2 > 0$ it cannot be at any $\mathfrak{q}$, and $\theta_\infty^{2^{s-2}} = \varepsilon_\infty$. Taking into account (3.6), it follows that $\theta^{2^{s-2}}$ is ramified at $\mathfrak{P}$ and $\mathfrak{P}'$. In other words, $e(\theta_p) = 2^{s-1}$. By Proposition 1.3,

(3.7) $$p \equiv 1 \pmod{2^{s-1}}.$$

Choose now a residue class character λ of $\mathfrak{o}_{L,\mathfrak{P}}^*$ mod $\mathfrak{P}$ of order 2^{s-1}. By replacing θ, if necessary, by an odd power θ^r, we may suppose, using again Proposition 1.3, that

(3.8) $$\theta_{\mathfrak{P}}(x) = \lambda(x) = \theta_{\mathfrak{P}'}(x)^{-1}, \qquad \text{for} \qquad x \in \mathbb{Z}_p^*.$$

Thus we have shown:

A. *The number of distinct fields N ($= N[p]$ in case I, $= N\{p\}$ in case II) is the number of characters θ of order 2^{s-1}, of quaternion type, satisfying* (3.6), (3.8). □

For such an idele class character θ we have of course $\theta(y) = 1$ if y is a global unit. Evaluating on the other hand $\theta(y)$ via the local components, we get from (3.4)–(3.6), that

$$\theta_{\mathfrak{P}}(y) \cdot \theta_{\mathfrak{P}'}(y) \cdot \prod_{\mathfrak{q} \mid D} \left(\frac{y}{\mathfrak{q}}\right)_L \cdot \theta_\infty(y) = 1. \tag{3.9}$$

But one verifies easily from (3.8), that

$$\theta_{\mathfrak{P}}(y)\theta_{\mathfrak{P}'}(y) = \lambda(y^2/N_{L/\mathbb{Q}}y).$$

Hence we obtain as a necessary condition for the existence of a θ, that

$$\lambda(y^2/N_{L/\mathbb{Q}}y) \cdot \prod_{\mathfrak{q} \mid D} \left(\frac{y}{\mathfrak{q}}\right)_L \cdot \rho_\infty(y) = 1, \tag{3.10}$$

for some $\rho_\infty = \varepsilon_\infty$ or $\rho_\infty = \text{sign}_\infty$ and for all global units. This condition is also sufficient, and moreover if it holds, then θ is uniquely determined, with the given choice of $\theta_\infty = \rho_\infty$. Indeed, we define a character θ on the group $\mathfrak{U}(L)$ of unit ideles, by

$$\theta(z) = \lambda(z^2/(N_{L/\mathbb{Q}}z)_p) \cdot \prod_{\mathfrak{q} \mid D} \left(\frac{z_{\mathfrak{q}}}{\mathfrak{q}}\right)_L \cdot \rho_\infty(z_\infty).$$

By (3.10) this extends to a character θ of $\mathfrak{U}(L)L^*$, trivial in L^*; moreover as the class number of L is odd, it also extends to a character θ of $\mathfrak{J}(L)$. θ is then easily seen to be an idele class character of order 2^{s-1}, of quaternion type satisfying (3.6) and (3.7). Moreover at each step the choice is forced on us, and so θ is unique. Finally observe that $y = -1$ will always satisfy (3.10). Hence we conclude

B. A necessary and sufficient condition for the existence of an idele class character θ of L, of order 2^{s-1}, of quaternion type, satisfying (3.6), (3.8), *and with some given $\theta_\infty = \rho_\infty$, as in* (3.5), *is that* (3.10) *holds for a fundamental unit of L. Moreover if θ exists then it is, for a given choice of ρ_∞, uniquely determined.* □

In case II we have, for any fundamental unit y of L,

$$\begin{cases} \theta_\infty(y) = 1, & \text{with } \theta_\infty \text{ as in (3.5),} \\ \left(\dfrac{y}{\mathfrak{q}_1}\right)_L \left(\dfrac{y}{\mathfrak{q}_2}\right)_L = -1, & \text{where } \mathfrak{q}_i \mid q_i. \end{cases} \tag{3.11}$$

As $N_{L/\mathbb{Q}}y = 1$ in this case, the first equation is obvious. Also one verifies that the value of $(y/\mathfrak{q}_1)_L(y/\mathfrak{q}_2)_L$ is independent of the choice of y. We choose $y < 0$. Then one can show that

$$\mathbb{Q}(\sqrt{-q_1}, \sqrt{-q_2}) = \mathbb{Q}(\sqrt{q_1 q_2}, \sqrt{y}).$$

Expressing the condition, for $\mathfrak{q}_2$ to split in this field, in two different ways, we get

$$\left(\frac{y}{\mathfrak{q}_2}\right)_L = \left(\frac{-q_1}{q_2}\right)_{\mathbb{Q}},$$

and analogously on interchange of q_1 and q_2. The second equation in (3.11) is now seen to be a consequence of the quadratic reciprocity law in $\mathbb{Q}$.

From the first equation in (3.11) we conclude that, in case II, the validity of (3.10) is independent of the choice of infinite component as in (3.5). Thus, if there is an idele class character θ satisfying our conditions, then there are exactly two – one with infinite component ε_∞, the other with sign_∞. The field corresponding to the first is real, that corresponding to the second is imaginary. By the second equation in (3.11) we now have (recalling that $N_{L/\mathbb{Q}}y = 1$), as existence criterion the condition

$$\lambda^2(y) = -1, \tag{3.12}$$

for a fundamental unit y – the choice of y being clearly irrelevant. In case I, on the other hand $\text{sign}_\infty(y) = -1$. Thus (3.10) will hold with at most one choice of ρ_∞. This gives the uniqueness of $N[p]$. Moreover the left hand side of (3.10) is of form $\lambda^2(y)\mu(y)$, where $\mu(y) = \pm 1$; both values occur, depending on the choice of ρ_∞. Thus in case I, the existence criterion is

$$\lambda^2(y) = \pm 1. \tag{3.13}$$

Again, once such an equation holds for some fundamental unit, it holds with the same value for all.

Now assume in case I that (3.13), and in case II that (3.12) is satisfied. Clearly the factor $(-1)^r$ in the formula of Proposition 2.2 is $\text{sign}(N)$. Thus by this Proposition we have

$$W(N^+\{p\})/W(N^-\{p\}) = -1$$

in case II, and

$$\begin{cases} W(N[p]) = \text{sign}(N[p])\left(\dfrac{2}{D}\right)_{\mathbb{Q}} v_s(p) & \text{(case I)}, \\ W(N^+\{p\}) = \left(\dfrac{2}{D}\right)_{\mathbb{Q}} v_s(p) & \text{(case II)}. \end{cases} \tag{3.14}$$

Moreover, in case I, $N_{L/\mathbb{Q}}y = -1$ and, $\lambda(-1) = (-1)^{p-1/2^{s-1}} = v_s(p)$. Thus by (3.10)

$$\lambda^2(y) = \left(\frac{y}{\mathfrak{q}}\right)_L \text{sign}(N[p])v_s(p). \tag{3.15}$$

In case I define, for $u, v = \pm 1$, $\mathcal{M}'_{u,v}$ as the set of odd primes p satisfying (3.3), for which $N[p]$ exists and so that $\lambda^2(y) = u$, $v_s(p) = v$. The four sets $\mathcal{M}'_{u,v}$ coincide in some order with the four sets $\mathcal{M}_{a,b}$. In case II, define, for $v = \pm 1$, $\mathcal{L}'_v$ as the set of odd primes p, satisfying (3.3), for which $N^+\{p\}$ exists and $v_s(p) = v$. The two sets $\mathcal{L}'_v$ coincide in some order with the two sets $\mathcal{L}_a$. We shall prove that each of these sets has positive density, using Tchebotareff's density theorem.

For any normal field $N/\mathbb{Q}$, in which p is not ramified, we write $(N/\mathbb{Q}, p)$ for the conjugacy class in $\mathrm{Gal}(N/\mathbb{Q})$ of Frobenius symbols of prime divisors of p. We consider in particular the fields $L\mathbb{Q}(2^{s-1})$, $L\mathbb{Q}(2^s)$ and the field M obtained by adjoining the 2^{s-2}-nd root of y to $L\mathbb{Q}(2^{s-1})$. (Recall $\mathbb{Q}(m)$ is the field of m-th roots of unity). We have the following facts:

$$(3.16)\quad \begin{cases} \text{(a)}\ \ \mathrm{Gal}(M/L\mathbb{Q}(2^{s-1})) \text{ is cyclic of order } 2^{s-2} \text{ (i.e., } \geqslant 2), \\ \text{(b)}\ \ M \cap L\mathbb{Q}(2^s) = L\mathbb{Q}(2^{s-1}), \\ \text{(c)}\ \ \mathrm{Gal}(L\mathbb{Q}/2^s)/L\mathbb{Q}(2^{s-1})) \text{ is of order 2.} \end{cases}$$

We shall indicate a proof below.

The primes p satisfying (3.3) and (3.7) are exactly the primes which split in $L\mathbb{Q}(2^{s-1})$. Therefore, for these primes, we have

$$(3.17)\qquad \begin{cases} (L\mathbb{Q}(2^s)/\mathbb{Q}, p) \subset \mathrm{Gal}(L\mathbb{Q}(2^s)/L\mathbb{Q}(2^{s-1})), \\ (M/\mathbb{Q}, p) \subset \mathrm{Gal}(M/L\mathbb{Q}(2^{s-1})). \end{cases}$$

By (3.16a), the additional condition (3.12) in case II, and any one of the two conditions $\lambda^2(y) = u$ for fixed u ($= +1$ or $= -1$) in case I, corresponds to fixing $(M/\mathbb{Q}, p)$. Similarly, by (3.16c), fixing the value of $v_s(p)$ amounts to fixing $(L\mathbb{Q}(2^s)/\mathbb{Q}, p)$. By (3.17) and (3.16b) the two classes $(M/\mathbb{Q}, p)$ and $(L\mathbb{Q}(2^s)/\mathbb{Q}, p)$ we have obtained come from a class $(M\mathbb{Q}(2^s)/\mathbb{Q}, p)$ in $\mathrm{Gal}(M\mathbb{Q}(2^s)/\mathbb{Q})$. Thus indeed $\mathscr{M}'_{u,v}$ and $\mathscr{L}'_v$ have positive densities for all choices of the parameters.

We return to (3.16). Assertion (c) and the inclusion $L\mathbb{Q}(2^{s-1}) \subset M \cap L\mathbb{Q}(2^s)$ are obvious. Everything else follows from the fact that $\sqrt{y} \notin L\mathbb{Q}(2^s)$. In case I one notes that the field $L(\sqrt{y})$ is clearly not Galois over $\mathbb{Q}$, while $L\mathbb{Q}(2^s)$ is in fact Abelian over $\mathbb{Q}$. In case II, we may take $y < 0$; then $L(\sqrt{y}) = \mathbb{Q}(\sqrt{-q_1}, \sqrt{-q_2})$ is non-ramified over L at all finite prime divisors, while the prime divisors of L above 2 are totally ramified in $L\mathbb{Q}(2^s)$. □

There is another family of quaternion groups with only one $\Omega_{\mathbb{Q}}$-orbit of irreducible symplectic characters, namely the groups H_{4l}, l an odd prime. For these a density result similar to those given here, can be established (cf. [F7]).

§4. The Distribution Theorem

We shall now look at root numbers from the point of view of a *fixed* finite group Γ and its representations as a Galois group of varying tame Galois extensions N/K of number fields. We then ask: What are the possible value distributions of root numbers of symplectic characters of this given group? This question indeed only makes sense, if we do not view Γ as the Galois group of one, given extension, as we have done mostly hitherto, but, for fixed Γ, vary the extensions. To make this new approach rigorous requires a complete change in our formal set up, which in any case is of wider relevance. We shall first of all have to define exactly what we mean by a "representation of Γ as Galois group" and by the root number distribution associated with such a representation. This is not at all difficult of course. As to the question itself we shall get a complete answer for Γ a quaternion group, and this will allow us to draw inferences for arbitrary groups. In particular we shall see that if K is big enough in a cyclotomic sense, then all tame symplectic root numbers "over K" have value 1.

With fixed K, we consider tame surjections

(4.1) $$\pi: \Omega_K \to \Gamma.$$

Denote by N_π the fixed field of the open subgroup $\operatorname{Ker}\pi$ of Ω_K. To say that π is tame is to mean that N_π/K is tame. If $T: \Gamma \to \mathrm{GL}_n(\mathbb{Q}^c)$ is a representation with character χ, then $\chi \circ \pi$ is the character of Ω_K associated with the representation $T \circ \pi$. As such it has a Artin root number $W(\chi \circ \pi)$, which we shall view as a function of the pair χ, π, writing

(4.2) $$W_\pi(\chi) = W(\chi \circ \pi).$$

Clearly

(4.3) $$W_\pi(\chi + \chi') = W_\pi(\chi) W_\pi(\chi')$$

and indeed W_π extends to an additive map on R_Γ. If g is an automorphism of Γ, acting on the left, then, from the definition,

(4.4) $$W_{g \circ \pi}(\chi) = W_\pi(\chi \circ g).$$

The importance of this formula lies in the fact that $\operatorname{Ker}\pi = \operatorname{Ker}\pi'$ if, and only if, $\pi' = g \circ \pi$ for some automorphism g.

Next let $h: \Gamma \to \Sigma$ be a map onto a quotient group. Then, again from the definition,

(4.5) $$W_{h \circ \pi}(\chi) = W_\pi(\operatorname{inf}_\Sigma^\Gamma \chi), \qquad \text{for} \qquad \chi \in R_\Sigma.$$

Now let Δ be a subgroup of Γ. The restriction of π to the inverse image of Δ defines a surjection $\pi[\Delta]$: $\Omega_L \to \Delta$, for some $L \subset N_\pi$. By the inductivity of W we have

(4.6) $$W_\pi(\operatorname{ind}_\Delta^\Gamma \chi) = W_{\pi[\Delta]}(\chi), \qquad \text{for} \qquad \chi \in R_\Delta.$$

Now we restrict to symplectic characters. We get a homomorphism W_π: $R_\Gamma^s \to \pm 1$ and of course $W(\mathrm{Tr}(\chi)) = 1$. By Galois invariance of symplectic root numbers, we may thus view W_π as an element of the group $\mathrm{Hom}_{\Omega_\mathbb{Q}}(R_\Gamma^s/\mathrm{Tr}(R_\Gamma), \pm 1)$, which has been a central object of our theory. The basic question is then: Which elements of this group are of form W_π? Let $\mathfrak{W}(K, \Gamma)$ be the set of such homomorphisms W_π. Our aim is to determine $\mathfrak{W}(K, \Gamma)$. The importance of this, in the context of Galois module structure of rings of integers, comes from Theorem 9, which we shall now restate in our new language. Given a tame surjection $\pi: \Omega_K \to \Gamma$ we get an isomorphism $\Gamma \cong \mathrm{Gal}(N_\pi/K)$ which makes the ring $\mathfrak{o}_{N,\pi}$ of algebraic integers in N_π into a locally free $\mathbb{Z}\Gamma$-module. As such it determines a class $U_\pi \in \mathrm{Cl}(\mathbb{Z}\Gamma)$. Then we have the new version of Theorem 9:

(4.7) $$t_\Gamma W_\pi = U_\pi.$$

Thus $t_\Gamma \mathfrak{W}(K, \Gamma)$ is the set of classes in $\mathrm{Cl}(\mathbb{Z}\Gamma)$ realizable over K as classes $(\mathfrak{o}_{N,\pi})_{\mathbb{Z}\Gamma} = U_\pi$.

Let now

(4.8) $\Gamma = H_{4m}$, $4m = 2^s h$, h odd; so $s \geqslant 3$, or $s = 2$ and $h > 1$.

For each positive divisor g of h we consider the Adams operation Ψ_g on R_Γ; if $\eta \in R_\Gamma$ then $\Psi_g \eta$, as a function on Γ, is defined by $\Psi_g \eta(\gamma) = \eta(\gamma^g)$. One knows that $\Psi_g \eta \in R_\Gamma$. From now on let ψ be the character of a faithful irreducible representation of Γ. Then the $\Psi_g \psi$, for $1 \leqslant g \mid h$, form a complete set of representatives for the $\Omega_{\mathbb{Q}}$-orbits of quaternion characters of H_{4m}, with the proviso that if, in (4.8), $s = 2$ then $\Psi_h \psi$ is degenerate. The character $\Psi_h \psi$ will be seen to play a special role; this is due to the fact that, module $\Omega_{\mathbb{Q}}$-action, it is the unique quaternion character inflated from a quaternion 2-group – but allowing degeneracy. Now define, via the Möbius function μ, virtual characters

(4.9) $$\xi_d = \sum_{g \mid d} (\psi - \Psi_g \psi)\mu(d/g), \qquad \text{for} \qquad 1 < d \mid h.$$

In other words we introduce an operator

$$\sum_{g \mid d} (1 - \Psi_g)\mu(d/g)$$

on R_Γ. ξ_d is the image of ψ.

Let $R_\Gamma^s I_\Omega$ be the subgroup of R_Γ^s generated by characters $\eta^\omega - \eta$ for $\eta \in R_\Gamma^s$ and $\omega \in \Omega_{\mathbb{Q}}$; this is the image of R_Γ^s under the action of the augmentation ideal of $\mathbb{Z}(\Omega_{\mathbb{Q}})$. We shall in the sequel view $\mathrm{Hom}_{\Omega_{\mathbb{Q}}}(R_\Gamma^s/\mathrm{Tr}(R_\Gamma), \pm 1)$ simply as the group of homomorphisms

(4.10) $$f\colon R_\Gamma^s \to \pm 1, \qquad f(\mathrm{Tr}(R_\Gamma)) = f(R_\Gamma^s I_\Omega) = 1.$$

The role the ξ_d can play in this context is indicated by the next Proposition.

4.1. Proposition. *The ξ_d for $1 < d \mid h$, together with $\Psi_h \psi$ in the case $s \geqslant 3$, form a basis of $R_\Gamma^s \bmod \mathrm{Tr}(R_\Gamma) + R_\Gamma^s I_\Omega$.*

Proof. We get such a basis in the form $\Psi_g \psi$, $1 \leqslant g \mid h$, with $g < h$ if $s = 2$ and $g \leqslant h$ if $s \geqslant 3$. For, these form a complete set of representatives for the $\Omega_{\mathbb{Q}}$-orbits of irreducible symplectic characters. The basis property applies then also to the $\psi - \Psi_g \psi$ (all g with $1 < g \mid h$), together with $\Psi_h \psi$ when $s \geqslant 3$. Now apply Möbius inversion. □

Recall some notation: $\mathbb{Q}(g)$ is the field of g-th roots of unity, $\mathbb{Q}^+(g)$ its maximal real subfield.

Theorem 38 (cf. [F26]). *Let* $\Gamma = H_{4m}$.
(i) *Suppose* $f \in \mathfrak{W}(K, \Gamma)$. *Then for* $1 < g \mid h$

$$\text{(a) } f(\xi_g) = 1 \text{ whenever } \mathbb{Q}(g) \subset K\mathbb{Q}^+(g),$$

and in the case $s \geqslant 3$

$$\text{(b) } f(\Psi_h \psi) = 1 \text{ whenever } \mathbb{Q}(2^s) \subset K.$$

(ii) *Suppose that* $f \in \mathrm{Hom}_{\Omega_\mathbb{Q}}(R_\Gamma^s/\mathrm{Tr}(R_\Gamma), \pm 1)$ *satisfies* (a), (b). *Then* $f \in \mathfrak{W}(K, \Gamma)$. *Moreover, given any finite set* $\mathscr{V}$ *of finite prime divisors of* K, *there exist infinitely many tame surjections* $\pi: \Omega_K \to \Gamma$, *such that the prime divisors of* $\mathscr{V}$ *are non-ramified in* N_π *and* $W_\pi = f$. □

Corollary 1. *The following conditions are equivalent for* $\Gamma = H_{4m}$.
(i) $\mathfrak{W}(K, \Gamma) = 1$.
(ii) *For all* g, $1 < g \mid h$, $\mathbb{Q}(g) \subset K\mathbb{Q}^+(g)$, *and in the case* $s \geqslant 3$ *also* $\mathbb{Q}(2^s) \subset K$.
(iii) *For all odd primes* l *dividing* h, $\mathbb{Q}(l) \subset K\mathbb{Q}^+(l)$, *and in case* $s \geqslant 3$ *also* $\mathbb{Q}(2^s) \subset K$.

Proof. By the theorem and Proposition 4.1, (i)⇔(ii). Clearly (ii)⇒(iii). But if $\mathbb{Q}(l) \subset K\mathbb{Q}^+(l)$ then $\mathbb{Q}(g) \subset K\mathbb{Q}^+(g)$ for any multiple g of l. Thus (iii)⇒(ii). □

Corollary 2. $\mathfrak{W}(K, H_{4m})$ *and* $t\mathfrak{W}(K, H_{4m})$ *are groups.* □

Next we consider an arbitrary finite group Γ together with its quaternion subquotients Δ/Σ; here Δ is a subgroup of Γ, Σ a normal subgroup of Δ with Δ/Σ quaternion. We define $s(\Gamma) = s \geqslant 3$, if 2^s is the maximal order of quaternion 2-groups which are subquotients of Γ; if there are no such subquotients we put $s(\Gamma) = 0$.

Corollary 3. *Let* Γ *be a finite group. Suppose that for every odd prime* l *which divides the order of some quaternion subquotient of* Γ, *we have* $\mathbb{Q}(l) \subset K\mathbb{Q}^+(l)$, *and that furthermore* $\mathbb{Q}(2^s) \subset K$ $(s = s(\Gamma))$. *Then* $\mathfrak{W}(K, \Gamma) = 1$. *Moreover for every tame surjection* $\pi: \Omega_K \to \Gamma$, $\mathfrak{o}_{N,\pi}$ *is a free* $\mathbb{Z}\Gamma$*-module.*

Proof. By Serre's induction theorem (cf. I note [1]), by the inductivity of root numbers (cf. (4.6)) and by Corollary 1, certainly $W_\pi = 1$. By (4.7), $U_\pi = 1$, i.e., $\mathfrak{o}_{N,\pi}$ is stably free over $\mathbb{Z}\Gamma$. If $K \neq \mathbb{Q}$, then $\mathfrak{o}_{N,\pi}$ is of rank > 1 over $\mathbb{Z}\Gamma$, and stably free modules of rank > 1 are free. If $K = \mathbb{Q}$, the hypothesis implies that Γ has no irreducible symplectic characters, and for such a Γ, stably free $\mathbb{Z}\Gamma$-modules are free. □

A weak, but easy to state consequence of the last Corollary is:

Corollary 4. *If* N/K *is a tame Galois extension, and* K *contains the* $[N:K]$*-th roots of unity, then* $\mathfrak{o}_N$ *is a free* $\mathbb{Z}\Gamma$*-module,* $(\Gamma \cong \mathrm{Gal}(N/K))$. □

Corollaries in the opposite direction are restricted to quaternion groups.

Corollary 5. *Let $\Gamma = H_{4m}$. Suppose that for all $g \mid h$, $\mathbb{Q}(g) \not\subset K\mathbb{Q}^+(g)$, and also in case $s \geqslant 3$, $\mathbb{Q}(2^s) \not\subset K$. Then*

$$\mathfrak{W}(K, \Gamma) = \mathrm{Hom}_{\Omega_\mathbb{Q}}(R_\Gamma^s/\mathrm{Tr}(R_\Gamma), \pm 1).$$

This is in particular the case if K is formally real. □

Remark. In the case of a quaternion group Γ, the action of its automorphisms leave the $\Omega_\mathbb{Q}$-orbits of irreducible symplectic characters fixed. Thus in this case W_π only depends on $\mathrm{Ker}\, \pi$. This is however not so in general, e.g., for $\Gamma = H_8 \times H_8$.

The questions which arise in this context concern the full extension of our results to arbitrary finite groups Γ. In particular

Question 1. Is $\mathfrak{W}(K, \Gamma)$ (or at least $t_\Gamma \mathfrak{W}(K, \Gamma)$) always a group?

We have seen that this is the case for Γ quaternion, and trivially for any group Γ which has no irreducible symplectic representations. By Corollary 3 it is also true for K "big enough". □

Question 2. If K is small enough, e.g., $K = \mathbb{Q}$, or better K just formally real, is $\mathfrak{W}(K, \Gamma) = \mathrm{Hom}_{\Omega_\mathbb{Q}}(R_\Gamma^s/\mathrm{Tr}(R_\Gamma), \pm 1)$? □

For the proof of the theorem we translate everything into the language of characters of quaternion type, used in §1 and §2. We have to consider pairs (L, θ), where L is a tame quadratic extension of K, and θ a tame idele class character of L, of quaternion type, of order $2m = 2^{s-1}h$. Write

(4.11) $$\theta = \theta_\zeta, \qquad \mathrm{ind}_L^K \zeta = \psi \circ \pi,$$

where $\pi\colon \Omega_K \to H_{4m}$, and ψ is faithful as above. In this way every pair (L, θ) yields a pair (ψ, π) modulo automorphisms of Γ and $\Omega_\mathbb{Q}$-action, and all pairs (ψ, π) are obtained in this way, each at most a finite number of times. Moreover from (4.11) we have, for $g \mid h$,

(4.11a) $$\theta^g = \theta_{\zeta^g}, \qquad \mathrm{ind}_L^K \zeta^g = \Psi_g \psi \circ \pi.$$

Let, for $g \mid h$,

(4.12) $$\mathfrak{g}_g(\theta) = \prod_{e(\theta_\mathfrak{p}) = g} \mathfrak{p}$$

(product over all prime divisors of K, for which $e(\theta_\mathfrak{p}) = g$ – see §1). Then we have, given (4.11),

4.2. Lemma. (i) $W_\pi(\Psi_h \psi) = W(\theta^h)$,
(ii) $W(\xi_g) = \phi_{L/K}(\mathfrak{g}_g(\theta))$.

(Note that by Proposition 1.3, the $\mathfrak{p}$ with $e(\theta_\mathfrak{p}) \mid h$ are non-ramified in L, whence $\phi_{L/K}(\mathfrak{g}_g(\theta))$ is defined.)

Proof. We already know (cf. §2), that

$$W(\theta^g) = W_\pi(\Psi_g \psi). \tag{4.13}$$

The case $g = h$ gives (i). Moreover by Proposition 2.1,

$$W(\theta)/W(\theta^d) = \prod_{e(\theta_\mathfrak{p})\,|\,d} \phi_{L/K}(\mathfrak{p}), \qquad \text{for } d\,|\,h.$$

In other words

$$W(\theta)/W(\theta^d) = \prod_{g\,|\,d} \phi_{L/K}(\mathfrak{g}_g(\theta)).$$

Now apply Möbius inversion and (4.13), to get (ii). □

Proof of Theorem 38 (i). First assume that $s \geqslant 3$, and that $\mathbb{Q}(2^s) \subset K$. By the last Lemma we have to show that $W(\theta^h) = 1$. As θ^h has order 2^{s-1}, we can apply the formula of Proposition 2.2. We shall show that each of the factors which appear in that formula must be $= 1$. Firstly $\sqrt{2} \in K$, whence $(2/\mathfrak{d})_K = 1$. K is totally imaginary, whence $(-1)^{r(\theta)} = 1$. Also $\sqrt{-1} \in K$, and so $(-1/\mathbf{N}_K\mathfrak{p})_\mathbb{Q} = 1$, for all odd prime divisors $\mathfrak{p}$ of K. This leaves us with factors of the form $v_t(\mathbf{N}_K\mathfrak{p})\phi_{L/K}(\mathfrak{p})$, where $e(\theta^h_\mathfrak{p}) = 2^{t-1}$, and $3 \leqslant t \leqslant s$. As K contains the primitive 2^t-th roots of unity, we must have $\mathbf{N}_K\mathfrak{p} \equiv 1 \pmod{2^t}$. But this implies firstly that $v_t(\mathbf{N}_K\mathfrak{p}) = 1$, and secondly, because of the congruence (1.6), that $\phi_{L/K}(\mathfrak{p}) = 1$.

Next assume that $\mathbb{Q}(g) \subset K\mathbb{Q}^+(g)$ for the divisor g of h, $g > 1$. We shall show that if $\mathfrak{p}\,|\,\mathfrak{g}_g(\theta)$, i.e., if $e(\theta_\mathfrak{p}) = g$, then $\phi_{L/K}(\mathfrak{p}) = 1$. In the first place, by the congruence (1.6), certainly $\mathbf{N}_K\mathfrak{p} \equiv \pm 1 \pmod{g}$. By cyclotomic classfield theory this implies that $\mathfrak{p}$ splits completely in $K\mathbb{Q}^+(g)$. But, by hypothesis, $K\mathbb{Q}^+(g) = K\mathbb{Q}(g)$, and the complete splitting of $\mathfrak{p}$ in $K\mathbb{Q}(g)$ implies in turn that $\mathbf{N}_K\mathfrak{p} \equiv 1 \pmod{g}$. By the congruence (1.6), $\phi_{L/K}(\mathfrak{p}) = 1$. □

Proof of Theorem 38 (ii). Here we shall only give the essential steps. Suppose again that L/K is a tame quadratic extension. The tame idele class characters θ of L, of quaternion type and of order $2^{s-1}h$ are precisely the characters

$$\begin{cases} \theta = \theta'\beta, & \theta' \text{ and } \beta \text{ tame idele class characters,} \\ \theta' \text{ of quaternion type of order } 2^{s-1}, & \beta \text{ of dihedral type of order } h. \end{cases} \tag{4.14}$$

Our first aim is to show that for any given pair (L, θ'), with θ' as in (4.14), with L satisfying Lemma 1.5 (ii) for $n = h$, and for any $f \in \operatorname{Hom}_{\Omega_\mathbb{Q}}(R^s_\Gamma/\operatorname{Tr}(R_\Gamma), \pm 1)$ $(\Gamma = H_{4m})$, which satisfies condition (i) (a) in Theorem 38, we can find a β in infinitely many ways so that

$$\phi_{L/K}(\mathfrak{g}_g(\theta'\beta)) = f(\xi_g),$$

and hence, by Lemma 4.2 and (4.11a),

$$W_\pi(\xi_g) = f(\xi_g).$$

For this we shall use the sets $\mathscr{T}_{n,e}$ of Lemma 1.5. Let $\mathscr{T}(f)$ be the subset of the product $\prod \mathscr{T}_{g,f(\xi_g)}$ $(1 < g \mid h)$ of "vectors" $\{\mathfrak{p}_g\}$ without repetition, i.e., so that $\mathfrak{p}_g \in \mathscr{T}_{g,f(\xi_g)}$ and $\mathfrak{p}_g \neq \mathfrak{p}_d$, if $g \neq d$. For each element of $\mathscr{T}(f)$ and each g, there is an idele class character $\beta[\mathfrak{p}_g]$ of order g of dihedral type, with $e(\beta[\mathfrak{p}_g]_{\mathfrak{p}_g}) = g$, and $\beta[\mathfrak{p}_g]$ non-ramified everywhere else. Let $\beta = \prod \beta[\mathfrak{p}_g]$ $(1 < g \mid h)$. Then indeed always $\phi_{L/K}(\mathfrak{g}_g(\theta'\beta)) = \phi_{L/K}(\mathfrak{p}_g) = f(\xi_g)$ as required.

We have not mentioned the set $\mathscr{V}$, i.e., we have given a proof for the case that $\mathscr{V}$ is empty. It is quite clear, however, that one can exclude any prime divisors of $\mathscr{V}$ from the chosen vectors in $\mathscr{T}(f)$ without affecting our result. Similar remarks apply in the remainder of the proof, where we shall continue to neglect the set $\mathscr{V}$.

From now on we may suppose that $\Gamma = H_{2^s}$, but we have to include the degenerate case $s = 2$. In this latter case, we let q be any rational prime $\equiv 1 \pmod 4$ which splits completely in K. There is a unique extension L' of $\mathbb{Q}$ cyclic of degree 4, in which only q is ramified. Now take $L = K(\sqrt{q})$, and θ the quadratic idele class character of L corresponding to the quadratic extension $L'K$ of L. Then indeed $L'K/K$ is cyclic of order 4, i.e., θ is degenerate quaternion. Now we apply the preceding construction to get characters $\theta\beta$ of quaternion type, of order $2h$ with prescribed root number distribution.

Next we consider the non-degenerate case $s \geqslant 3$. Let $2^{t(K)-1}$ be the order of the 2-primary part of the group of roots of unity in K. If $t(K) \geqslant 3$, we choose for L any field

$$(4.15) \qquad \begin{cases} L = K(\sqrt{q}), & q \text{ a rational prime splitting in } K, \\ \text{and} \quad q \equiv 1 + 2^{t-1} & (\text{mod } 2^t), \quad \text{where} \quad t = t(K). \end{cases}$$

If $t(K) = 2$, we choose for L any field

$$(4.16) \qquad \begin{cases} L = K(\sqrt{q_1 q_2}), & q_1, q_2 \text{ rational primes} \\ \text{splitting in } K, & q_1 \equiv q_2 \equiv -1 \pmod 4. \end{cases}$$

To obtain characters of quaternion type of order 2^{s-1}, we first apply Propositions 3.1, or 3.2, to get a prime p, such that, in the respective cases (4.15), or (4.16), the appropriate field $N = N[p]$, or $N = N\{p\}$ respectively, exists – of course with $\mathrm{Gal}(N/\mathbb{Q}) \cong H_{2^s}$. If we also assume that p splits in K, then we get indeed a cyclic extension NK/L of degree 2^{s-1}, so that any associated idele class character θ of L of order 2^{s-1} is of quaternion type. Note, by the way, that we have infinitely many choices for L, and then for p – finitely many of these may have subsequently to be excluded.

The final step in the proof is then to show, for given θ and with $s \geqslant t(K)$, that there are infinitely many idele class characters α of L of dihedral type, of order $2^{t(K)-1}$ with the following two properties: (a) $\theta\alpha$ is still of order 2^{s-1} – it certainly will be of quaternion type, and (b) $W(\theta\alpha)/W(\theta) = -1$.

To obtain this result one needs a further density result, like that in Lemma 1.5, but rather more sophisticated. One looks at finite prime divisors $\mathfrak{s}$ of K with the following two properties.

(a) There is an idele class character α of L of order 2^{t-1}, of dihedral type, with $e(\alpha_{\mathfrak{s}}) = 2^{t-1}$, and α non-ramified everywhere else.

(b) In case of L of type (4.15): $\phi_{L/K}(\mathfrak{s}) = 1$, $v_t(\mathbf{N}_K\mathfrak{s}) = -1$; in case of L of type (4.16) ($t(K) = 2$): $(-1/\mathbf{N}_K\mathfrak{s})_{\mathbb{Q}} = -1$. One then establishes that the set of such prime divisors has positive density, excepting finitely many choices for L and p. Now one chooses an $\mathfrak{s}$ for which $e(\theta_{\mathfrak{s}}) = 1$. Then indeed $\theta\alpha$ has the required properties. □

VI. Relative Structure

§1. The Background

In this chapter we continue to look at tame Galois extensions N/K of number fields, and to study the module structure of $\mathfrak{o}_N$ over the Galois group Γ. But now we are interested in $\mathfrak{o}_N$, not as a $\mathbb{Z}\Gamma$-module, but as an $\mathfrak{o}\Gamma$-module, where $\mathfrak{o}$ is the ring of integers in an intermediary field – usually $\mathfrak{o} = \mathfrak{o}_K$. We shall refer to this as *relative Galois module structure.*

Our general theory, developed in the preceding chapters provides us with two basic pieces of information: Firstly of course, by Theorem 9 and Theorem 13 we have determined $\mathcal{N}_{K/\mathbb{Q}}((\mathfrak{o}_N)_{\mathfrak{o}_K\Gamma})$, which is exactly $tW_{N/K}$. Secondly by Theorems 7 and 8 we have a description of $(\mathfrak{o}_N)_{\mathfrak{o}_K\Gamma} \cdot (\hat{\mathfrak{o}}_N)^{-1}_{\mathfrak{o}_K\Gamma}$, essentially in terms of conductors. In general, this does not take us very far, and indeed one cannot expect such precise results on relative structure, as one could over $\mathbb{Z}\Gamma$. Thus $\mathfrak{o}_N$ need not even be free over $\mathfrak{o}_K$. Moreover at present there is no arithmetical invariant which could take over the role that the Galois Gauss sums are playing over $\mathbb{Q}$ – although there may possibly be some parallel object for a theory over imaginary quadratic fields. Of course the stated conductor interpretation allows us to show that $\mathfrak{o}_N$ may be far removed from being free over $\mathfrak{o}_K\Gamma$.

The type of question one would look at is: Given K and Γ what are the classes in $\mathrm{Cl}(\mathfrak{o}_K\Gamma)$ which are realizable in the form $(\mathfrak{o}_N)_{\mathfrak{o}_K\Gamma}$ for $\mathrm{Gal}(N/K) \cong \Gamma$, N tame – possibly imposing further restrictions on the allowable extensions N. There are of course always sufficient conditions for $\mathfrak{o}_N$ to be a free $\mathfrak{o}_K\Gamma$-module: If N' is a tame Galois extension of $\mathbb{Q}$ with Galois group $\cong \Gamma$, and $\mathfrak{o}_{N'}$ is free over $\mathbb{Z}\Gamma$, and if furthermore the discriminants over $\mathbb{Q}$ of N' and of K are coprime then the composite field $N'K$ has the structure of the tensor product $N' \otimes_{\mathbb{Q}} K$, and its ring of integers the structure of the tensor product $\mathfrak{o}_{N'} \otimes_{\mathbb{Z}} \mathfrak{o}_K$. It follows that $\mathrm{Gal}(N'K/K) \cong \Gamma$ and that $\mathfrak{o}_{N'K}$ is free over $\mathfrak{o}_K\Gamma$, by "transport of structure". What one wants are of course somewhat less trivial conditions!

There are two aspects, where work of some generality has been done and where one can now hope for further interesting developments. The first connects the problem of relative Galois module structure with the classical embedding problem of algebraic number theory; we shall give a survey of this in §2 and §3. The second deals with certain Abelian extensions and has produced significant connections with the Galois module structure of class groups – in other words with the Stickelberger relations, and beyond that also with other features of Iwasawa theory. A brief account of this will form the contents of §4 and §5.

We shall conclude this section by describing briefly some work and some problems dealing with special situations.

(a) *Module and root number ratios*

This (cf. [U3]) is the only known result on relative structure which involves a root number criterion. We consider the quaternion 2-group H_{2^s} $(s \geqslant 3)$ and its quotient $D_{2^{s-1}}$ modulo its centre. For $s > 3$, $D_{2^{s-1}}$ is the dihedral group and for $s = 3$ it is the Vierer group. We say that two tame representations (cf. (V.4.1)) π, π': $\Omega_K \to H_{2^s}$ are *related* if the composita $\Omega_K \to H_{2^s} \to D_{2^{s-1}}$ coincide. As H_{2^s} has only one quaternion character ψ, modulo $\Omega_{\mathbb{Q}}$-action, we can simply write

$$W_\pi = W_\pi(\psi) \qquad (= W(\psi \circ \pi))$$

(cf. (V.4.2)).

If π and π' are related then the fixed subfield N_π and $N_{\pi'}$ of $\operatorname{Ker}\pi$ and $\operatorname{Ker}\pi'$, respectively, are both cyclic of degree 2^{s-1} over the same quadratic extension L of K. We indicate the situation in a diagram giving the degrees:

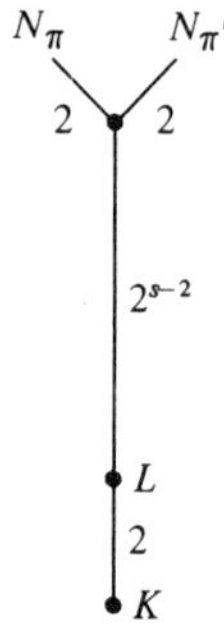

π and π' then give rise to surjections $\pi_1, \pi_1' : \Omega_L \to \Sigma$, where here Σ is the cyclic group of order 2^{s-1}. Via π_1, π_1' the rings of integers $\mathfrak{o}_{N,\pi}, \mathfrak{o}_{N,\pi'}$ become locally free $\mathfrak{o}_L\Sigma$-modules. We write V_π, $V_{\pi'}$ for their classes in $\mathrm{Cl}(\mathfrak{o}_L\Sigma)$. We then want to compare $V_\pi V_{\pi'}^{-1}$ with $W_\pi W_{\pi'}^{-1}$. Under certain further hypotheses this becomes possible. The main condition is:

A. There exist tame quadratic extensions $L_i = K(\sqrt{D_i})$ $(i = 1, 2)$ of K, so that for the relative discriminants we have

$$\mathfrak{d}(L/K) = \mathfrak{d}(L_1/K)\mathfrak{d}(L_2/K),$$

and so that for all prime divisors $\mathfrak{p}$ of K

$$\left(\frac{D_1, D_2}{\mathfrak{p}}\right)\left(\frac{-1, D_1}{\mathfrak{p}}\right)\left(\frac{-1, D_2}{\mathfrak{p}}\right) = 1.$$

Now we can state the result

Theorem 39 (cf. [U3], [U4]). *Suppose that A holds, that 2 is non-ramified in $L/\mathbb{Q}$, and that the map $\mathrm{Cl}(\mathfrak{o}_L) \to \mathrm{Cl}(\mathfrak{o}_M)$ is injective, where $M = L(2^{s-1})$. Then there* *is a subgroup Y of* $\mathrm{Cl}(\mathfrak{o}_L\Sigma)$, *containing all ratios* $V_\pi V_{\pi'}^{-1}$, *whenever* π, π' *are related, with* $N_\pi, N_{\pi'}$ *cyclic over* L, *and further there is a homomorphism* $l\colon Y \to \pm 1$ *such that* $l(V_\pi V_{\pi'}^{-1}) = W_\pi W_{\pi'}^{-1}$. □

Remark. It is easy to find examples, when

$$\mathrm{Cl}(\mathfrak{o}_L) \to \mathrm{Cl}(\mathfrak{o}_M)$$

is injective.

(b) *Norm resolvent – Gauss sum ratio for cyclotomic fields*

We shall start with a general observation, and will end up with the well known product representation of the classical Gauss sum – modulo a sign however.

Let N/K be a tame Galois extension of number fields with Galois group Γ and let a be an algebraic integer, generating a normal basis of N. Then we clearly have divisibility relations $P(\chi) \mid (a \mid \chi)$ and $\mathscr{N}_{K/\mathbb{Q}}P(\chi) \mid \mathscr{N}_{K/\mathbb{Q}}(a \mid \chi)$ for all χ (see III.1.5, 1.6 for the definitions). Hence we also have the

Criterion. $\mathscr{N}_{K/\mathbb{Q}}(a \mid \chi)/\tau(\chi)$ *is always an algebraic integer in* $\mathbb{Q}(\chi)$. *It is a unit at a prime* q, *for all* χ, *if, and only if,* $a\mathfrak{o}_{K,q}\Gamma = \mathfrak{o}_{N,q}$. *It is a global unit, for all* χ, *if and only if* $a\mathfrak{o}_K\Gamma = \mathfrak{o}_N$.

Now let p be an odd prime, l a prime with $p \equiv 1 \pmod{l}$. (We could make things more general, but this example will suffice to give the picture). We take $N = \mathbb{Q}(p)$ and K its subfield with $[N:K] = l$. We moreover take $a = y_p$, a primitive p-th root of unity, and χ an Abelian character of $\mathrm{Gal}(N/K)$ of order l. χ can also be viewed as a character of the group $\mu_l(p)$ of l-th roots of unity in $\mathbb{F}_p$, and the resolvent is then given by

$$(y_p \mid \chi) = \sum_{r \in \mu_l(p)} y_p^r \chi(r).$$

Any particular right transversal of Ω_K in $\Omega_{\mathbb{Q}}$ is given by a set of representatives $\{s\}$ of $\mathbb{F}_p^*$ mod $\mu_l(p)$. For the norm resolvent we then get

$$\mathscr{N}_{K/\mathbb{Q}}(y_p \mid \chi) = \prod_s \sum_r y_p^{rs} \chi(r).$$

We have an isomorphism $\mathbb{Z}/(p) \cong \mathfrak{o}_K/\mathfrak{p}$ of residue class fields, where $\mathfrak{p}$ is the unique prime divisor of K above p. Therefore modulo on l-th root of unity, $\tau(\chi) \equiv G_\chi$, the classical Gauss sum of the residue class character mod p, given by χ. Thus we now have as a special case from our criterion:

$\prod_s \sum_r y_p^{rs}\chi(r)/G_\chi$ *is an algebraic integer in* $\mathbb{Q}(\chi)$. *It is a unit precisely at all* q *at which* y_p *generates a local normal integral basis of* $\mathbb{Q}(p)$ *over* K. *In particular it is a unit at* p.

The aim is then to study the precise arithmetic nature of these quotients and their relationship to other arithmetic invariants.

Example. $l = 2$. Clearly y_p does generate a normal integral basis of $\mathbb{Q}(p)$ over $\mathbb{Q}^+(p)$, its maximal real subfield. Therefore indeed

$$\sum \left(\frac{x}{p}\right) e^{2\pi i x/p} = \pm \prod_{\substack{s \bmod \pm 1 \\ \bmod p}} (e^{2\pi i s/p} - e^{-2\pi i s/p}).$$

(c) *Complex Multiplication*

This will be left to the imagination of the reader. Compare with the situation in cyclotomic fields!

§2. Galois Module Structure and the Embedding Problem

The contents of this and the next section comes from J. Brinkhuis's doctoral thesis (cf. [Bri]). One knows a certain classical necessary condition for the solution of the embedding problem which will here emerge quite trivially from some easy observations about Galois module structure of fields. Going over to rings of integers instead, a problem arises which generalizes the embedding problem, connecting it in a natural way with the problem of Galois module structure. We shall only give a brief survey of some aspects of this developing theory, and in the next section describe applications to a concrete arithmetic situation.

Let S be a commutative ring, Σ a finite group acting as a group on S by ring automorphisms, Δ a finite group (not at this stage necessarily Abelian) and Γ a group extension of Δ by Σ (i.e., Δ is the normal subgroup, and $\Sigma \cong \Gamma/\Delta$). We let Γ act on the group rings $S\Delta$, by acting on S via the surjection $\Gamma \to \Sigma$ and the action of Σ, and by acting simultaneously on Δ by conjugation. An $S\Delta$-Γ-module X is a $S\Delta$-module with the additional structure of Γ-module, satisfying two conditions: (i) The action of Δ on X via Γ is the same as the action via $S\Delta$. (ii) Writing the action of Γ exponentially, we have

$$(xa)^\gamma = x^\gamma a^\gamma, \qquad \text{for} \quad x \in X, \quad a \in S\Delta, \quad \gamma \in \Gamma.$$

We shall now connect these module theoretic concepts with (non-Abelian) cohomology. Note that the map $c\colon \Delta \to S\Delta^*$ with $c(\delta) = \delta^{-1}$ is a 1-cocycle, whose cohomology class we denote by $c_{\Delta,S}$.

2.1. Proposition. *There exists a $S\Delta$-Γ-module which is free of rank one over $S\Delta$ if, and only if, $c_{\Delta,S}$ lies in the image of the restriction map*

$$\mathrm{res}_\Delta^\Gamma\colon H^1(\Gamma, S\Delta^*) \to H^1(\Delta, S\Delta^*).$$

Proof. If X is an $S\Delta$-Γ-module, free over $S\Delta$ on a generator x_0, then for each $\gamma \in \Gamma$, $\exists a(\gamma) \in S\Delta^*$, with $(x_0^\gamma)a(\gamma) = x_0$. One verifies that a is a 1-cocycle $\Gamma \to S\Delta^*$, and for all $\delta \in \Delta$, $a(\delta) = \delta^{-1}$. Conversely given such a 1-cocycle and an $S\Delta$-module X, free on a generator x_0, then the rule $(x_0 a)^\gamma = x_0 a(\gamma)^{-1} a^\gamma$ (for all $a \in S\Delta$) makes X into a $S\Delta$-Γ-module. □

We now turn to the embedding problem, with Γ, Δ and Σ as above. Let moreover L/K be a given Galois extension of fields of finite degree, with Galois group Σ. A solution of the "embedding problem $[L/K, \Gamma]$" is a Galois extension N of K, containing L, together with a commutative diagram with exact rows and columns

$$\begin{array}{ccccccccc} & & & & & & 1 & & \\ & & & & & & \downarrow & & \\ 1 & \to & \Omega_L & \to & \Omega_K & \to & \Sigma & \to & 1 \\ & & \downarrow{\scriptstyle \pi} & & \downarrow{\scriptstyle \pi'} & & \downarrow{\scriptstyle 1} & & \\ 1 & \to & \Delta & \to & \Gamma & \to & \Sigma & \to & 1 \\ & & \downarrow & & \downarrow & & \downarrow & & \\ & & 1 & & 1 & & 1 & & \end{array} \tag{2.1}$$

with $\Omega_N = \operatorname{Ker} \pi = \operatorname{Ker} \pi'$. By abuse of language we shall often speak of N as a solution.

We have immediately

Corollary 1 (*to Proposition 2.1*). *If the embedding problem for $[L/K, \Gamma]$ has a solution, then*

$$c_{\Delta,L} \in \operatorname{Im} \operatorname{res}_{\Delta}^{\Gamma} \colon H^1(\Gamma, L\Delta^*) \to H^1(\Delta, L\Delta^*).$$

For, if N is a solution, then N is a $L\Delta$-Γ-module, free of rank one over $L\Delta$. □

The condition given in Corollary 1 was first established in [Wo], by a rather lengthier argument.

It is now already obvious how one can formulate a similar condition for rings of algebraic integers.

From now on we take L, K etc. to be algebraic number fields.

Corollary 2. *Suppose that the embedding problem $[L/K, \Gamma]$ has a solution, given by a field N, and so that $\mathfrak{o}_N$ is free as an $\mathfrak{o}_L\Delta$-module. Then*

$$c_{\Delta,\mathfrak{o}_L} \in \operatorname{Im} \operatorname{res}_{\Delta}^{\Gamma} \colon H^1(\Gamma, \mathfrak{o}_L\Delta^*) \to H^1(\Delta, \mathfrak{o}_L\Delta^*). \qquad \square$$

Remark. Using Noether's criterion (cf. Theorem 3) one can obtain a result, similar to Corollary 1, for the tame embedding problem of number fields, but we shall not go into this here.

From now on we assume Δ to the Abelian. Then Δ and $S\Delta^*$ become Σ-modules (via the action of Σ on both S and on Δ), and the sequence of groups

$$H^1(\Gamma, S\Delta^*) \xrightarrow{\text{res}} H^1(\Delta, S\Delta^*)^{\Sigma} \xrightarrow{d_e} H^2(\Sigma, S\Delta^*) \tag{2.2}$$

is exact, where $e \in H^2(\Sigma, \Delta)$ corresponds to the group extension Γ, and d_e is the transgression coming from the Hochschild-Serre spectral sequence given by Γ. Moreover the class $c_{\Delta,S}$ is clearly fixed under the action of Σ. We thus get a well known reformulation of the criterion of Corollary 1, for Abelian Δ. We state the analogous consequence of Corollary 2.

Corollary 3. *Under the hypothesis of Corollary 2,*

$$d_e(c_{\Delta,\mathfrak{o}_L}) = 1. \qquad \square$$

(We write the group operation multiplicatively.)

Next one can show that – at least to within sign –

$$j_{S,\Delta}(e) = d_e(c_{\Delta,S}),$$

where $j_{S,\Delta}\colon H^2(\Sigma,\Delta) \to H^2(\Sigma, S\Delta^*)$ is induced from $\Delta \subset S\Delta^*$. Hence we get

Corollary 4. *Under the hypothesis of Corollary 2,*

$$j_{\mathfrak{o}_L,\Delta}(e) = 1. \qquad \square$$

The question is now how far this criterion is also sufficient. For a proper treatment we need however a more general approach, which should also give us some insight into all the possible structures of $\mathfrak{o}_N$ as an $\mathfrak{o}_L\Delta$-module, or as an $\mathfrak{o}_L\Delta$-Γ-module in the context of tame solutions N of the embedding problem. From now on L/K is always a tame extension of number fields with Galois group Σ. We shall now only fix the Σ-module Δ, but not any particular group extension of Δ by Σ. Instead we shall consider all such group extensions, and for each of these all the potential Galois module structures for rings of integers in tame solutions. Because of the restriction to tame ramification, we shall also replace the groups Ω_L and Ω_K in (2.1) by the Galois groups $\tilde{\Omega}_L = \mathrm{Gal}(\tilde{L}/L)$, and $\tilde{\Omega}_K = \mathrm{Gal}(\tilde{L}/K)$, respectively, where $\tilde{L}$ is the maximal tame Abelian extension of L. Thus, in (2.1), π and π' are now tame representations, in the sense of V §4. We moreover shall have to extend the type of diagrams (2.1) considered. The map π in (2.1) is a homomorphism $\tilde{\Omega}_L \to \Delta$ of Σ-modules, i.e., $\pi \in \mathrm{Hom}_\Sigma(\tilde{\Omega}_L, \Delta) = H^1(\tilde{\Omega}_L, \Delta)^\Sigma$. As we wish to work with this group, we shall have, for the time being, to drop the further hypothesis in (2.1), namely that π is surjective, i.e., we have to allow what are known as *improper* solutions of the embedding problem. We shall later see that we shall actually gain information on *proper* solutions, i.e, solutions with π surjective.

We associate with any element $\pi \in \mathrm{Hom}_\Sigma(\tilde{\Omega}_L, \Delta)$ the L-algebra

$$\mathrm{Map}_{\pi,\tilde{\Omega}_L}(\Delta, \tilde{L}) = A_\pi. \tag{2.3}$$

This is the set of maps $f\colon \Delta \to \tilde{L}$, with the property that $f(\delta \cdot \pi(\omega)) = f(\delta)^\omega$. It is a ring, with addition and multiplication defined pointwise, i.e., by

$$(f + g)(\delta) = f(\delta) + g(\delta).$$

This makes it into an L-algebra. Moreover Δ acts on A_π by algebra automorphisms:

$$f^\delta(\delta_1) = f(\delta\delta_1),$$

and one can show that with respect to this action $A_\pi \cong L\Delta$. (A_π is a "Galois algebra".) Its algebraic structure is that of a product of fields, each isomorphic over L to the fixed field N'_π of $\operatorname{Ker}\pi \subset \tilde{\Omega}_L$. If in particular π is surjective then the map $f \mapsto f(1)$ actually is an isomorphism $A_\pi \cong N'_\pi$, i.e., A_π is – to within isomorphism – the field $N = N'_\pi$, giving the associated proper solution. Conversely if A_π is a field, then π is surjective.

As $L\Delta$ is commutative, the tensor product $A_{\pi_1} \otimes_{L\Delta} A_{\pi_2}$ still has the structure of an L-algebra and an $L\Delta$-module, and one can show that with respect to both structures

(2.4) $$A_{\pi_1} \otimes_{L\Delta} A_{\pi_2} \cong A_{\pi_1\pi_2}.$$

Finally A_π admits for each $\sigma \in \Sigma$, σ-semilinear automorphisms s, i.e., automorphisms s of A_π as a ring, so that $(ab)^s = a^s b^\sigma$ for $a \in A_\pi$, $b \in L\Delta$. (Recall that Σ acts on $L\Delta$ both via L and via Δ). Indeed choose an element v of $\tilde{\Omega}_K = \operatorname{Gal}(\tilde{L}/K)$, which maps onto σ, and then set

(2.5) $$f^s(\delta) = f(\delta^{\sigma^{-1}})^v.$$

Now let for each π, $\mathfrak{a}_\pi$ be the integral closure of $\mathfrak{o}_L$ in A_π. The $\mathfrak{a}_\pi$ are locally free rank one $\mathfrak{o}_L\Delta$-modules, thus giving us a map

(2.6) $$\begin{cases} i\colon \operatorname{Hom}_\Sigma(\tilde{\Omega}_L, \Delta) \to \operatorname{Cl}(\mathfrak{o}_L\Delta)^\Sigma, \\ i(\pi) = (\mathfrak{a}_\pi)_{\mathfrak{o}_L\Delta}. \end{cases}$$

The same argument which led us to the existence of σ-linear automorphisms of A_π (see (2.5)), gives such automorphisms of $\mathfrak{a}_\pi$. Thus indeed the class $(\mathfrak{a}_\pi)_{\mathfrak{o}_L\Delta}$ is fixed under Σ.

Now we observe that in the case of a commutative ring, such as $\mathfrak{o}_L\Delta$, the locally free class group is the Picard group, i.e., can be viewed as the group of isomorphism classes of rank one projectives under tensor product. In other words, in our case

(2.7) $$(\mathfrak{a}_{\pi_1})_{\mathfrak{o}_L\Delta} \cdot (\mathfrak{a}_{\pi_2})_{\mathfrak{o}_L\Delta} = (\mathfrak{a}_{\pi_1} \otimes_{\mathfrak{o}_L\Delta} \mathfrak{a}_{\pi_2}).$$

In general we can however not expect an analogue to the isomorphism (2.4). To define a useful criterion for the validity of such an isomorphism we shall say that a finite prime divisor $\mathfrak{p}$ of L is ramified in A_π if it is ramified in the associated field N'_π. Then A_{π_1} and A_{π_2} are said to have *coprime ramification*, if no finite prime divisor of L is ramified in both A_{π_1} and A_{π_2}. We get then the criterion

(2.8) $$\mathfrak{a}_{\pi_1\pi_2} \cong \mathfrak{a}_{\pi_1} \otimes_{\mathfrak{o}_L\Delta} \mathfrak{a}_{\pi_2} \quad \text{if} \quad A_{\pi_1} \text{ and } A_{\pi_2} \text{ have coprime ramification.}$$

Next we introduce the equivariant class group $\operatorname{Cl}(\mathfrak{o}_L\Delta, \Sigma)$ (cf. [FW1, FW2]). For this we consider rank one projective $\mathfrak{o}_L\Delta$-modules X, which are also Σ-modules, with the property that for $x \in X$, $a \in \mathfrak{o}_L\Delta$, $\sigma \in \Sigma$, $(xa)^\sigma = x^\sigma a^\sigma$. $\operatorname{Cl}(\mathfrak{o}_L\Delta, \Sigma)$ is then the group of isomorphism classes of such modules, with respect to the tensor product over $\mathfrak{o}_L\Delta$.

We can now introduce the basic diagram:

$$
\begin{array}{ccccccccc}
1 \to & H^1(\Sigma, \Delta) & \to & H^1(\tilde{\Omega}_K, \Delta) & \to & H^1(\tilde{\Omega}_L, \Delta)^\Sigma & \xrightarrow{d} & H^2(\Sigma, \Delta) \\
 & \downarrow j & & \downarrow i' & & \downarrow i & & \downarrow j \\
1 \to & H^1(\Sigma, \mathfrak{o}_L\Delta^*) & \to & \mathrm{Cl}(\mathfrak{o}_L\Delta, \Sigma) & \to & \mathrm{Cl}(\mathfrak{o}_L\Delta)^\Sigma & \xrightarrow{g} & H^2(\Sigma, \mathfrak{o}_L\Delta^*)
\end{array} \tag{2.9}
$$

The top row is the exact restriction – inflation sequence, coming from the Hochschild-Serre spectral sequence for the module Δ. Of course $H^1(\tilde{\Omega}_L, \Delta) = \mathrm{Hom}(\tilde{\Omega}_L, \Delta)$. The map i was defined in (2.6); i' has a similar definition, which we shall not give here explicitly, as it will not be required. Neither i nor i' need to be homomorphisms of groups. The maps j are homomorphisms induced by the embedding $\Delta \to \mathfrak{o}_L\Delta^*$. The lower row is a special case of part of a long exact sequence of homomorphisms (cf. [FW1], [FW2]). We shall describe the two relevant maps. Firstly $\mathrm{Cl}(\mathfrak{o}\Delta, \Sigma) \to \mathrm{Cl}(\mathfrak{o}\Delta)^\Sigma$ is the forgetful map – forgetting the Σ-structure of modules. To define g, consider the semilinear automorphisms s of a rank one locally free $\mathfrak{o}_L\Delta$-module X; in other words such an s is an automorphism of X as additive group, so that, for some $\sigma \in \Sigma$, $(xa)^s = x^s a^\sigma$ $(x \in X,\ a \in \mathfrak{o}_L\Delta)$. These semilinear automorphisms form a group G, and the map $s \mapsto \sigma = \sigma_s$ yields a homomorphism $G \to \Sigma$. If $(X) \in \mathrm{Cl}(\mathfrak{o}\Delta)^\Sigma$, this homomorphism is surjective. Its kernel is $\mathrm{Aut}_{\mathfrak{o}_L\Delta}(X) = \mathfrak{o}_L\Delta^*$. Thus we have an exact sequence

$$1 \to \mathfrak{o}_L\Delta^* \to G \to \Sigma \to 1,$$

giving rise to an element of $H^2(\Sigma, \mathfrak{o}_L\Delta^*)$, which by definition is $g((X))$.

Theorem 40. (i) *The diagram* (2.9) *commutes and has exact rows.*
(ii) *If A_{π_1} and A_{π_2} have coprime ramification, then $i(\pi_1)i(\pi_2) = i(\pi_1\pi_2)$.*

We shall not give a proof here. The exactness of the top row is a well known result on spectral sequences, that of the bottom row is in [FW1] of [FW2]. (ii) is a consequence of (2.8). □

Corollary 4 to Proposition 2.1 has now become quite obvious from our new point of view. For, if there is a solution N of the embedding problem $[L/K, \Gamma]$, with $\mathfrak{o}_N$ free over $\mathfrak{o}_L\Delta$, this means that there exists a $\pi \in H^1(\tilde{\Omega}_L, \Delta)^\Sigma$, with $i(\pi) = 1$, and $d(\pi) = e$, where e is the cohomology class of the group extension Γ. But then by the commutativity of (2.9), $j(e) = 1$. As we shall see, the diagram (2.9) can be used to get results on the sufficiency of the condition in Corollary 4.

The next theorem tells us that we have not lost anything by allowing improper solutions.

Theorem 41. *Given $\pi^* \in H^1(\tilde{\Omega}_L, \Delta)^\Sigma$, there exist infinitely many $\pi' \in H^1(\tilde{\Omega}_L, \Delta)^\Sigma$, which are surjections, and so that*

$$d(\pi') = d(\pi^*), \qquad i(\pi') = i(\pi^*).$$

We give the idea of the proof. For any π, let $\mathscr{R}(\pi)$ be the set of prime divisors $\mathfrak{p}$ in K, so that if $\mathfrak{P}$ is a prime divisor of L, above $\mathfrak{p}$, then $\mathfrak{P}$ is ramified in A_π. Standard techniques for the embedding problem, similar to the ones used in Chapter V, but more refined, allow us to construct an infinite sequence $\{\pi_n\}$ of elements of $H^1(\tilde{\Omega}_L, \Delta)^\Sigma$ with the following properties: (i) For all n, π_n is surjective, and A_{π_n} contains no proper non-ramified extension of L. (ii) The $\mathscr{R}(\pi_n)$ are mutually disjoint, and disjoint from $\mathscr{R}(\pi^*)$. It follows that any product $\pi_{i_1}\pi_{i_2}\cdots\pi_{i_r}$ with $i_1 < i_2 < \cdots < i_r$ is surjective, and for distinct sequences $i_1, \ldots, i_r$ these are all distinct. As, however, the group $H^2(\Sigma, \Delta) \times \mathrm{Cl}(\mathfrak{o}\Delta)^\Sigma$ is finite, there is a pair n, m of natural numbers, with $n < m$, so that

$$d(\pi_1\pi_2\cdots\pi_n) = d(\pi_1\pi_2\cdots\pi_m),$$

$$i(\pi_1\pi_2\cdots\pi_n) = i(\pi_1\pi_2\cdots\pi_m).$$

Our conditions imply that $\mathscr{R}(\pi_1\pi_2\cdots\pi_n)$ and $\mathscr{R}(\pi_{n+1}\cdots\pi_m)$ are disjoint. Therefore we conclude that

$$d(\pi_{n+1}\cdots\pi_m) = 1, \qquad i(\pi_{n+1}\cdots\pi_m) = 1.$$

Put $\pi_{n+1}\cdots\pi_m = \pi^{(1)}$. Repeat the process. We get an infinite sequence $\{\pi^{(n)}\}$ with the following properties: (i) The $\pi^{(n)}$ are surjective and $A_{\pi^{(n)}}$ contains no proper non-ramified extension of L. (ii) The $\mathscr{R}(\pi^{(n)})$ are mutually disjoint, and disjoint from $\mathscr{R}(\pi^*)$. (iii) $d(\pi^{(n)}) = 1$, $i(\pi^{(n)}) = 1$. These properties imply that, for all n, $\pi^{(n)}\pi^*$ is surjective, that the $\pi^{(n)}\pi^*$ are all distinct, and that – because of coprime ramification –

$$i(\pi^{(n)}\pi^*) = i(\pi^{(n)})i(\pi^*) = i(\pi^*),$$

and trivially also

$$d(\pi^{(n)}\pi^*) = d(\pi^*). \qquad \square$$

Corollary. *In the embedding problem $[L/K, \Gamma]$ assume that $\Gamma \cong \Delta \rtimes \Sigma$. Then there exist infinitely many fields N in L^c giving a solution of the embedding problem, and so that always $\mathfrak{o}_N$ is free over $\mathfrak{o}_L\Delta$.*

Proof. Take π^* the identity element of $H^1(\tilde{\Omega}_L, \Delta)^\Sigma$, i.e., the null map $\tilde{\Omega}_L \to \Sigma$. $\square$

More surprising than this result, are those in the opposite direction, excluding the possibility of $\mathfrak{o}_N$ to be free over $\mathfrak{o}_L\Delta$. We give one example:

2.2. Proposition. *Suppose Δ is of odd order m, Σ acts trivially on Δ and $L \subset \mathbb{Q}(h)$, where h is square free and prime to m. Then the map j: $H^2(\Sigma, \Delta) \to H^2(\Sigma, \mathfrak{o}_L\Delta^*)$ in* (2.9) *is injective.* $\square$

Suppose then $\pi \in H^1(\tilde{\Omega}_L, \Delta)^\Sigma$ has image $d(\pi) \neq 1$. Then it follows that we must have $i(\pi) \neq 1$, under the hypotheses of the proposition.

Example: Let $N/\mathbb{Q}$ be a tame extension with cyclic Galois group of odd prime power order l^n. Let L be an intermediary field, $Q \subsetneqq L \subsetneqq N$. Then $\mathfrak{o}_N$ cannot be free over $\mathfrak{o}_L \operatorname{Gal}(N/L)$. We shall later see that this is not so when $l = 2$.

§3. An Example

We shall now apply the theory described in §2 in a concrete arithmetic situation: $K = \mathbb{Q}$, L is a tame Abelian extension of $\mathbb{Q}$, whose Galois group Σ is an elementary Abelian l-group, l a prime, and Δ is of order l. Thus Σ must act trivially on Δ. The basic diagram (2.9) now simplifies very considerably, as we shall briefly show. Firstly, we observe that quite generally, i.e., in the original situation in which (2.9) was defined, we can replace $\mathrm{Cl}(\mathfrak{o}_L\Delta)$ by the kernel $\mathrm{Cl}(\mathfrak{o}_L\Delta)_0$ of the map $\mathrm{Cl}(\mathfrak{o}_L\Delta) \to \mathrm{Cl}(\mathfrak{o}_L)$, which associates with the class of a module X over $\mathfrak{o}_L\Delta$ that of the module X^Δ of fixed points, over $\mathfrak{o}_L$. For, the fixed module in our case is always $\mathfrak{o}_L$ itself. An analogous observation applies also to the equivariant version $\mathrm{Cl}(\mathfrak{o}_L\Delta, \Sigma)$. In the particular case we are now considering let

$$M = L(\zeta),$$

ζ a primitive l-th root of unity. Then we get an isomorphism of Σ-modules, from the group $\mathrm{Cl}(\mathfrak{o}_L\Delta)_0$ onto the ray class group C of $\mathfrak{o}_M \bmod (1-\zeta)$, with Σ acting as Galois group of $M/\mathbb{Q}(\zeta)$, i.e., keeping ζ fixed. Moreover on taking Σ-fixed points the map $C^\Sigma \to \mathrm{Cl}(\mathfrak{o}_M)^\Sigma$ becomes an isomorphism. Thus in our diagram $\mathrm{Cl}(\mathfrak{o}\Delta)^\Sigma$ is now replaced by $\mathrm{Cl}(\mathfrak{o}_M)^\Sigma$. In fact, as can be shown, in all entries of the bottom row of (2.9), $\mathfrak{o}_L\Delta$ can be replaced by $\mathfrak{o}_M$.

We next come to the top row. By results in [F27], we can split off a trivial infinite part, to be left with an exact sequence involving only canonically defined finite groups. Let $\mathscr{R}_L$ be the set of finite rational prime divisors ramified in L. Denote by $L^{(1)}$ the maximal tame Abelian extension of $\mathbb{Q}$, non-ramified at all $p \in \mathscr{R}_L$, and $\tilde{L}^{(2)}$ the maximal tame Abelian extension of L, non-ramified outside $\mathscr{R}_L$. Write furthermore

$$\tilde{\Omega}_{\mathbb{Q}}^{(1)} = \operatorname{Gal}(L^{(1)}/\mathbb{Q}), \qquad \tilde{\Omega}_L^{(1)} = \operatorname{Gal}(L^{(1)}L/L),$$
$$\tilde{\Omega}_{\mathbb{Q}}^{(2)} = \operatorname{Gal}(\tilde{L}^{(2)}/\mathbb{Q}), \qquad \tilde{\Omega}_L^{(2)} = \operatorname{Gal}(\tilde{L}^{(2)}/L).$$

Translating results in [F27] into our present language, we have

3.1. Proposition.

$$H^1(\tilde{\Omega}_{\mathbb{Q}}, \Delta) = H^1(\tilde{\Omega}_{\mathbb{Q}}^{(1)}, \Delta) \oplus H^1(\tilde{\Omega}_{\mathbb{Q}}^{(2)}, \Delta),$$
$$H^1(\tilde{\Omega}_L, \Delta)^\Sigma = H^1(\tilde{\Omega}_L^{(1)}, \Delta) \oplus H^1(\tilde{\Omega}_L^{(2)}, \Delta)^\Sigma,$$

and the map $H^1(\tilde{\Omega}_{\mathbb{Q}}, \Delta) \to H^1(\tilde{\Omega}_L, \Delta)^\Sigma$ *of* (2.9) *is the direct sum of an isomorphism* $H^1(\tilde{\Omega}_{\mathbb{Q}}^{(1)}, \Delta) \cong H^1(\tilde{\Omega}_L^{(1)}, \Delta)$ *and a homomorphism* $H^1(\tilde{\Omega}_{\mathbb{Q}}^{(2)}, \Delta) \to H^1(\tilde{\Omega}_L^{(2)}, \Delta)^\Sigma$. □

Next consider an element π of $\mathrm{Hom}(\tilde{\Omega}_{\mathbb{Q}}^{(1)}, \Delta) = H^1(\tilde{\Omega}_{\mathbb{Q}}^{(1)}, \Delta)$. We assume π to be surjective – as Δ is of prime order, this will be the case, except when $\pi = 1$. The fixed field N'_π of $\mathrm{Ker}\,\pi$ is then a tame cyclic extension of degree l of $\mathbb{Q}$, and so its ring $\mathfrak{o}_\pi$ of integers, viewed as a $\mathbb{Z}\Delta$-module is free. This already implies that $i'(\pi) = 1$. Next let π' be the image of π in $\mathrm{Hom}(\tilde{\Omega}_L^{(1)}, \Delta) = H^1(\tilde{\Omega}_L^{(1)}, \Delta)$. The fixed field of $\mathrm{Ker}\,\pi'$ is the composite $N'_\pi L$. As the ramification of $N'_\pi/\mathbb{Q}$ is coprime to that of $L/\mathbb{Q}$, $N'_\pi L \cong N'_\pi \otimes_{\mathbb{Q}} L$, and for the same reason the ring of algebraic integers in the tensor product is actually $\mathfrak{o}_\pi \otimes_{\mathbb{Z}} \mathfrak{o}_L$. It follows that the ring of integers in $N'_\pi L$ is free over $\mathfrak{o}_L\Delta$, by "lifting structure". Hence $i(\pi') = 1$. We conclude:

3.2. Proposition.

$$H^1(\tilde{\Omega}_{\mathbb{Q}}^{(1)}, \Delta) \subset \mathrm{Ker}\, i'$$
$$H^1(\tilde{\Omega}_L^{(1)}, \Delta) \subset \mathrm{Ker}\, i. \qquad \square$$

By the two preceding Propositions, and taking account of the changes which can be made in the bottom row of (2.9), we now get a new commutative diagram, with exact rows

$$\begin{array}{ccccccccc}
1 \to & H^1(\Sigma, \Delta) & \to & H^1(\tilde{\Omega}_{\mathbb{Q}}^{(2)}, \Delta) & \to & H^1(\tilde{\Omega}_L^{(2)}, \Delta)^\Sigma & \xrightarrow{d} & H^2(\Sigma, \Delta) \\
& \downarrow & & \downarrow{\scriptstyle i'} & & \downarrow{\scriptstyle i} & & \downarrow{\scriptstyle j} \\
1 \to & H^1(\Sigma, \mathfrak{o}_M^*) & \to & \mathrm{Cl}(\mathfrak{o}_M, \Sigma) & \to & \mathrm{Cl}(\mathfrak{o}_M)^\Sigma & \xrightarrow{g} & H^2(\Sigma, \mathfrak{o}_M^*).
\end{array} \qquad (3.1)$$

All groups in this diagram are finite and can in every given case be computed.

We shall describe $\mathrm{Ker}\, d$, using classical genus theory. Let now $\pi \in \mathrm{Hom}(\tilde{\Omega}_{\mathbb{Q}}^{(2)}, \Delta) = H^1(\tilde{\Omega}_{\mathbb{Q}}^{(2)}, \Delta)$, and let N'_π be the fixed field of $\mathrm{Ker}\,\pi$. Excluding the case $\pi = 1$, we see that N'_π is cyclic of degree l over $\mathbb{Q}$, and ramified only at primes which ramify in L. From this one deduces that $N'_\pi L/L$ is non-ramified. Thus if π' is the image of π in $H^1(\tilde{\Omega}_L^{(2)}, \Delta)^\Sigma$, then by Theorem 40 (ii), $i(\pi'\pi^*) = i(\pi')i(\pi^*)$, for every π^*. In particular we now have

3.3. Proposition. *The map i defines a homomorphism* $\mathrm{Ker}\, d \to \mathrm{Cl}(\mathfrak{o}_M)^\Sigma$, *and thus* $i(\mathrm{Ker}\, d)$ *is a group. Moreover if* $e \in \mathrm{Im}\, d$, *then* $i(\mathrm{Ker}\, d)$ *acts freely and transitively on* $i(d^{-1}e)$. $\square$

In [F27] the group $\mathrm{Im}\, d$ was determined in the present situation. In conjunction with Proposition 3.3, this gives us almost complete information. The group $\mathrm{Ker}\, d$ can certainly be described in terms of the so called genus group of L – it is its dual. To be more precise, let L' be the maximal Abelian extension of $\mathbb{Q}$ containing L, which is non-ramified (at all finite prime divisors) over L, (i.e., L' is the "narrow genus field" of L). Then our previous discussion shows that $\mathrm{Ker}\, d \cong \mathrm{Hom}(\mathrm{Gal}(L'/L), \Delta)$. On the other hand, we can also describe L' as the maximal Abelian extension of $\mathbb{Q}$, whose Galois group is an elementary l-group and which is ramified only in $\mathscr{R}_L$. This implies that $[L' : L] = l^{t-m}$, where $t = \mathrm{card}(\mathscr{R}_L)$, and $[L : \mathbb{Q}] = l^m$. Hence $\mathrm{Ker}\, d$ is an elementary Abelian l-group of order l^{t-m}. If in particular L is its own genus field, then d is injective.

If l is odd, then, by Proposition 2.2, j is injective, hence trivially the converse of Corollary 4 to Proposition 2.1 now holds. But in fact one can show this also for $l = 2$, i.e., we have for all l

3.4. Proposition. *If* $jd\pi = 1$ *then* $\exists \pi' \in H^1(\tilde{\Omega}_L^{(2)}, \Delta)^\Sigma$ *with* $d\pi' = d\pi$ *and* $i\pi' = 1$. □

Now we take $l = 2$. Then $M = L$ and $\mathrm{Cl}(\mathfrak{o}_L)^\Sigma$ is the subgroup of so called "ambivalent" ideal classes. One can also show that $\mathrm{Cl}(\mathfrak{o}_L, \Sigma)$ is the group of Σ-fixed ("ambivalent") fractional ideals of $\mathfrak{o}_L$, modulo the subgroup of rational fractional ideals. This group is trivial to compute: it is of order 2^t, t the number of rational primes ramified in L. It follows that $\mathrm{Ker}\, g$ is the group of ambivalent ideal classes modulo the classes of ambivalent ideals – a very classical object. The map i on $\mathrm{Ker}\, d$ can now be described more directly. Let $\pi \in \mathrm{Ker}\, d$, $\pi \neq 1$. The fixed field of $\mathrm{Ker}\, \pi$ is of form $L(\sqrt{b})$, $b \equiv 1 \pmod 4$ and square free in $\mathbb{Q}^*$, and the ideal $b\mathfrak{o}_L$ the square of an ideal $\mathfrak{b}$ in L. We then have

$$i(\pi) = \text{class of } \mathfrak{b}.$$

Finally we specialize to a quadratic field

$$L = \mathbb{Q}(\sqrt{D}), \qquad D \equiv 1 \pmod 4. \tag{3.2}$$

Then trivially $H^2(\Sigma, \Delta)$ is of order 2, the non-trivial element corresponding to the cyclic group of order 4. Classically one knows that d is surjective if, and only if, in (3.2),

$$D = \prod p_k, \qquad p_k \text{ distinct primes} \equiv 1 \pmod 4).$$

One can then show

3.5. Proposition. *If $D < 0$, i.e., L is imaginary, then i is injective.*
If $D > 0$, i.e., L is real, then there are exactly two elements of $H^1(\tilde{\Omega}_L^{(2)}, \Delta)^\Sigma$, *which map under i into* 1. *The non-identity element among these lies in* Ker *d if, and only if, the norm of the fundamental unit of L is* 1. □

We translate this into more concrete terms. We consider the set of fields N, with

$$\begin{cases} N \text{ quadratic over } L,\ N \text{ Abelian over } \mathbb{Q}, \\ \text{all rational primes } p \text{ which ramify in } N \text{ already ramify in } L. \end{cases} \tag{3.3}$$

These fields N are thus either cyclic over $\mathbb{Q}$, or the composites of two distinct quadratic extensions of $\mathbb{Q}$.

3.5a. Proposition. *If L is imaginary, then none of the fields N has a normal integral basis over L. If L is real then there is exactly one such field N with normal integral basis over L. This field is the composite of two quadratic extensions, or is cyclic, according to whether the norm of the fundamental unit of L is* $+1$, *or* -1, *respectively.* □

Remark. Let t be the number of rational primes ramified in L, and suppose the norm of the fundamental unit is -1. Then there are exactly 2^{t-1} fields N of type (3.3), which are cyclic over $\mathbb{Q}$. Class field theoretically there is nothing to distinguish a canonical one among these – but as we have seen there is one canonical such field from the standpoint of Galois module structure.

§4. Generalized Kummer Theory

Recent advances, dealing with relative extensions with certain Abelian Galois groups, and due to Leon McCulloh (see [MC2], [MC3], [MC4]), form the contents of this and of the next section. They arise out of a combination of ideas and methods, coming from two distinct, but related earlier developments. We shall begin by describing these.

The first goes back to the *Stickelberger relations*. These concern the Galois module structure of the ideal class group in absolutely Abelian fields. Here we shall restrict ourselves to a cyclotomic field $\mathbb{Q}(f)$. We denote its ring of integers now by $\mathbb{Z}[f]$. We write

$$G = G_f = \mathrm{Gal}(\mathbb{Q}(f)/\mathbb{Q}) \tag{4.1}$$

for the Galois group. It acts on the group μ_f of f-th roots of unity in $\mathbb{Q}(f)$. We define a map

$$t = t_f : G \to \mathbb{Z}$$

by

$$0 \leqslant t(g) < f, \qquad y^g = y^{t(g)} \qquad \text{for all} \qquad y \in \mu_f. \tag{4.2}$$

(If $f = l^m$, l a prime, then $t(g) \equiv u_l(\omega)^{-1} \pmod{l^m}$ for $\omega \in \Omega_{\mathbb{Q}}$, $\omega \mapsto g$ under $\Omega_{\mathbb{Q}} \to G_f$. Here $u_l : \Omega_{\mathbb{Q}} \to \mathbb{Z}_l^*$ is the homomorphism which we used to describe Galois action on Gauss sums, cf. III, §3.) Now write

$$\Theta = \Theta_f = \sum_{g \in G} t(g) g^{-1} \in \mathbb{Z}(G). \tag{4.3}$$

The *Stickelberger ideal* of $\mathbb{Z}G$ is defined to be

$$J_f = \mathbb{Z}G \cap (\Theta_f/f)\mathbb{Z}G, \tag{4.4}$$

and the Stickelberger theorem asserts that J_f annihilates the class group $\mathrm{Cl}(\mathbb{Z}[f])$, i.e., that

$$\mathrm{Cl}(\mathbb{Z}[f])^{J_f} = 1. \tag{4.5}$$

All the proofs until 1975, including the original one of Stickelberger, were based on the prime factorization of the classical, local, tame Abelian Gauss-sums (see e.g., [Le2], or [Ct] (Theorem 3.1)). One exception was a proof of Hilbert (cf. [Hi], Theorem 136) for the special case when f is a prime. This latter foreshadows the idea of the new proof to which we shall come presently.

Recall that the prime decomposition of Gauss sums has also played an important role in the proof of the relation between Gauss sums and resolvents (see Theorem 27). Moreover the Stickelberger relations themselves have been used directly to obtain information on the Galois module structure of rings of integers (see [Ma1], [Co1], [Co2], [Ty8]). To complete this picture of interdependence of these several aspects, the author showed in [F19] that the Stickelberger relations (4.5) can be derived directly from our explicit knowledge of the structure of cyclotomic rings $\mathbb{Z}[p]$ as modules over $\mathrm{Gal}(\mathbb{Q}(p)/\mathbb{Q})$, for all $p \equiv 1 \pmod{f}$. All one needs in this context is the obvious observation that any primitive p-th root of unity generates a normal integral basis. In fact for these primes p, the same considerations led in [F19] to a new proof for the prime decomposition of Gauss sums, for residue class characters mod p of order f.

The second development arose out of the problem of realizable classes (realizable by rings of integers). In [MC1] McCulloh succeeded in solving this problem in the following "Kummer" situation: l is a prime, K is a number field containing the primitive l-th roots of unity, and Γ is of order l. He showed that the classes $(\mathfrak{o}_N)_{\mathfrak{o}_K\Gamma}$, for N/K tame with $\mathrm{Gal}(N/K) \cong \Gamma$, form a subgroup of $\mathrm{Cl}(\mathfrak{o}_K\Gamma)$, again determined by a certain Stickelberger ideal. Lindsay Childs extended one half of this criterion to arbitrary basefields K (cf. [Chi2]). As we shall state later a generalization of these results, we shall not give a detailed statement at this stage. In this context other work of Lindsay Childs should also be mentioned. This is ring theoretic, but has applications to extensions of the type considered here, provided that one restricts oneself to non-ramified ones (cf. [Chi1]). For a more general ring theoretic approach with arithmetic applications see also [Mr].

The more recent work of McCulloh's is based on a combination of the techniques of [MC1] and of [F19]. It leads to a considerable generalization of previous results concerning realizable classes, and to new Stickelberger type relations on class groups of group rings. The original Stickelberger relations of course have been one of the main ingredients of Iwasawa theory. There are some areas of McCulloh's work which point into this direction as well. Thus in [MC2] an index relation is derived, quite similar to a class number formula of Iwasawa's. Most of this will be subject matter of §5. Here we shall introduce the underlying formal theory in our own way, emphasizing that aspect which seems to be the crucial one, namely something which may be called a new type of Kummer theory. In this the group of roots of unity is replaced by a given finite Abelian group Γ, the group which is to be represented as Galois group. Abelian characters are replaced by homomorphisms into Γ, the cyclotomic field by a group ring, and the Galois group of the basefield $\mathbb{Q}(f)$ for Kummer theory by a certain "natural" group G of automorphisms of Γ. Some of these ideas also occur in [Q2].

Let then K be a number field, K^t its maximal tame extension, $\mathfrak{O}$ the integral closure of $\mathfrak{o}_K$ in K^t, $\Omega = \mathrm{Gal}(K^t/K)$. We shall use the group $\mathrm{Hom}(\Omega, \Gamma)$ of (continuous) homomorphisms to define certain "eigenspaces" in the group ring $K^t\Gamma$. We let Ω act on $K^t\Gamma$ by ring automorphisms, via its action on K^t. Then $K\Gamma$

is the fixed subring. Given $\pi \in \mathrm{Hom}(\Omega, \Gamma)$, define

$$X_\pi = (x \in K^t\Gamma \mid x^\omega = x\pi(\omega) \text{ for all } \omega \in \Omega). \tag{4.6}$$

4.1. Proposition. *X_π contains an element x of $(K^t\Gamma)^*$, and every such x is a free generator of X_π as a $K\Gamma$-module.*

Proof. It is clear that X_π is a $K\Gamma$-module. Let then y be a free generator of the fixed field N_π of $\operatorname{Ker}\pi$ over $K\Delta$, where $\Delta = \mathrm{Gal}(N_\pi/K)$. Via π, Δ is identified with the subgroup $\operatorname{Im}\pi$ of Γ, and $x = \sum y^\omega \pi(\omega)^{-1}$ (sum mod $\operatorname{Ker}\pi$) is the corresponding resolvent element. By Proposition I.4.1, it is an invertible element of $K^t\Delta$, i.e., lies in $K^t\Gamma^*$. Moreover, as one verifies, $x \in X_\pi$. If now $z \in X_\pi$, then $(zx^{-1})^\omega = zx^{-1}$ for all $\omega \in \Omega$, and hence $z = xa$, $a \in K\Gamma$. It is now clear that x is a free generator of X_π over $K\Gamma$. □

Remark 1. If π is surjective i.e., $\Gamma \cong \mathrm{Gal}(N_\pi/K)$, then the elements z of X_π are precisely the resolvent elements $z = \sum b^\omega \pi(\omega)^{-1}$ (sum mod $\operatorname{Ker}\pi$), for all $b \in N_\pi$. The map $b \mapsto z$ is an isomorphism $N_\pi \cong X_\pi$ of $K\Gamma$-modules.

Remark 2. It is clear that multiplication in $K^t\Gamma$ yields a map $X_\pi \times X_{\pi'} \to X_{\pi\pi'}$.

Let

$$H(K\Gamma) = H = K^t\Gamma^* \cap \left(\bigcup_\pi X_\pi\right). \tag{4.7a}$$

Thus an element x of $K^t\Gamma^*$ lies in H, if for some $\pi = \pi_x \in \mathrm{Hom}(\Omega, \Gamma)$, we have

$$x^\omega = x\pi_x(\omega) \qquad \text{for all} \qquad \omega \in \Omega. \tag{4.7b}$$

We have evidently

4.2. Proposition. *H is a subgroup of $K^t\Gamma^*$, containing $K\Gamma^*$ and the map $x \mapsto \pi_x$ gives rise to an exact sequence*

$$1 \to K\Gamma^* \to H(K\Gamma) \to \mathrm{Hom}(\Omega, \Gamma) \to 1.$$ □

Next we consider, for variable π, $\mathfrak{o}_K\Gamma$ submodules of X_π which are finitely generated, locally free of rank one. We shall call them *radical modules* (of $\mathfrak{o}_K\Gamma$). If Y, Y' are two such modules define $Y \cdot Y'$ to be the $\mathfrak{o}_K\Gamma$-submodule of $K^t\Gamma$, generated by the products yy', $y \in Y$, $y' \in Y'$. This is clearly again a radical module. In fact we have

4.3. Proposition. *The radical modules form an Abelian group $\mathfrak{H}(\mathfrak{o}_K\Gamma)$ under multiplication, containing the group $\mathfrak{I}(\mathfrak{o}_K\Gamma)$ of invertible fractional ideals of $\mathfrak{o}_K\Gamma$. The map $Y \mapsto \pi$, where $Y \subset X_\pi$ gives rise to an exact sequence*

$$1 \to \mathfrak{I}(\mathfrak{o}_K\Gamma) \to \mathfrak{H}(\mathfrak{o}_K\Gamma) \to \mathrm{Hom}(\Omega, \Gamma) \to 1.$$ □

Remark. One can define local or idelic analogues to the group $H(K\Gamma)$ and obtain a representation of $\mathfrak{H}(\mathfrak{o}_K\Gamma)$ as a quotient of the idelic version.

Mapping a radical module Y into its class $(Y)_{\mathfrak{o}_K\Gamma}$ in the class group, gives rise to a homomorphism

$$\mathfrak{H}(\mathfrak{o}_K\Gamma) \to \mathrm{Cl}(\mathfrak{o}_K\Gamma). \tag{4.8}$$

We shall be interested in the classes of certain special modules. For each π, define

$$\mathfrak{b}_\pi = [x \in X_\pi \cap \mathfrak{O}\Gamma], \tag{4.9}$$

(recall that $\mathfrak{O}$ is the ring of integers in K^t). On localizing one verifies that $\mathfrak{b}_\pi$ is a radical module. If in particular π is surjective, then the isomorphism $N_\pi \cong X_\pi$ (Remark 1) yields an isomorphism

$$\mathfrak{o}_{N,\pi} \cong \mathfrak{b}_\pi \tag{4.10}$$

of $\mathfrak{o}_K\Gamma$-modules ($\mathfrak{o}_{N,\pi}$ the ring of integers in N_π).

Now let G be a group of automorphisms of Γ (written exponentially). Then, for any commutative ring B, the group $(B\Gamma)^*$ is a $\mathbb{Z}G$-module, G acting trivially on B. Thus $K^t\Gamma^*$ is a $\mathbb{Z}G$-module, and the action of $\mathbb{Z}G$ commutes with that of Ω. Clearly G and $\mathbb{Z}G$ act on $\mathrm{Hom}(\Omega, \Gamma)$ via their action on Γ. It follows that $H(\mathfrak{o}_K\Gamma)$ and $\mathfrak{H}(\mathfrak{o}_K\Gamma)$, as well as $(K\Gamma)^*$ and $\mathfrak{J}(\mathfrak{o}_K\Gamma)$ are $\mathbb{Z}G$-modules and the sequences of Propositions 4.2 and 4.3 are sequences of $\mathbb{Z}G$-modules. Similarly (4.8) admits $\mathbb{Z}G$-action.

Let now A be the annihilating ideal of Γ in $\mathbb{Z}G$. Then we have

4.4. Proposition. *A is the annihilating ideal of* $\mathrm{Hom}(\Omega, \Gamma)$, *and so*

$$H(K\Gamma)^A \subset K\Gamma^*,$$

$$\mathfrak{H}(\mathfrak{o}_K\Gamma)^A \subset \mathfrak{J}(\mathfrak{o}_K\Gamma).$$

Indeed, it is clear that A annihilates $\mathrm{Hom}(\Omega, \Gamma)$. As Hom (Ω, Γ) contains surjections, A is indeed its full annihilating ideal. For the remainder of the proposition use the exact sequences of Propositions 4.2 and 4.3. □

Let now l be a prime number, and let Γ from now on be an l-group. Write $\mathfrak{J}^{(l)}(\mathfrak{o}_K\Gamma)$ for the subgroup of $\mathfrak{J}(\mathfrak{o}_K\Gamma)$ consisting of fractional ideals $\mathfrak{m}$ with $\mathfrak{m}_l = \mathfrak{o}_{K,l}\Gamma$. This group is free Abelian and can naturally be identified with a product of groups of fractional ideals of Dedekind domains.

Let l^n be the exponent of Γ. Then $l^n \in A$, hence $H(K\Gamma)^{l^n} \subset K\Gamma^*$, $\mathfrak{H}(\mathfrak{o}_K\Gamma)^{l^n} \subset \mathfrak{J}(\mathfrak{o}_K\Gamma)$. Assume from now on that π is surjective and let b be a generator of X_π over $K\Gamma$, and a unit of $K^t\Gamma$. As N_π/K is tame, hence non-ramified above l, we can also assume that b is a unit above l. Moreover l^n is the exact order of $b \bmod K\Gamma^*$. Put $b^{l^n} = a \in K\Gamma^*$. Then $a \in \mathfrak{o}_{K,l}\Gamma^*$. Write

$$\begin{cases} a\mathfrak{o}_K\Gamma = \mathfrak{a}\mathfrak{m}^{l^n}, & \mathfrak{a}, \mathfrak{m} \in \mathfrak{J}^{(l)}(\mathfrak{o}_K\Gamma) \\ \mathfrak{a}\ l^n\text{-th power free}, & \mathfrak{a} \text{ integral i.e., } \mathfrak{a} \subset \mathfrak{o}_K\Gamma. \end{cases} \tag{4.11}$$

This decomposition is possible, in view of our remarks on the structure of $\mathfrak{J}^{(l)}(\mathfrak{o}_K\Gamma)$. Moreover, $\mathfrak{a}$ and the class of $\mathfrak{m}$ are uniquely determined by π, via (4.11).

4.5. Proposition. *With* $\mathfrak{m}$ *as above,*

$$(\mathfrak{o}_{N,\pi})_{\mathfrak{o}_K\Gamma} = (\mathfrak{m})^{-1}_{\mathfrak{o}_K\Gamma}.$$

Proof. Put $\mathfrak{m}_1 = \mathfrak{b}_\pi b^{-1}$, b as above. Then $\mathfrak{m}_1$ and $\mathfrak{b}_\pi^{l^n} \in \mathfrak{J}^{(l)}(\mathfrak{o}_K\Gamma)$, and trivially $(\mathfrak{o}_{N,\pi})_{\mathfrak{o}_K\Gamma} = (\mathfrak{b}_\pi)_{\mathfrak{o}_K\Gamma} = (\mathfrak{m}_1)_{\mathfrak{o}_K\Gamma}$. Also $\mathfrak{m}_1^{l^n} = \mathfrak{b}_\pi^{l^n} a^{-1}$. But from the definition of $\mathfrak{b}_\pi$ it follows that $\mathfrak{b}_\pi^{l^n}$ is l^n-th power free in $\mathfrak{J}^{(l)}(\mathfrak{o}_K\Gamma)$ and integral. Hence $\mathfrak{m}_1 = \mathfrak{m}^{-1}$. □

From now on we specialize further. Γ is now to be homogeneous, i.e., the direct product of say k cyclic factors, each of order l^n. We shall now follow closely McCulloh's own account (cf. [MC3], [MC4]). It will be useful to view Γ in a particular way. Let F_k be the non-ramified extension of $\mathbb{Q}_l$ of degree k and

$$S = S_{n,k}$$

be the residue class ring of $\mathfrak{o}_{F_k} \bmod l^n$. Then

$$\Gamma = S_{n,k}^+ = S^+ \tag{4.12a}$$

is the additive group of $S_{n,k}$, and we take

$$G = S_{n,k}^* = S^* \tag{4.12b}$$

to be its multiplicative group, acting on Γ in the obvious way.

A special case of (4.12) is that of an elementary Abelian group Γ, when we have

$$\begin{cases} \Gamma = \mathbb{F}_{l^k}^+, \\ G = \mathbb{F}_{l^k}^*. \end{cases} \tag{4.13}$$

There is a question of notation. Γ acts in the first place as a Galois group over K, and as such it has been written as a multiplicative group. But in the present context it appears as an additive group. To avoid notational complication we shall accordingly, and for the remainder of the present section only, write Γ additively. In fact we shall be working in two rings, namely $S = S_{n,k}$ and $\mathbb{Z}G$. We shall write

$$A = A_\Gamma = A_{n,k}$$

for the annihilating ideal of Γ in $\mathbb{Z}G$.

The ring $\mathbb{Z}G$ acts on Γ via the ring homomorphism $\mathbb{Z}G \to S$, induced by $g \mapsto g$ (for $g \in G$). We have in fact clearly

4.6. Proposition. *The ring homomorphism* $\mathbb{Z}G \to S$ *gives rise to an exact sequence*

$$0 \to A \to \mathbb{Z}G \to \Gamma \to 0$$

of Abelian groups. □

Let here

$$T: S_{n,k} \to \mathbb{Z}/l^n\mathbb{Z} = S_{n,1}$$

be the trace. As $F_k/\mathbb{Q}_l$ is non-ramified, T defines a non-singular pairing $S \times S \to \mathbb{Z}/l^n\mathbb{Z}$. Let γ_i be a basis of S over $\mathbb{Z}/l^n\mathbb{Z}$, and γ_i' the dual basis. Then we have the standard formula

(4.14) $$\sum_i T(\delta\gamma_i')\gamma_i = \delta, \qquad \text{for all} \qquad \delta \in \Gamma.$$

(Note that actually $\gamma_i, \gamma_i' \in G$).

Now define a map

$$t = t_{n,k}: S \to \mathbb{Z},$$

by

(4.15) $$1 \leqslant t(\gamma) < l^n, \qquad T(\gamma) \equiv t(\gamma) \bmod l^n.$$

This gives by restriction to G the natural generalization of the map t of (4.2).

From (4.14) and Proposition 4.6 we now deduce

4.7. Lemma. *For all $g \in G$, the elements $l^n g$, and*

$$d(g) = g - \sum_i t(g\gamma_i')\gamma_i,$$

lie in A. Moreover the elements

$$l^n\gamma_i \text{ and } d(g) \qquad (g \neq \gamma_i, \text{ all } i)$$

form a $\mathbb{Z}$-basis of A. □

We next define the *generalized Stickelberger element*

$$\Theta = \Theta_{n,k} \in \mathbb{Z}G,$$

by

(4.16) $$\Theta = \sum_{g \in G} t(g)g^{-1},$$

and the *generalized Stickelberger ideal* in $\mathbb{Z}G$ by

(4.17) $$J = J_{n,k} = J_\Gamma = \mathbb{Z}G \cap (\Theta/l^n)\mathbb{Z}G.$$

More generally we define, for any $\gamma \in \Gamma$,

(4.16a) $$\Theta(\gamma) = \sum_{g \in G} t(g^{-1}\gamma)g \in \mathbb{Z}G$$

and we verify that for $h \in G$

(4.18) $$\Theta(\gamma h) = \Theta(\gamma)h.$$

4.8. Proposition. (i) *The map* $\gamma \mapsto \Theta(\gamma)/l^n \bmod \mathbb{Z}G$ *is an isomorphism*

$$\Gamma \cong (\mathbb{Z}G(\Theta/l^n) + \mathbb{Z}G)/\mathbb{Z}G$$

of $\mathbb{Z}G$*-modules.*

(ii) $$A = [a \in \mathbb{Z}G \mid a(\Theta/l^n) \in \mathbb{Z}G].$$

(iii) $$J = A(\Theta/l^n).$$

Proof. (i) The given map is clearly additive, by the additivity of the trace T. It preserves G action by (4.18). Moreover, writing an element γ as a $\mathbb{Z}$-linear combination of basis elements γ_i, where $\gamma_i \in G$, we see that $\Theta(\gamma)/l^n \bmod \mathbb{Z}G$ does indeed lie in $(\mathbb{Z}G(\Theta/l^n) + \mathbb{Z}G)/\mathbb{Z}G$. The map is clearly surjective – just look at the image of $\gamma = 1$, which is $\Theta/l^n \bmod \mathbb{Z}G$. Moreover if γ lies in the kernel, then $\Theta(\gamma) \equiv 0 \pmod{l^n\mathbb{Z}G}$, i.e., for all $g \in G$, $\mathrm{Tr}(\gamma g) = 0$. This implies that $\gamma = 0$. Thus the map is injective.

(ii) is an immediate consequence of (i) and of the definition of A. Similarly for (iii), using also the definition of J. □

4.9. Proposition. *Let* $n = 1$. *Then*

$$J + A = \mathbb{Z}G.$$

Proof. We work with the images $\tilde{J}, \tilde{A}$ of J and A, respectively, in $\mathbb{F}_l G$, and show that the image e of $-\Theta$ is an idempotent, which fixes Γ elementwise. As $e \in \tilde{J}$, and $\tilde{A}$ is the annihilating ideal of Γ in $\mathbb{F}_l G$, this gives the result.

Now G acts freely and transitively on the set of non-zero elements of Γ. Hence Γ is a simple $\mathbb{F}_l G$-module. Moreover the order of G is prime to l. Thus Γ corresponds to an idempotent e, where by the standard formula

$$(\text{order } G)e = \sum T(g)g^{-1}.$$

As order $G \equiv -1 \pmod{l}$, the assertion follows. □

Remark. For certain aspects of the theory, e.g., the index formula, the ideal $J_{n,k}$ is the wrong generalization of $J_{1,k}$. Instead one has to consider the $\mathbb{Z}G$-module generated by all the $\Theta(\gamma)l^{-n}$ $(\gamma \in \Gamma)$ (see (4.16a)), and take its intersection $J'_{n,k}$ with $\mathbb{Z}G$. Of couse for $n = 1$ the two ideals coincide.

§5. The Generalized Class Number Formula and the Generalized Stickelberger Relation

We continue to employ the notation of the preceding section. But we shall mostly restrict ourselves to an elementary Abelian l-group Γ, i.e., to the case $S = S_{1,k}$. Following [MC3], we shall give a description of the "realizable classes" in $\mathrm{Cl}(\mathfrak{o}_K\Gamma)$

in terms of the Stickelberger ideal, where K is always a given number field. Specializing to $K = \mathbb{Q}$ this leads to "Stickelberger relations" on the $\mathbb{Z}G$-module $\mathrm{Cl}(\mathbb{Z}\Gamma)$. This result is complemented by another, independent one, namely a class number formula for $\mathrm{Cl}(\mathbb{Z}\Gamma)$ in terms of a module index in $\mathbb{Z}G$, coming from the Stickelberger ideal – thus also called the "index relation" (cf. [MC2]). This is quite analogous to a formula due to Iwasawa (cf. [I]) for cyclotomic fields. In fact the salient features in the two theories exhibit such a remarkable similarity, that one suspects there is a common generalization.

Let then here

$$\Gamma = \mathbb{F}_{l^k}^+, \qquad G = \mathbb{F}_{l^k}^*.$$

The group G contains an element, which acts on Γ by multiplication by -1. We shall denote this by j, reserving the symbol -1 for the negative identity of $\mathbb{Z}$. For any $\mathbb{Z}G$-module M we denote by M^- the kernel in M under multiplication by $1+j$. Thus $\Gamma = \Gamma^-$; j induces on $\mathbb{Z}\Gamma$ the standard involution $^-$, and $\mathrm{Cl}(\mathbb{Z}\Gamma)^-$ is the subgroup of classes c with $c\bar{c} = 1$. We write again $J = J_{1,k}$ for the Stickelberger ideal. We have

Theorem 42 (cf. [MC2]). *Let l be an odd prime. Then*

$$\mathrm{order}(\mathrm{Cl}(\mathbb{Z}\Gamma)^-) = [\mathbb{Z}G^- : J^-]. \qquad \square$$

Remark 1. In this form the result does not generalize to $\Gamma = S_{n,k}^+$ for $n > 1$. Instead one uses the more general Stickelberger elements, defined in (4.16a), and the corresponding "second" Stickelberger ideal $J'_{n,k}$.

McCulloh's generalization is suggestive in its close analogy to cyclotomic fields. Let again for the moment $\Gamma = S_{n,k}^+$, with arbitrary n. Let Δ be the maximal subgroup of Γ of exponent l, and write

$$\sigma = \sum_{\delta \in \Delta} \delta.$$

Then for all $n \geqslant 1$, and all k (with l odd),

$$\mathrm{order}(\mathrm{Cl}(\mathbb{Z}\Gamma/\sigma\mathbb{Z}\Gamma)^-) = [\mathbb{Z}G^- : J'^-].$$

Note that when $k = 1$, then $\mathbb{Z}\Gamma/\sigma\mathbb{Z}\Gamma$ is the ring of integers in $\mathbb{Q}(l^n)$.

Remark 2. For $n = 1$ and $l = 2$ one can instead prove the equation

$$\mathrm{order}(\mathrm{Cl}(\mathbb{Z}\Gamma)) = [\mathbb{Z}G : J].$$

Remark 3. As McCulloh points out, the statement of Theorem 42 exhibits features pointing to a connection with the work of Kubert-Lang (cf. [KL]) on the group of cuspidal divisor classes in fields of modular functions of an appropriate level.

We shall only briefly outline some aspects of the proof of Theorem 42.

Let $\mathfrak{M}$ be the maximal order in $\mathbb{Q}\Gamma$. Then (cf. (I.2.20)), we have an exact sequence of $\mathbb{Z}G$-modules

$$1 \to D(\mathbb{Z}\Gamma) \to \mathrm{Cl}(\mathbb{Z}\Gamma) \to \mathrm{Cl}(\mathfrak{M}) \to 1, \tag{5.1}$$

giving rise to a sequence

$$1 \to D(\mathbb{Z}\Gamma)^- \to \mathrm{Cl}(\mathbb{Z}\Gamma)^- \to \mathrm{Cl}(\mathfrak{M})^- \to 1. \tag{5.2}$$

One then has

5.1. Lemma. (5.2) *is exact.*

Proof. The sequences (5.1) and (5.2) split up into products of sequences, one for each q-primary part, with q running through the rational primes. For q odd, one clearly gets an exact sequence, viewing all groups as modules over the group of order 2, generated by j. Moreover we know (cf. [RU1]) that $D(\mathbb{Z}\Gamma)$ is an l-group, and as l is odd, we have an isomorphism

$$\mathrm{Cl}(\mathbb{Z}\Gamma)_2 \cong \mathrm{Cl}(\mathfrak{M})_2$$

of 2-primary parts, and thus an isomorphism

$$\mathrm{Cl}(\mathbb{Z}\Gamma)_2^- \cong \mathrm{Cl}(\mathfrak{M})_2^-.$$

Putting everything together we get the exactness of (5.2). □

We continue to write $J = J_{1,k}$, $G = \mathbb{F}_{l^k}$, and also abbreviate for the moment $G_1 = \mathbb{F}_l^*$, $J_1 = J_{1,1}$. The embedding $\mathbb{F}_l \subset \mathbb{F}_{l^k}$ gives an embedding $J_1 \subset \mathbb{Z}G_1 \subset \mathbb{Z}G$. Theorem 42 now follows clearly from Lemma 5.1 and the following two propositions.

5.2. Proposition. *For l odd,*

$$\mathrm{order}(\mathrm{Cl}(\mathfrak{M})^-) = [\mathbb{Z}G^- : (J_1\mathbb{Z}G)^-].$$

5.3. Proposition. *For l odd,*

$$J_1\mathbb{Z}G \supset J,$$

and

$$\mathrm{order}(D(\mathbb{Z}\Gamma)^-) = [(J_1\mathbb{Z}G)^- : J^-]. \tag{5.3}$$

Proof of 5.2. $\mathfrak{M}$ is the product of $\mathbb{Z}$ and of $[G:G_1]$ copies of the ring $\mathbb{Z}[l]$ of integers in $\mathbb{Q}(l)$. (G_1 as defined prior to 5.2.) As $j \in G_1$, we conclude that $\mathrm{Cl}(\mathfrak{M})^-$ is the product of $[G:G_1]$ copies of $\mathrm{Cl}(\mathbb{Z}[l])^-$. On the other hand, as $\mathbb{Z}G$ is free over $\mathbb{Z}G_1$, of rank $[G:G_1]$, we also have

$$[\mathbb{Z}G^- : (J_1\mathbb{Z}G)^-] = [\mathbb{Z}G_1^- : J_1^-]^{[G:G_1]}.$$

Thus the proof of 5.2 reduces to the proof of the relation

$$\operatorname{order}(\mathrm{Cl}(\mathbb{Z}[l])^-) = [\mathbb{Z}G_1^- : J_1^-].$$

This however is Iwasawa's formula (cf. [I]).

Outline of proof of 5.3. The proof consists essentially of separate explicit computations of the two sides of (5.3), each ending up with the same expression

$$(5.4) \qquad l^M, \qquad M = k(l^k + 3) - [G:G_1](l+3).$$

The computation for order $(D(\mathbb{Z}\Gamma)^-)$ was done by the author in [F29], that for $[(J_1\mathbb{Z}G)^- : J^-]$ by McCulloh in [MC2].

The inclusion relation in Proposition 5.3 is a consequence of the equation

$$\Theta_{1,k} = \Theta_{1,1}\beta$$

with

$$\beta = \sum_{\substack{g \in G \\ t(g)=1}} g^{-1}.$$

The essential step in the proof of the equation

$$(5.5) \qquad [(J_1\mathbb{Z}G)^- : J^-] = l^M, \qquad M \text{ as in } (5.4),$$

consists of showing that

$$(5.6) \qquad |\phi(\beta)|^2 = l^{s(\phi)}, \qquad s(\phi) \text{ a non-negative integer},$$

for each Abelian character ϕ of G, extended by linearity to $\mathbb{Z}G$. This will then imply that

$$[\Theta_1\mathbb{Z}G : \Theta\mathbb{Z}G]^2 = l^s, \quad s = \sum s(\phi), \quad \text{sum over those } \phi \text{ with } \phi(\Theta_1) \neq 0.$$

One also knows that if ϕ_1 is an odd character mod l, i.e., an Abelian character of G_1 with $\phi_1(j) = -1$, then $\phi_1(\Theta_{1,1}) \neq 0$. This is a fairly deep classical result, which together with (5.6) lies at the arithmetic core of the proof of (5.5). The rest is a straightforward module theoretic argument, using also Proposition 4.8 (iii).

To prove (5.6) one uses classical Gauss sums for finite fields. If ϕ is a character of G, denote in the sequel by ϕ_1 its restriction to G_1. Also write

$$\beta_0 = \sum_{\substack{g \in G \\ t(g)=0}} g^{-1}.$$

Then, with β as above, one obtains

$$(5.7) \qquad \sum_{g \in G} \phi(g) = \phi^{-1}(\beta_0) + \phi^{-1}(\beta) \sum_{h \in G_1} \phi_1(h).$$

Similarly for the Gauss sums we get

(5.8) $$\sum_{g\in G} \phi(g)e^{2\pi i t(g)/l} = \phi^{-1}(\beta_0) + \phi^{-1}(\beta) \sum_{h\in G_1} \phi(h)e^{2\pi i t_1(h)/l},$$

where we recall that for $h \in G_1 = \mathbb{F}_l^*$, $t_1(h)$ is the least non-negative integer in the residue class h.

If ϕ_1 is non-trivial (i.e., not the identity character) then both sums in (5.7) vanish, thus $\phi^{-1}(\beta_0) = 0$, and we get an expression for $\phi(\beta)$ from (5.8). If ϕ_1 is trivial, but ϕ is not, then the left hand of (5.7) vanishes, but none of the other sums in (5.7) and (5.8) do. We can thus eliminate $\phi^{-1}(\beta_0)$ and again get an expression for $\phi(\beta)$. Finally if ϕ is trivial, then the computation is trivial. Using the fact that, for any $k \geqslant 1$,

$$\left|\sum_{g\in G} \phi(g)e^{2\pi i t(g)/l}\right| = l^{k/2} \qquad \text{(absolute value of Gauss sums)}$$

if ϕ is non-trivial, we end up with

$$|\phi(\beta)|^2 = \begin{cases} l^{k-1}, & \text{if } \phi_1 \text{ is non-trivial,} \\ l^{k-2}, & \text{if } \phi_1 \text{ is trivial, but } \phi \text{ is not,} \\ l^{2(k-1)}, & \text{if } \phi \text{ is trivial.} \end{cases}$$

This is the precise form of (5.6).

Next we come to the equation

(5.9) $$\operatorname{order}(D(\mathbb{Z}\Gamma)^-) = l^M, \qquad M \text{ as in (5.4).}$$

As Γ is Abelian, we can omit the "Det" in the formulae for $D(\mathbb{Z}\Gamma)$ (cf. (I.2.18), (I.2.19)) and thus have

$$D(\mathbb{Z}\Gamma) \cong \mathfrak{U}(\mathfrak{M})/\mathfrak{M}^*\mathfrak{U}(\mathbb{Z}\Gamma)$$

($\mathfrak{U}$ = unit ideles). But, for all $p \neq l$, $\mathfrak{M}_{\mathfrak{p}}^* = \mathbb{Z}_p\Gamma^*$. Hence we get in fact an exact sequence

(5.10) $$1 \to \mathfrak{M}^*/\mathbb{Z}\Gamma^* \to \mathfrak{M}_l^*/\mathbb{Z}_l\Gamma^* \to D(\mathbb{Z}\Gamma) \to 1,$$

coming from the localization map $\mathfrak{M}^* \to \mathfrak{M}_l^*$. Let $\mathfrak{r}_l$ be the radical of $\mathfrak{M}_l$ and write for any subring $\mathfrak{A}$ of $\mathfrak{M}_l$, $Y^{(1)}(\mathfrak{A}) = \mathfrak{A} \cap 1 + \mathfrak{r}_l$. As $D(\mathbb{Z}\Gamma)$ is an l-group (cf. [RU1]), or more directly as $\mathfrak{M}^*(1 + \mathfrak{r}_l) = \mathfrak{M}_l^*$ (Hilbert's cyclotomic units!), one deduces from (5.10), on taking l-primary components, that one also gets an exact sequence

$$1 \to Y^{(1)}(\mathfrak{M})/Y^{(1)}(\mathbb{Z}\Gamma) \to Y^{(1)}(\mathfrak{M}_l)/Y^{(1)}(\mathbb{Z}_l\Gamma) \to D(\mathbb{Z}\Gamma) \to 1.$$

As all the groups in this sequence have l-power order, the sequence is still exact on applying the functor $M \mapsto M^-$. Moreover $Y^{(1)}(\mathfrak{M})^-$ and $Y^{(1)}(\mathbb{Z}\Gamma)^-$ are finite groups, in fact $Y^{(1)}(\mathbb{Z}\Gamma)^- = \Gamma$ and $Y^{(1)}(\mathfrak{M})^-$ is the product of the copies of μ_l (l-th roots of unity) in $\mathfrak{M}$. We thus have now the equation

$$\operatorname{order}(D(\mathbb{Z}\Gamma)^-) = [\operatorname{order}(Y^{(1)}(\mathfrak{M}_l)^-/Y^{(1)}(\mathbb{Z}_l\Gamma)^-)\ \operatorname{order}(\Gamma)]/\operatorname{order}(Y^{(1)}(\mathfrak{M})^-).$$

It remains to evaluate the index

$$[(1+\mathfrak{r}_l)^- : ((1+\mathfrak{r}_l)^- \cap \mathbb{Z}_l\Gamma)].$$

Via the filtration of $1+\mathfrak{r}_l$ defined by the subgroups $1+\mathfrak{r}_l^k$, and the usual isomorphisms

$$\mathfrak{r}_l^k/\mathfrak{r}_l^{k+1} \text{ (additive)} \cong 1+\mathfrak{r}_l^k/1+\mathfrak{r}_l^{k+1} \text{ (multiplicative)},$$

one shows that the above index is also an index

$$[\mathfrak{r}_l^- : (\mathfrak{r}_l \cap \mathbb{Z}_l\Gamma)^-]$$

of additive groups. Its evaluation is now fairly straightforward. □

We next come to the main theorem of the present section. Unless otherwise mentioned, Γ is again $\mathbb{F}_{l^k}^+$, G is $\mathbb{F}_{l^k}^*$ and J is the Stickelberger ideal of $\mathbb{Z}G$. The functor $X \mapsto X^\Gamma$, or the functor $X \mapsto X \otimes_{\mathfrak{o}_K\Gamma} \mathfrak{o}_K$ (they are equivalent on locally free $\mathfrak{o}_K\Gamma$-modules) defines a homomorphism $\mathrm{Cl}(\mathfrak{o}_K\Gamma) \to \mathrm{Cl}(\mathfrak{o}_K)$ whose kernel we denote by $\mathrm{Cl}(\mathfrak{o}_K\Gamma)_0$. This homomorphism is split by the homomorphism $\mathrm{Cl}(\mathfrak{o}_K) \to \mathrm{Cl}(\mathfrak{o}_K\Gamma)$ coming from ring extension, i.e., from the functor $Y \mapsto Y \otimes_{\mathfrak{o}_K} \mathfrak{o}_K\Gamma$. Both homomorphisms admit $\mathbb{Z}G$-action. It follows in particular that

$$\text{(5.11)} \qquad \mathrm{Cl}(\mathfrak{o}_K\Gamma)_0^J = \mathrm{Cl}(\mathfrak{o}_K\Gamma)_0 \cap \mathrm{Cl}(\mathfrak{o}_K\Gamma)^J.$$

We shall now consider tame surjections $\pi\colon \Omega_K \to \Gamma$. We shall use the notation of §4, i.e., rewrite these as surjections from $\mathrm{Gal}(K^t/K)$ onto Γ, where K^t is the maximal tame extension of K. Denote by $\mathfrak{R}(K,\Gamma)$ the set of *realizable classes*, i.e., the subset of $\mathrm{Cl}(\mathfrak{o}_K\Gamma)$ of classes of form $(\mathfrak{o}_{N,\pi})_{\mathfrak{o}_K\Gamma}$, π as above.

Theorem 43 (cf. [MC3]).

$$\mathfrak{R}(K,\Gamma) = \mathrm{Cl}(\mathfrak{o}_K\Gamma)_0^J.$$ □

Before turning to the proof we note an important Corollary.

Corollary (Stickelberger relation).

$$\mathrm{Cl}(\mathbb{Z}\Gamma)^J = 1.$$

By the Theorem we have to show that

$$\mathfrak{R}(\mathbb{Q},\Gamma) = 1.$$

This is of course an immediate consequence of Theorem 9, since for an Abelian group we have $R_\Gamma^s = \mathrm{Tr}(R_\Gamma)$. It can however be established directly without any effort. For, every tame Abelian extension N of $\mathbb{Q}$ is subfield of a field $\mathbb{Q}(m)$ of m-th roots of unity, m square free. Any primitive m-th root y of 1 will generate a normal integral basis of $\mathbb{Q}(m)$ and its trace $t_{\mathbb{Q}(m)/N}y$ will thus generate a normal integral basis of N. Thus indeed always $(\mathfrak{o}_{N,\pi})_{\mathbb{Z}\Gamma} = 1$. □

The proof of Theorem 43 falls naturally into two parts, namely the proofs of the inclusion relations

(5.12) $$\mathfrak{R}(K,\Gamma) \subset \mathrm{Cl}(\mathfrak{o}_K\Gamma)^J,$$

and

(5.13) $$\mathrm{Cl}(\mathfrak{o}_K\Gamma)_0^J \subset \mathfrak{R}(K,\Gamma),$$

respectively. Note here in passing, that always $\mathfrak{o}_{N,\pi}^{\Gamma} = \mathfrak{o}_K$, therefore $(\mathfrak{o}_{N,\pi})_{\mathfrak{o}_K\Gamma} \in \mathrm{Cl}(\mathfrak{o}_K\Gamma)_0$. By (5.11), the relation (5.12) thus does imply that

$$\mathfrak{R}(K,\Gamma) \subset \mathrm{Cl}(\mathfrak{o}_K\Gamma)_0^J,$$

and this in conjunction with (5.13) yields the Theorem.

In preparation for the proof we shall first need a good description of $\mathfrak{J}^{(l)}(\mathfrak{o}_K\Gamma)$, and of its subgroup $\mathfrak{J}^{(l)}(\mathfrak{o}_K\Gamma)_0$ of modules M, which under the augmentation $K\Gamma \to K$ get mapped onto $\mathfrak{o}_K$. As Γ is Abelian, it is now more convenient to work with the set (and multiplicative group) $\hat{\Gamma}$ of its Abelian characters, i.e., replace the Hom groups $\mathrm{Hom}(R_\Gamma, \cdot)$ by the equivalent Map groups $\mathrm{Map}(\hat{\Gamma},\ )$. Thus via

$$c \mapsto [\phi \mapsto \phi(c)] \qquad (\phi \in \hat{\Gamma} \text{ extended linearly to } K\Gamma \to \mathbb{Q}^c),$$

we get an isomorphism

$$K\Gamma^* \cong \mathrm{Map}_{\Omega_K}(\hat{\Gamma}, \mathbb{Q}^{c*}).$$

However, $\hat{\Gamma}$ is fixed under $\Omega_{K(l)}$, ($K(l)$ the field of l-th roots of unity over K). Hence, writing from now on

(5.14) $$D = \mathrm{Gal}(K(l)/K),$$

we also have

(5.15) $$K\Gamma^* \cong \mathrm{Map}_D(\hat{\Gamma}, K(l)^*).$$

We shall now apply a similar description to $\mathfrak{J}^{(l)}(\mathfrak{o}_K\Gamma)$. Ring extension from $\mathfrak{o}_K\Gamma$ to the maximal order $\mathfrak{M}$ gives rise to an isomorphism $\mathfrak{J}^{(l)}(\mathfrak{o}_K\Gamma) \cong \mathfrak{J}^{(l)}(\mathfrak{M})$. On the other hand $\mathfrak{M}$ is a direct product of Dedekind domains, more precisely of one copy of $\mathfrak{o}_K$, corresponding to the identity character (or augmentation) ε, and of copies of $\mathfrak{o}_{K(l)}$. As all prime ideals of $\mathfrak{o}_K$, not above l, are non-ramified in $K(l)$, it follows that

$$\mathfrak{J}^{(l)}(\mathfrak{o}_{K(l)})^D = \mathfrak{J}^{(l)}(\mathfrak{o}_K).$$

We conclude from this discussion:

(i) The map

$$\mathfrak{c} \mapsto [\phi \mapsto \phi(\mathfrak{c})] \qquad (\mathfrak{c} \in \mathfrak{J}^{(l)}(\mathfrak{o}_K\Gamma))$$

gives rise to an isomorphism

(5.16) $$\mathfrak{J}^{(l)}(\mathfrak{o}_K\Gamma) \cong \mathrm{Map}_D(\hat{\Gamma}, \mathfrak{J}^{(l)}(\mathfrak{o}_{K(l)})).$$

Evaluation at ε defines a homomorphism

$$\hat{\varepsilon}\colon \mathrm{Map}_D(\hat{\Gamma}, \mathfrak{J}^{(l)}(\mathfrak{o}_{K(l)})) \to \mathfrak{J}^{(l)}(\mathfrak{o}_K),$$

and (5.16) yields an isomorphism

(5.17) $$\mathfrak{J}^{(l)}(\mathfrak{o}_K\Gamma)_0 \cong \mathrm{Ker}\,\hat{\varepsilon}.$$

One also needs a more explicit – although non-canonical – description for computational purposes. As D acts freely on the set $\hat{\Gamma}$ with ε removed, we have

(5.17a) $$\mathfrak{J}^{(l)}(\mathfrak{o}_K\Gamma)_0 \cong \mathrm{Map}(\Phi, \mathfrak{J}^{(l)}(\mathfrak{o}_{K(l)}))$$

where Φ is a set of representatives of the orbits of $\hat{\Gamma}$, with ε removed, under D.

(ii) $\mathfrak{J}^{(l)}(\mathfrak{o}_K\Gamma)$ and $\mathfrak{J}^{(l)}(\mathfrak{o}_K\Gamma)_0$ are free on prime ideals. Writing a fractional ideal as product of prime ideals

$$\mathfrak{c} = \prod_{\mathfrak{p}} \mathfrak{p}^{v_{\mathfrak{p}}(\mathfrak{c})},$$

we know that $\mathfrak{c}$ is integral, i.e., $\mathfrak{c} \subset \mathfrak{o}_K\Gamma$, precisely if the $v_{\mathfrak{p}}(\mathfrak{c})$ are all non-negative.

If $x \in K\Gamma^*$ and x a unit above l, we shall also write

$$v_{\mathfrak{p}}(x\mathfrak{o}_K\Gamma) = v_{\mathfrak{p}}(x).$$

(iii) The groups $\mathfrak{J}^{(l)}(\mathfrak{o}_K\Gamma)$, $\mathfrak{J}^{(l)}(\mathfrak{o}_K\Gamma)_0$ are $\mathbb{Z}G$-modules. Also $\hat{\Gamma}$ is a $\mathbb{Z}G$-module, via the action of $\mathbb{Z}G$ on Γ, and so is $\mathrm{Map}_D(\hat{\Gamma}, \mathfrak{J}^{(l)}(\mathfrak{o}_{K(l)}))$, via the action of $\mathbb{Z}G$ on $\hat{\Gamma}$. With respect to these structures, (5.16), (5.17) are isomorphisms of $\mathbb{Z}G$-modules.

We now turn to the proof of (5.12). Let $\pi\colon \mathrm{Gal}(K^t/K) \to \Gamma$ be a surjection. Go back to the notation introduced in §4. Let $b \in K^t\Gamma^*$ be a generator of X_π, and a unit above l, and write $a = b^l$. Thus $a \in K\Gamma^*$. By (4.11)

$$a\mathfrak{o}_K\Gamma = \mathfrak{a}\mathfrak{m}^l, \qquad \mathfrak{a}, \mathfrak{m} \in \mathfrak{J}^{(l)}(\mathfrak{o}_K\Gamma),$$

$\mathfrak{a}$ integral and l-power free.

5.4. Proposition (cf. [MC3]). *There is a square free integral ideal* $\mathfrak{n}$ *of* $\mathfrak{o}_K\Gamma$, $\mathfrak{n} \in \mathfrak{J}^{(l)}(\mathfrak{o}_K\Gamma)$, *with*

$$\mathfrak{a} = \mathfrak{n}^{\Theta}. \qquad \square$$

Remark. This proposition, and consequently (5.12), generalize in appropriate form to $n > 1$.

We shall first deduce (5.12) from the proposition. Let A be again the annihilating ideal of Γ in $\mathbb{Z}G$. With b as above, we have $b^\lambda \in K\Gamma^*$ for all $\lambda \in A$, by

Proposition 4.4. Therefore, and recalling that $\lambda\Theta/l \in \mathbb{Z}G$ (by Proposition 4.8) we get, from Proposition 5.4,

$$b^{\lambda l}\mathfrak{o}_K\Gamma = (a\mathfrak{o}_K\Gamma)^\lambda = (\mathfrak{n}^{\lambda\Theta/l}\mathfrak{m}^\lambda)^l.$$

As $\mathfrak{J}^{(l)}(\mathfrak{o}_K\Gamma)$ is torsion free, we deduce that

$$b^\lambda\mathfrak{o}_K\Gamma = \mathfrak{n}^{\lambda\Theta/l}\mathfrak{m}^\lambda.$$

Going over to module classes we now have

$$(\mathfrak{m})_{\mathfrak{o}_K\Gamma}^{-\lambda} = (\mathfrak{n})_{\mathfrak{o}_K\Gamma}^{\lambda\Theta/l},$$

and this holds for all $\lambda \in A$. By Proposition 4.9, there is a $\lambda \in A$ and a $v \in J$ so that

$$(\mathfrak{m})_{\mathfrak{o}_K\Gamma}^{\lambda+v} = (\mathfrak{m})_{\mathfrak{o}_K\Gamma}.$$

Thus, as $\lambda\Theta/l \in J$, we get

$$(\mathfrak{m})_{\mathfrak{o}_K\Gamma} = (\mathfrak{m})_{\mathfrak{o}_K\Gamma}^{v}(\mathfrak{n})_{\mathfrak{o}_K\Gamma}^{-\lambda\Theta/l} \in \mathrm{Cl}(\mathfrak{o}_K\Gamma)^J.$$

By Proposition 4.5, $(\mathfrak{o}_{N,\pi})_{\mathfrak{o}_K\Gamma} \in \mathrm{Cl}(\mathfrak{o}_K\Gamma)^J$, as we had to show. □

Proof of Proposition 5.4. We shall define $\mathfrak{n}$ in terms of the element a above, by giving the values $v_\mathfrak{p}(\mathfrak{n})$ for all $\mathfrak{p}$, as follows:

$$\text{(5.18)} \qquad \begin{cases} v_\mathfrak{p}(\mathfrak{n}) = 1, & \text{if for all } g \in G, \quad v_\mathfrak{p}(a^g) \equiv t(g) \pmod{l}, \\ v_\mathfrak{p}(\mathfrak{n}) = 0, & \text{otherwise.} \end{cases}$$

As $v_\mathfrak{p}(a^g) = 0$ for all g, with at most finitely many exceptions $\mathfrak{p}$, we also have $v_\mathfrak{p}(\mathfrak{n}) = 0$, with at most finitely many exceptions. Thus (5.18) does indeed define an $\mathfrak{n} \in \mathfrak{J}^{(l)}(\mathfrak{o}_K\Gamma)$, and $\mathfrak{n}$ is clearly integral and square free. Using the obvious equation $v_\mathfrak{p}(\mathfrak{n}^{h^{-1}}) = v_{\mathfrak{p}^h}(\mathfrak{n})$, for $h \in G$, we deduce

$$\text{(5.18a)} \qquad \begin{cases} v_\mathfrak{p}(\mathfrak{n}^{h^{-1}}) = 1 & \text{if for all } g \in G, \quad v_\mathfrak{p}(a^g) \equiv t(gh) \pmod{l}, \\ v_\mathfrak{p}(\mathfrak{n}^{h^{-1}}) = 0, & \text{otherwise.} \end{cases}$$

The $\mathfrak{n}^{h^{-1}}$ are again integral and square free. Moreover for $h, f \in G$, $h \neq f$, the ideals $\mathfrak{n}^h$ and $\mathfrak{n}^f$ are coprime, i.e., $v_\mathfrak{p}(\mathfrak{n}^h)v_\mathfrak{p}(\mathfrak{n}^f) = 0$, for all $\mathfrak{p}$. Otherwise we would have for all $g \in G$, $t(gh^{-1}) = t(fh^{-1})$. Going back to the trace $T\colon \mathbb{F}_{l^k} \to \mathbb{F}_l$, this would imply that $T(g(h^{-1} - f^{-1})) = 0$ for all $g \in G$, whence $h^{-1} - f^{-1} = 0$.

The fact that the $\mathfrak{n}^{h^{-1}}$ are mutually coprime now implies that $\mathfrak{n}^\Theta$ is an l-power free integral ideal in $\mathfrak{J}^{(l)}(\mathfrak{o}_K\Gamma)$. It remains for us to show that, for all $\mathfrak{p}$,

$$\text{(5.19)} \qquad v_\mathfrak{p}(\mathfrak{n}^\Theta) \equiv v_\mathfrak{p}(a) \pmod{l}.$$

By what we have seen above, we have $v_\mathfrak{p}(\mathfrak{n}^\Theta) > 0$ if and only if the following condition is satisfied: $\exists h \in G$, with $t(h) = v_\mathfrak{p}(\mathfrak{n}^\Theta)$, so that for all $g \in G$, $v_\mathfrak{p}(a^g) \equiv$

$t(gh) \pmod{l}$. Putting here $g = 1$, we get (5.19) for such a prime ideal $\mathfrak{p}$. To complete the proof of (5.19) we thus have to show that if, for a given prime ideal $\mathfrak{p}$, $v_\mathfrak{p}(a) \equiv v$ $\pmod{l}$, $0 < v < l$, then $v_\mathfrak{p}(\mathfrak{n}^{h^{-1}}) = 1$ for some $h \in G$, or equivalently, by (5.18a)

$$(5.20) \qquad v_\mathfrak{p}(a^g) \equiv t(gh) \pmod{l}, \qquad \text{for all} \qquad g \in G.$$

This is the crux of the whole proof of Proposition 5.4. We shall show that (5.20) does hold for

$$h = \sum_i \bar{v}_\mathfrak{p}(a^{\gamma_i})\gamma_i'$$

where $\bar{v}_\mathfrak{p}(a^{\gamma_i})$ is the class of $v_\mathfrak{p}(a^{\gamma_i}) \bmod l$, $\{\gamma_i\}$ is a basis of Γ over $\mathbb{F}_l$, with $\gamma_1 = 1$, and $\{\gamma_i'\}$ the dual basis. As $\bar{v}_\mathfrak{p}(a^{\gamma_1}) = \bar{v}_\mathfrak{p}(a) \neq 0$, the element h of $\Gamma = \mathbb{F}_{l^k}$ does indeed belong to $G = \mathbb{F}_{l^k}^*$. Multiplying the defining equation of h by γ_j and applying t, we get

$$(5.21) \qquad t(h\gamma_j) \equiv v_\mathfrak{p}(a^{\gamma_j}) \pmod{l}.$$

By Lemma 4.7 and Proposition 4.4,

$$b^{d(g)} \in K\Gamma^*, \qquad \text{for all} \qquad g \in G,$$

where $d(g) = g - \sum_j t(g\gamma_j')\gamma_j$. Therefore

$$v_\mathfrak{p}(a^g) \equiv \sum_j t(g\gamma_j')v_\mathfrak{p}(a^{\gamma_j}) \pmod{l}.$$

Hence, by (5.21),

$$v_\mathfrak{p}(a^g) \equiv \sum_j t(g\gamma_j')t(h\gamma_j) \pmod{l}.$$

The right hand side in this congruence is $\equiv t(gh)$. Thus we get (5.20), as required.

□

We now turn to the proof of (5.13). Here we shall only give an outline, and omit some rather technical details. In any case this proof is probably not yet in its definitive form.

The map $\mathfrak{I}^{(l)}(\mathfrak{o}_K\Gamma) \to \mathrm{Cl}(\mathfrak{o}_K\Gamma)$ is surjective, by weak approximation. Similarly $\mathfrak{I}^{(l)}(\mathfrak{o}_K\Gamma)_0 \to \mathrm{Cl}(\mathfrak{o}_K\Gamma)_0$ is surjective. It is also a $\mathbb{Z}G$-homomorphism. Thus any given class in $\mathrm{Cl}(\mathfrak{o}_K\Gamma)_0^J$ is of form $(\mathfrak{s}')_{\mathfrak{o}_K\Gamma}$, with $\mathfrak{s}' \in \mathfrak{I}^{(l)}(\mathfrak{o}_K\Gamma)_0^J$. We want to represent this class in the form $(\mathfrak{o}_{N,\pi})_{\mathfrak{o}_K\Gamma}$. By Proposition 4.8 (iii)

$$\mathfrak{s}' = \prod_i \mathfrak{s}_i^{(v_i\Theta/l)}, \qquad \mathfrak{s}_i \in \mathfrak{I}^{(l)}(\mathfrak{o}_K\Gamma)_0, \qquad v_i \in A.$$

Hence for any $v \in A$

$$(5.22) \qquad \mathfrak{s}'^v = \mathfrak{s}^{v\Theta/l}, \qquad \mathfrak{s} = \prod_i \mathfrak{s}_i^{v_i}.$$

The strategy will be to find a square free integral ideal $\mathfrak{n}$ in $\mathfrak{I}^{(l)}(\mathfrak{o}_K\Gamma)_0$, in the same class as $\mathfrak{s}$ so that $\mathfrak{n}^{\Theta}$ is (integral and) l-power free. Then we shall have

(5.23a) $$y\mathfrak{s} = \mathfrak{n}, \qquad y \in K\Gamma^*, \qquad y \text{ a unit at } l$$

and so

(5.23b) $$\mathfrak{s}'^{l} y^{\Theta} = \mathfrak{n}^{\Theta}.$$

The aim at this final stage is to find a surjection π, such that

(5.24) $$y^{\Theta} = b^{l}, \qquad b \in K^{t}\Gamma^*, \qquad b \text{ a generator of } X_{\pi}.$$

By Proposition 4.5, it then follows that $(\mathfrak{s}')_{\mathfrak{o}_K\Gamma} = (\mathfrak{o}_{N,\pi})_{\mathfrak{o}_K\Gamma}$.

We shall give a construction of $\mathfrak{n}$, so that moreover $\mathfrak{n}^{\Theta}$ is not divisible by the prime ideals in a finite preassigned set. This guarantees that there are an infinity of such π. We shall proceed via the isomorphism (5.17a), using the same symbol for the ideals as for the associated maps. We impose the following requirements on $\mathfrak{n}$.

(i) For each $\phi \in \Phi$, $\mathfrak{n}(\phi)$ is a prime ideal of $\mathfrak{o}_{K(l)}$, of degree 1 over $\mathfrak{o}_K$.

(ii) The $\mathfrak{n}(\phi)$ for distinct ϕ lie over distinct prime ideals of $\mathfrak{o}_K$.

(iii) $\mathfrak{n}(\phi)$ lies in the same ray class as $\mathfrak{s}(\phi)$ modulo l^{k+1}.

(iv) The $\mathfrak{n}(\phi)$ and their conjugates are prime to any preassigned ideal of $\mathfrak{o}_K$ and are distinct from any prime ideals occurring in $\mathfrak{s}(\phi')$ for all ϕ'.

It is clear that such an $\mathfrak{n}$ will exist. It will be integral and square free. By (iii) it will lie in the same class as $\mathfrak{s}$. Moreover, now reverting to the canonical description (5.17) of $\mathfrak{I}^{(l)}(\mathfrak{o}_K\Gamma)_0$, the $\mathfrak{n}(\chi)$ (i.e., the $\chi(\mathfrak{n})$) for $\chi \in \hat{\Gamma}$ are distinct prime ideals of $\mathfrak{o}_{K(l)}$. For, every χ is of form ϕ^{d}, $d \in D$, $\phi \in \Phi$, and the assertion thus follows from conditions (i) and (ii) above.

Next the $\mathfrak{n}^{g}$ are integral and square free, as $\mathfrak{n}$ is. Moreover, for $g \neq h$, $\mathfrak{n}^{g}$ and $\mathfrak{n}^{h}$ are coprime. Otherwise for some $\chi \in \hat{\Gamma}$, $\chi(\mathfrak{n}^{g}) = \chi(\mathfrak{n}^{h})$, or $\chi^{g^{-1}}(\mathfrak{n}) = \chi^{h^{-1}}(\mathfrak{n})$, i.e., $\mathfrak{n}(\chi^{g^{-1}}) = \mathfrak{n}(\chi^{h^{-1}})$. But this, as we have seen, implies $\chi^{g^{-1}} = \chi^{h^{-1}}$. As G acts freely on $\hat{\Gamma}$ with ε removed, it follows that $g = h$.

From the above we can finally conclude that $\mathfrak{n}^{\Theta}$ is integral and l-power free. We shall in conclusion outline the construction of an element b, as in (5.24), without however giving proofs. These use the particular properties of $\mathfrak{n}$ we have postulated and are at present based on classical Kummer theory.

We fix once and for all an element $\psi \in \hat{\Gamma}$, different from ε. We then let Y be the $\mathbb{Z}G$-submodule of $K\Gamma^*$ generated by y. We put $\psi(Y) = Y' \subset K(l)^*$ and write $M_1 = K(l)(Y'^{1/l})$. This is a Kummer extension of $K(l)$. By the defining condition (iii) on $\mathfrak{n}$, $M_1/K(l)$ is tame. As Y' is seen to be D-invariant, M_1/K is Galois, and as $([M_1 : K(l)], \text{order } D) = 1$, the exact sequence

$$1 \to \mathrm{Gal}(M_1/K(l)) \to \mathrm{Gal}(M_1/K) \to D \to 1$$

splits. Fix a splitting and put $M = M_1^{D}$. Identifying $D = \mathrm{Gal}(M_1/M) = \mathrm{Gal}(M(l)/M)$ we get (just as in (5.15)) an isomorphism

$$M\Gamma^* \cong \mathrm{Map}_D(\hat{\Gamma}, M(l)^*).$$

By the construction of $M(l) = M_1$, there is an element $\tilde{y} \in M\Gamma^*$ with $\varepsilon(\tilde{y}) = 1$, and

$\chi(\tilde{y})^l = \chi(y)$ for all $\chi \in \hat{\Gamma}$. Thus $\tilde{y}^l = y$. Now put $b = \tilde{y}^\Theta$. Then b has the required properties. It is already obvious of course that for all $v \in A$, $b^v = \tilde{y}^{\Theta v} = y^{\Theta/lv} \in K\Gamma^*$. What we have not shown here is that $b^\omega = b\pi(\omega)$, for some surjection π: $\mathrm{Gal}(K^t/K) \to \Gamma$ and all $\omega \in \mathrm{Gal}(K^t/K)$. This is where classical Kummer theory enters. An alternative approach, more along the lines of the theory of §4, is as follows: As $\tilde{y}^l = \tilde{y}^{\omega l}$, for all ω, we have $\tilde{y}^\omega = \tilde{y}\rho(\omega)$, $\rho(\omega)$ an l-th root of unity in $K^t\Gamma^*$, depending on ω. But then $\rho(\omega)^\Theta = \pi(\omega) \in \Gamma$, with π: $\mathrm{Gal}(K^t/K) \to \Gamma$ a homomorphism. It follows that, for all ω, $b^\omega = b\pi(\omega)$. The fact that π is surjective should be deduced from the special structure of the $\mathbb{Z}G$-module Y. □

Appendix

Here we shall add a brief outline of some recent developments in areas which are not covered in our survey, but are related to it.

A. Hermitian Theory for Tame Extensions

(See also the remark at the end of II §4, and note [3] to Chapter III.) Literature: [F20], [F24], [F25], [Ri2], [CN-T1], [CN-T2]. The motivation came in the first place from the fact that the basic equation $tW_{N/K} = U_{N/K}$ of Theorem 9 determines the (stable) module structure, over $\mathbb{Z}\Gamma$, of the ring $\mathfrak{o}_N$ of integers in a tame Galois extension N/K of number fields with group Γ, in terms of the symplectic root numbers, but not the other way round. Thus one needs additional algebraic structure to get more information on root numbers, after the pattern of Theorem 9. Such additional structure is provided by the trace form, and indeed this allows one to describe all symplectic root numbers in terms of this Hermitian structure, not only on the global, but also on the local level, i.e., for Langlands' constants. This had been conjectured by the author and was recently proved in [CN-T1].

It was necessary first to lay the algebraic foundations. For, the theory of non-singular forms over integral group rings which had been developed by K-theorists, mainly for topological applications, is not appropriate: the ring of integers with a trace form is usually singular in the sense of this latter theory. The basic notions which had to be developed were that of a generalized discriminant and a generalized Pfaffian, say for group rings. These are related to the corresponding classical notions in roughly the same way as our determinant for group rings is to the classical determinant. For instance for every virtual symplectic character χ of the underlying group Γ and for every invertible element a of a group ring $F\Gamma$, symmetric under the basic involution of $F\Gamma$, we get a Pfaffian $Pf_\chi(a)$. In the applications to arithmetic, the relevant Pfaffian is then given again in terms of resolvents – yet another role of these fundamental objects of our theory (cf. [F20], [F24], [F25]).

The theory proceeds essentially on a local level, and there is a close parallel between it and the theory of global Galois module structure as described in this book.

B. Norm Properties of Normal Bases

The theorems on the existence of normal integral bases are almost wholly existence theorems, not involving any actual construction and not yielding any information on other arithmetic properties of such bases. Specifically one might look at the norms of generators of normal integral bases, and indeed there is a wealth of problems and there are some results in this direction (cf. [Bu1], [Bu2], [Bu3]).

Let then N be a Galois extension of $\mathbb{Q}$ with Galois group Γ. Supposing that one does have a free generator a of $\mathfrak{o}_N$ over $\mathbb{Z}\Gamma$, one wants to study the norms of such generators. These form precisely an orbit $a(\mathbb{Z}\Gamma)^*$, and for fixed a the norms of these elements a^λ may then be investigated as functions of the variable $\lambda \in \mathbb{Z}\Gamma^*$. The procedure immediately generalizes in a natural manner. One need not restrict oneself to normal integral bases, and in fact need not assume their existence, but may simply study the norm properties of the orbit $a(\mathbb{Z}\Gamma)^*$, with a being a free generator of N over $\mathbb{Q}\Gamma$, hence also a free generator of $\mathfrak{o}_{N,p}$ over $\mathbb{Z}_p\Gamma$, for all p outside a finite exceptional set $\mathcal{T}_a$ of primes. Of course if $b \in a(\mathbb{Z}\Gamma)^*$ then $\mathcal{T}_b = \mathcal{T}_a$.

An obvious fundamental question is whether the following statement is true:

I. *Given $M > 0$, and $a \in N$ so that $a\mathbb{Q}\Gamma = N$, the inequality $|N_{N/\mathbb{Q}}(a^\lambda)| < M$ has only finitely many solutions $\lambda \in \mathbb{Z}\Gamma^*$.*

This has been proved for certain groups Γ, in particular for Γ of prime order. In this case it is a consequence of the following theorem:

II. *Let Γ be of prime order l. Let $\| \ \|$ be a Euclidean norm on $\mathbb{R}\Gamma$. Then, given $\varepsilon > 0$, $\exists c = c(x, \varepsilon) > 0$ with*

$$|N_{N/\mathbb{Q}}(x^\lambda)| \geqslant c\|\lambda\|^{l-\varepsilon}.$$

In general it is unknown whether I is valid. (Late Footnote: Bushnell tells me that G. R. Everest has now proved I.) There are a number of other results on the properties of the function $\lambda \mapsto |N_{N/\mathbb{Q}}(a^\lambda)|$.

In a different direction, one can also look at divisibility properties of norms $N_{N/\mathbb{Q}}(a^\lambda)$. Thus, one can, for non-Abelian Γ, find elements b in $a(\mathbb{Z}\Gamma)^*$ for which $N_{N/\mathbb{Q}}(b)$ is divisible by the primes in a given finite set $\mathcal{S}$ and not divisible by those in another finite $\mathcal{R}$ both disjoint from $\mathcal{T}_a$ – provided that $\mathcal{S}$ and $\mathcal{R}$ satisfy certain, not too stringent conditions.

C. The Wild Case

The obvious question which arises from the tame theory is of course: What happens in the "wild case", i.e., for extensions which are not necessarily tame? More precisely, which results can be generalized? We shall give here a partial, far from complete, survey of several strands which make up the "wild" theory as it stands at present. The first thing to observe is that while, under the restriction to tame ramification, the local structure is trivial, this is not so in general. In fact a major part of the theory is by its nature essentially local.

(i) *Module Conductors and Module Resolvents*

This topic occurred already briefly in I §1 and then in some detail in III §8, but in both instances only in its relevance within the tame theory. However its real significance lies outside of this. Module conductors and module resolvents are invariants which reflect ramification, specifically wild ramification, on the one hand, and Galois module structure on the other. They thus form an important tool to describe the connection between these two aspects. The author's paper on Kummer extensions (cf. [F2]) was a predecessor of this theory, which he then developed in the early 1960s (cf. [F3]) with a final account appearing in [F14]. At this stage there were still certain restrictions, which could subsequently be removed, and the role of the Galois Gauss sums was not at that time appreciated. A later more general and complete formulation, with applications, is contained in [Ne1] and [Ne2].

Going back to the results in III §8, it is of interest to note that in general the deviation of Artin conductors from module conductors, or of Galois Gauss sums from norm module resolvents which occurs in the wild case, has an interesting module theoretic interpretation, leading to specific information on Galois module structure. Moreover this deviation collapses, i.e., Artin conductors and module conductors "coincide", with an analogous result for Galois Gauss sums, when we work over local fields of residue class characteristic p and restrict ourselves to characters which are "lifted" from projective $\mathbb{F}_p^c\Gamma$-modules, i.e., characters in the image of the so called e-map (cf. [F14], [Ne1], [Ne2], [Q2]).

(ii) *The Order of* $\mathfrak{o}_N$

Let N/K be a Galois extension with Galois group Γ, either global or local non-Archimedean, and let K be an extension of finite degree of a field F. We define the order of $\mathfrak{o}_N$ in $F\Gamma$ by

$$\mathfrak{A}_{N,F}(\Gamma) = [\lambda \in F\Gamma \mid \mathfrak{o}_N \lambda \subset \mathfrak{o}_N].$$

This is indeed on $\mathfrak{o}_F$-order in $F\Gamma$, containing $\mathfrak{o}_F\Gamma$. The determination of $\mathfrak{A}_{N,F}$ is a local problem, and does involve ramification. $\mathfrak{A}_{N,F}$ has been determined in a number of situations, but no general results are known. (In the tame case of course always $\mathfrak{A}_{N,F}(\Gamma) = \mathfrak{o}_F\Gamma$.)

To develop a global theory, one of the approaches might be first to prove that $\mathfrak{o}_N$ is always locally free, or at least projective over $\mathfrak{A}_{N,F}$, and then to study $\mathfrak{o}_N$ via the locally free or projective class group of $\mathfrak{A}_{N,F}$. Unfortunately this is not true. (See [Be2], [Be3].) The "reason" behind this failure is the bad behaviour of orders with respect to change of group or basefield, in particular with respect to induction. In fact even "obvious" localization procedures may turn out to be useless in this context: We give an example. Let $\mathfrak{p}$ be a prime divisor of N above a rational prime p. Then, although $\mathfrak{o}_{N,\mathfrak{p}}$ may be free over $\mathfrak{A}_{N_\mathfrak{p},\mathbb{Q}_p}(\Gamma_\mathfrak{p})$, $\Gamma_\mathfrak{p}$ the local Galois group, $\mathfrak{o}_{N,p}$

need not be free over $\mathfrak{A}_{N,\mathbb{Q}}(\Gamma)_p$. Example (S. M. J. Wilson):

$$p = 2, \qquad N = \mathbb{Q}\left(\sqrt{-7}, \sqrt{\frac{\pm 1 + \sqrt{-7}}{2}}\right).$$

On the other hand for certain groups Γ as Galois groups over $\mathbb{Q}$, $\mathfrak{o}_N$ is indeed locally free over $\mathfrak{A}_{N,\mathbb{Q}}$, e.g., for $\Gamma = D_{2p}$ (cf. [Be1]). Then there is the outstanding such example: Leopoldt (cf. [Le1]) has, for all Abelian Galois groups Γ over $\mathbb{Q}$, determined the order $\mathfrak{A}_{N,\mathbb{Q}}$ explicitly in terms of the ramification structure and has proved that $\mathfrak{o}_N$ is indeed always free over $\mathfrak{A}_{N,\mathbb{Q}}$. He moreover described certain canonical free generators. This is done in terms of the classical Abelian Gauss sums, which in this case are essentially the resolvents.

(iii) *Martinet's Conjecture*

Let again N/K be a Galois extension of number fields with Galois group Γ. Martinet conjectured, in the case $K = \mathbb{Q}$, that if $\mathfrak{M}$ is a maximal order in $\mathbb{Q}\Gamma$, containing $\mathbb{Z}\Gamma$, then in $\mathrm{Cl}(\mathfrak{M})$ we have $(\mathfrak{o}_N\mathfrak{M}) = 1$. This was discussed in Chapt. I §1, and the proof of the conjecture in the tame case, even for $K \neq \mathbb{Q}$, was one of the early successes of the author's approach (see the Corollary to Theorem 6). The conjecture was also proved for certain Galois groups over $\mathbb{Q}$, in particular p-groups (cf. [Co1]). Subsequently, in [F28], the methods developed originally for the tame theory were extended and applied to this problem, and positive results similar to Cougnard's were proved for arbitrary basefields.

The conjecture is however false – the first counterexample (over $\mathbb{Q}$) being given in [Co2]. For certain metacyclic fields over $\mathbb{Q}$, such as that occurring in this counterexample, a revised and more natural version of the conjecture was established (cf. [Co3], and also for Abelian fields [Cha]). This deals with the class of the module $\mathfrak{o}_N \otimes_{\mathbb{Z}\Gamma} \mathfrak{M}$ in the Grothendieck group $\mathfrak{G}_0(\mathfrak{M})$ of finitely generated $\mathfrak{M}$-modules. However even this type of result fails in general. In [Wi2], Wilson in fact proved that the classes $(\mathfrak{o}_N \otimes_{\mathbb{Z}\Gamma} \mathfrak{M})$ and $(\mathfrak{o}_N\mathfrak{M})$ will genuinely depend on the choice of $\mathfrak{M}$. He gave examples where these classes do indeed vary. (This makes sense, as the groups $\mathfrak{G}_0(\mathfrak{M})$ for varying $\mathfrak{M}$ can be identified with each other in a natural manner).

(iv) *Queyrut's Approach*

See [Q2], [Q3], [Q4]. The impasse in which the global theory for wild extensions found itself, due to the failure of Martinet's conjecture, on the one hand, and the failure of rings of integers $\mathfrak{o}_N$ to be locally free over their orders $\mathfrak{A}_N$, on the other hand, was overcome by a new approach, due to Queyrut, whose essential ingredient is a new variant of the $\mathfrak{K}$-theory of orders. We briefly describe this in the particular case relevant to our applications.

We let $\mathcal{S}$ be a finite set of finite rational prime divisors and consider two Grothendieck groups associated with the integral group ring $\mathbb{Z}\Gamma$, Γ a finite group. The first is $\mathfrak{G}_{\oplus}^{\mathcal{S}}(\mathbb{Z}\Gamma)$, the Grothendieck group of $\mathbb{Z}\Gamma$-modules, which are finitely generated and $\mathbb{Z}$-free, modulo exact sequences, whose localization splits at all prime divisors not in $\mathcal{S}$. The second is $\mathfrak{K}_0^{\mathcal{S}}(\mathbb{Z}\Gamma)$, the Grothendieck group of finitely generated and $\mathbb{Z}$-free $\mathbb{Z}\Gamma$-modules, whose localizations are free over $\mathbb{Z}_p\Gamma$, for all p

outside $\mathscr{S}$. Associated with these Grothendieck groups one defines corresponding classgroups $\mathfrak{K}_0^{\mathscr{S}}(\mathbb{Z}\Gamma)$ and $\mathfrak{G}_{\oplus}^{\mathscr{S}}(\mathbb{Z}\Gamma)$, and for these one obtains a Hom description, following the pattern of our theory but involving new ideas.

For any Galois extension N/K of number fields the ring $\mathfrak{o}_N$ of integers now defines a class $(\mathfrak{o}_N)_{\mathbb{Z}\Gamma}$ in $\mathfrak{K}_0^{\mathscr{S}}(\mathbb{Z}\Gamma)$, provided that – as we shall assume here – $\mathscr{S}$ contains all prime divisors above which there is wild ramification in N/K. One thus regains a genuine global object, which reduces to $U_{N/K}$ when the ramification is tame and $\mathscr{S}$ is empty. Moreover for this class a description corresponding to that in Theorem 6 is derived. This can then be used to extend certain tame results to the wild case. In particular Queyrut obtains and proves a correct general form of Martinet's conjecture, replacing $\mathrm{Cl}(\mathfrak{M})$ by $\mathfrak{G}_{\oplus}^{\mathscr{S}}(\mathbb{Z}\Gamma)$. Indeed if $\mathscr{S}$ is empty, then these two groups coincide.

Queyrut's theory is thus a genuine generalization of this aspect of the tame theory. Naturally the results have to be weaker, as our original question concerning the existence of a normal integral basis has a negative answer already for local reasons. The principal new tool is the $\mathfrak{K}$-theory, with an exceptional set $\mathscr{S}$, something which is of independent interest. As far as Galois Gauss sums are concerned, no really new results are involved – just a rather clever use of what was already known from our original theory. The reason for this is that the only local Galois Gauss sums at wildly ramified primes which are needed here are those of characters "lifted from projective modules" in the residue class characteristic. These characters are however all induced from tame ones.

(v) *Further Work*

There are some further global results, dealing with the wild case, which do at present not fit into the Queyrut theory. With the usual meaning of N/K and Γ, the general pattern of these results can be described as follows: In some component of $\mathbb{Q}\Gamma$ or of $K\Gamma$ one defines an order, call it $\mathfrak{B}$, and one then associates with the Γ-module $\mathfrak{o}_N$ a $\mathfrak{B}$-module X_N in a canonical manner, and this is then proved to be free. There are three pieces of work of this nature to be mentioned. In [Ty8] Kummer extensions N/K are dealt with on the basis of earlier results in [F2]. In [Ty9] certain monomial representations are used to define pairs $(\mathfrak{B}, X_N)$; the method is applicable to any group Γ whenever it has such representations. In [Co4], $K = \mathbb{Q}$, and Γ is of a special metacyclic type. There are two aspects of special interest to be pointed out. Firstly in [Co4] Gauss sums for wild characters play a role, and these are not of the special type considered by Queyrut. Secondly the results in [Ty8] and [Ty9] point to yet another possible form of Martinet's conjecture. This involves not the minimal $\mathfrak{M}$-module containing $\mathfrak{o}_N$, but the maximal $\mathfrak{M}$-module contained in $\mathfrak{o}_N$ and first introduced in this context in [F2].

Final Remark. Wild local Galois Gauss sums occur in the latest versions of the theory of module conductors and module resolvents (see (i)), in Queyrut's approach (see (iv)) and in Cougnard's work (see (v)). It is also clear that the known properties certainly characterize them to within roots of unity, but there is not yet a complete theory on the lines of [FT].

Literature List

[Arm1] Armitage, J. V.: On a theorem of Hecke in number fields and function fields. Inv. Math. *2* (1967), 238–246

[Arm2] Armitage, J. V.: Zeta-functions with zero at $s = \frac{1}{2}$. Inv. Math. *15* (1972), 199–205

[Art1] Artin, E.: Die gruppentheoretische Struktur der Diskriminanten algebraischer Zahlkörper. Crelle *164* (1931), 1–11, or Collected Papers, 180–194

[Art2] Artin, E.: Zur Theorie der L-Reihen mit allgemeinen Gruppencharakteren. Hamb. Abh. *8* (1930), 292–306, or Collected Papers, 165–179

[Ba] Bass, H.: Algebraic K-theory. Benjamin, New York 1968

[Be1] Bergé, A. M.: Sur l'arithmetique d'une extension diédrale. Ann. Inst. Fourier *22*, 2 (1972), 31–59

[Be2] Bergé, A. M.: Arithmetique d'une extension galoisienne à groupe d'inertie cyclique. Ann. Inst. Fourier *28*, 4 (1978), 17–44

[Be3] Bergé, A. M: Projectivité des anneaux d'entiers sur leur ordres associés. In Thèse, Bordeaux 1979

[Br] Brauer, R.: On Artin's L-series with general group characters. Ann. of Math. *18* (1947), 502–514

[Bri] Brinkhuis, J.: Embedding problems and Galois modules, Thesis. University of Leiden 1981

[Bu1] Bushnell, C. J.: Norms of normal integral generators. J. London Math. Soc. (2), *15* (1977), 199–209

[Bu2] Bushnell, C. J.: Norm distribution in Galois orbits. Crelle *310* (1979), 81–99

[Bu3] Bushnell, C. J.: Diophantine approximation and norm distribution in Galois orbits, to appear in Illin. Journ. Math.

[Cha] Chatelain, D.: Étude du $\mathfrak{O}$ module $\mathfrak{O} \otimes_{\mathbb{Z}_g} \mathfrak{O}_N$ pour une extension $N/\mathbb{Q}$ abelienne. Sém. de Théorie des nombres, Besançon 1976/77

[Chb] Chinburg, T.: On the Galois structure of algebraic integers and units, to appear

[Chi1] Childs, L. N.: The group of unramified Kummer extensions of prime degree. Proc. London Math. Soc. (3), *35* (1977), 407–422

[Chi2] Childs, L. N.: Stickelberger relations and tame extensions of prime degree. Illin. Journ. Math. *25* (1981), 258–266

[Chs] Chase, S.: Ramification Invariants and Torsion Galois Module Structure in Number Fields, to appear

[CN1] Cassou-Noguès, Ph.: Théorèmes de base normale, Sem. de théorie de nombres. Bordeaux, exp. 27 (1976/77)

[CN2] Cassou-Noguès, Ph.: Quelques theorèmes de base normale d'entiers. Ann. Inst. Fourier *28* (1978), 1–33

[CN3] Cassou-Noguès, Ph.: Structure galoisienne des anneaux d'entiers. Proc. London Math. Soc. *38* (1979), 545–576

[CN4] Cassou-Noguès, Ph.: Module de Frobenius et structure galoisienne des anneaux entiers. J. of Algebra *71* (1981), 268–289

[CN-T1] Cassou-Noguès, Ph., Taylor, M. J.: Local rootnumbers and Hermitian Galois module structure of rings of integers, to appear

[CN-T2] Cassou-Noguès, Ph., Taylor, M. J.: Constante de l'equation fonctionelle de la fonction L d'Artin d'une représentation symplectique et modérée, to appear in Ann. Inst. Fourier

[Co1] Cougnard, J.: Propriétés galoisiennes des anneaux d'entiers des p-extensions. Comp. Math. *33*, 3 (1976), 303–336

[Co2] Cougnard, J.: Un contre exemple à une conjecture de J. Martinet, in "Algebraic Number Fields", Proceedings of the Durham Symposium 1975, Ac. Press London 1977, 539–560

[Co3] Cougnard, J.: Une propriété de l'anneau des entiers des extensions galoisiennes non abeliennes de degré pq des rationnels. Comp. Math. *40*, 3 (1980), 407–415

[Co4] Cougnard, J.: Propriétés locales et globales de certaines extensions métacycliques, to appear

[Ct] Coates, J.: p-adic L-functions and Iwasawa theory, in "Algebraic Number Fields", Proceedings of the Durham Symposium 1975, Ac. Press London, 1977.

[De1] Deligne, P.: Les constantes des équations fonctionelles des fonctions L. Springer Lecture Notes 349 (1974), 501–597

[De2] Deligne, P.: Les constantes locales de l'équation fonctionelle de la fonction L d'Artin d'une répresentation orthogonale. Invent. Math. *35* (1976), 299–316

[DH] Davenport, H., Hasse, H.: Die Nullstellen der Kongruenzzetafunktionen in gewissen zyklischen Fällen. Crelle *172* (1935), 151–182

[DM] Damey, P., Martinet, J.: Plongement d'une extension quadratique dans une extension quaternionienne. Crelle *262/263* (1973), 323–338

[Dw] Dwork, B.: On the Artin root number. Amer. J. Math. *78* (1956), 444–472

[F1] Fröhlich, A.: Discriminants of algebraic number fields. Math. Z. *74* (1960), 18–28

[F2] Fröhlich, A.: The module structure of Kummer extensions over Dedekind domains. Crelle *209* (1962), 39–53

[F3] Fröhlich, A.: Some topics in the theory of module conductors, Oberwolfach Berichte *2* (1966), 59–83

[F4] Fröhlich, A.: Resolvents, discriminants and trace invariants. J. of Algebra *4* (1966), 643–662

[F5] Fröhlich, A.: Local fields, in Algebraic Number Theory, Proceedings of the Brighton Conference of 1965. Academic Press 1967, 1–41

[F6] Fröhlich, A.: Artin root numbers and normal integral bases for quaternion fields. Invent. Math. *17* (1972), 143–166

[F7] Fröhlich, A.: Artin root numbers, conductors and representations for generalized quaternion groups. Proc. London Math. Soc. *28* (1974), 402–438

[F8] Fröhlich, A.: Module invariants and root numbers for quaternion fields of degree $4l^r$. Proc. Camb. Phil. Soc. *76* (1974), 393–399

[F9] Fröhlich, A.: The Galois module structure of algebraic integer rings in fields with generalized quaternion group. Bull. Soc. Math. France, Memoire *37* (1974), 31–86

[F10] Fröhlich, A.: Locally free modules over arithmetic orders. Crelle *274/75* (1975), 112–138

[F11] Fröhlich, A.: Galois module structure and Artin L-functions. Société Mathematique de France, Asterisque *24–25* (1975), 9–13

[F12] Fröhlich, A.: Artin root numbers for quaternion characters. Symposia Mathematica *15* (1975), 353–363

[F13] Fröhlich, A.: Resolvents and trace form. Math. Proc. Cambridge Phil. Soc. *78* (1975), 185–210

[F14] Fröhlich, A.: Module conductors and module resolvents. Proc. London Math. Soc. *32* (1976), 279–321

[F15] Fröhlich, A.: A normal integral basis theorem. J. of Algebra *39* (1976), 131–137

[F16] Fröhlich, A.: Galois module structure and Artin L-functions. Proc. Int. Congress 1974, Vancouver, Vol 1, 351–356

[F17] Fröhlich, A.: Arithmetic and Galois module structure for tame extensions. Crelle *286/287* (1976), 380–440

[F18] Fröhlich, A.: Galois module structure, in "Algebraic Number Fields". Proceedings of the Durham Symposium 1975. Ac. Press London 1977, 133–191

[F19] Fröhlich, A.: Stickelberger without Gauss sums, in "Algebraic Number Fields". Proceedings of the Durham Symposium 1975. Ac. Press London 1977, 589–607

[F20] Fröhlich, A.: Symplectic local constants and Hermitian Galois module structure. International Symposium, Kyoto 1976 (ed. S. Iyanaga), Japan Society for the Promotion of Science, Tokyo 1977, 25–42

[F21] Fröhlich, A.: Non-Abelian Jacobi sums, Number theory and algebra. Academic Press 1977, 71–75

[F22] Fröhlich, A.: On parity problems, Sem. de. Th. des Nombres, Bordeaux (1978–79), 21

[F23] Fröhlich, A.: Galois module structure and root numbers for quaternion extensions of degree 2^n. J. of Number Theory *12* (1980), 499–518

[F24] Fröhlich, A.: Class groups, in particular Hermitian class groups. To appear

[F25] Fröhlich, A.: The Hermitian class group. In "Integral Representations and Applications", Proc. Oberwolfach 1980 (Ed. K. W. Roggenkamp), Springer Lecture Notes *882*, 1981, 191–206

[F26] Fröhlich, A.: Value distributions of symplectic root numbers. To appear in Proc. London Math. Soc.

[F27] Fröhlich, A.: The rational characterization of certain sets of relatively Abelian extensions. Phil. Trans. Royal Soc. London A (1959), 385–425

[F28] Fröhlich, A.: Some problems of Galois module structure for wild extensions. Proc. London Math. Soc. *27* (1978), 193–212

[F29] Fröhlich, A.: On the classgroup of integral group rings of finite Abelian groups II. Mathematika *19* (1972), 51–56

[FKW] Fröhlich, A., Keating, M., Wilson, S.: The class group of quaternion and dihedral 2-groups. Mathematika *21* (1974), 64–71

[FM] Fröhlich, A., McEvett, A. M.: The representation of groups by automorphisms of forms. J. of Algebra *12* (1969), 114–133

[FQ] Fröhlich, A., Queyrut, J.: On the functional equation of the Artin *L*-function for characters of real representations. Inv. Math. *20* (1973), 125–138

[FST] Fröhlich, A., Serre, J.-P., Tate, J.: A different with an odd class. Crelle *209* (1962), 6–7

[FT] Fröhlich, A., Taylor, M. J.: The arithmetic theory of local Galois Gauss sums for tame characters. Trans. Royal Soc. A *298* (1980), 141–181

[FW1] Fröhlich, A., Wall, C. T. C.: Equivariant Brauer groups in algebraic number theory. Bull. Soc. Math. France, Mémoire *25* (1971), 91–96

[FW2] Fröhlich, A., Wall, C. T. C.: Graded monoidal categories, Compositio Mathematica *28* (1974), 229–285

[Ge] Gechter, J.: Artin rootnumbers for real characters. Proc. Am. Math. Soc. *57* (1976), 35–38

[Ha] Hasse, H.: Artinsche Führer, Artinsche *L*-Funktionen und Gaussche Summen über endlich-algebraischen Zahlkörpern. Acta Salmanticensa 1954, or: Mathematische Abhandlungen, 3. Band, 35–151

[He1] Hecke, E.: Vorlesungen über die Theorie der algebraischen Zahlen. Akad. Verl. Leipzig 1923

[He2] Hecke, E.: Über die *L*-Funktionen und den Dirichletischen Primzahlsatz für einen beliebigen Zahlkörper. Nachr. der kön. Ges. der Wissenschaften, Göttingen 1917, 299–318, or: Mathematische Werke, 178–197

[He3] Hecke, E.: Eine neue Art von Zetafunktionen und ihre Beziehungen zur Verteilung der Primzahlen. Math. Z. *1* (1918), 357–376, and *6* (1920), 11–51, or: Mathematische Werke, 249–289

[Hi] Hilbert, D.: Die Theorie der algebraischen Zahlkörper ("Zahlbericht"), Jahr. Ber. der deutschen Math. Ver. *4* (1897), 175–546, or: Gesammelte Abhandlungen, 63–363

[HS] Hasse, H., Schilling, O.: Die Normen aus einer normalen Divisionsalgebra über einem algebraischen Zahlkörper, Crelle *174* (1936), 248–252, or: H. Hasse, Mathematische Abhandlungen I, 531–535

[I] Iwasawa, K.: A class number formula for cyclotomic fields, Ann. of Math. *76* (1962), 171–179

[J1] Jacobinski, H.: Über die Geschlechter von Gittern über Ordnungen, Crelle *230* (1968), 29–39

[J2] Jacobinski, H.: Genera and decomposition of lattices over orders. Acta Math. *121* (1968), 1–29

[JL] Jacquet, H., Langlands, R. P.: Automorphic forms on GL(2), Springer Lecture Notes 114, 1970

[KL] Kubert, D., Lang, S.: Cartan-Bernoulli numbers on values of *L*-series, Math. Ann. *240* (1979), 21–26

[Le1] Leopoldt, H. W.: Über die Hauptordnung der ganzen Elemente eines abelschen Zahlkörpers, Crelle *201* (1959), 119–149

[Le2] Leopoldt, H. W.: Zur Arithmetik in abelschen Zahlkörpern, Crelle *209* (1962), 54–71

[Lg] Lang, S.: Algebraic number theory. Addison-Wesley, 1970

[Lm] Lam, T. Y.: Induction theorems for Grothendieck groups and Whitehead groups of finite groups. Ann. Scien. Ecole Norm. Sup. *1* (1968), 91–148

[Ls] Langlands, R. P.: On Artin's L-function, in "Complex Analysis". Rice Univ. Studies *56* (1970), 23–28

[Ma1] Martinet, J.: Sur l'arithmétique d'une extension galoisiennes à groupe de Galois diédral d'ordre $2p$, Ann. Inst. Fourier *19* (1969), 1–80

[Ma2] Martinet, J.: Modules sur l'algèbra du group quaternionien. Ann. Sci. Ecole Norm. Sup. *4* (1971), 399–408

[Ma3] Martinet, J.: Character theory and Artin L-functions, in "Algebraic Number Fields". Proceedings of the Durham Symposium 1975, Ac. Press London 1977, 1–87

[Ma4] Martinet, J.: H_8, in "Algebraic Number Fields". Proceedings of the Durham Symposium 1975. Ac. Press London 1977, 525–538

[MC1] McCulloh, L. R.: A Stickelberger condition on Galois module structure for Kummer extensions of prime degree, in "Algebraic Number Fields", Proceedings of the Durham Symposium 1975. Ac. Press London 1977, 561–588

[MC2] McCulloh, L. R.: A class number formula for elementary Abelian group rings. J. of Algebra *68* (1981), 443–452

[MC3] McCulloh, L. R.: Galois module structure of elementary Abelian extensions, to appear in J. of Algebra

[MC4] McCulloh, L. R.: Stickelberger relations in classgroups and Galois module structure, Proc. Journées Arithm. 1980 (ed. J. V. Armitage), Cambridge University Press

[Mi] Miyata, Y.: On the characterization of the first ramification group as the vertex of the ring of integers. Nagoya Math. J. *43* (1971), 151–156

[Mr] Maurer, D.: The semigroup of Galois algebras with applications to algebraic number theory. To appear

[Ne1] Nelson, A.: Monomial representations and Galois module structure, Thesis. University of London 1979

[Ne2] Nelson, A.: Modules over group rings and Abelian subquotients, to appear in J. of Algebra

[No] Noether, E.: Normalbasis bei Körpern ohne höhere Verzweigung. Crelle *167* (1932), 147–152

[Q1] Queyrut, J.: Extensions quaternioniennes généralisées et constente de l'équation fonctionelle des séries L d'Artin. Publ. Math. Univ. Bordeaux I, *4* (1972/73), 91–113

[Q2] Queyrut, J.: Structure Galoisienne des anneaux d'entiers des extension sauvagement ramifiées. Ann. Inst. Fourier *31*, 3 (1981) 1–35

[Q3] Queyrut, J.: S-groupes des classes d'un ordre arithmétique. J. of Algebra *76* (1982), 234–260

[Q4] Queyrut, J.: Modules radicaux sur des ordres arithmétiques, to appear in J. of Algebra

[Re] Reiner, I.: Class groups and Picard groups of group rings and orders. AMS reg. conf. series 26 (1975)

[Ri1] Ritter, J.: The class group à la Fröhlich, in "Integral Representations and Applications", Proc. Oberwolfach 1980 (Ed. K. W. Roggenkamp), Springer Lecture Notes, 882 (1981), 174–190

[Ri2] Ritter, J.: On orthogonal and orthonormal characters. J. of Algebra *76* (1982), 519–531

[RU1] Reiner, I., Ullom, S.: Class groups of integral group rings. Trans. Am. Math. Soc *179* (1972), 1–30

[RU2] Reiner, I., Ullom, S.: A Mayer-Vietoris sequence for class groups. J. of Algebra *31* (1974), 305–342

[Se1] Serre, J-P.: Corps locaux. Hermann, Paris 1962

[Se2] Serre, J-P.: Représentations linéaires des groupes finis, 2me ed. Hermann, Paris 1977

[Se3] Serre, J-P.: Conducteurs d'Artin des caractères réels. Inv. Math. *14* (1971), 173–183

[Se4] Serre, J-P.: Local class field theory (prepared by J. V. Armitage and J. Neggers), in "Algebraic Number Theory", Proceedings of the Brighton Conference 1965. Ac. Press, London 1967, 129–161

[Se5] Serre, J-P.: Modular forms and Galois representations (prepared in collaboration with C. J. Bushnell), in "Algebraic Number Fields", Proceedings of the Durham Symposium 1975. Ac. Press, London 1977, 193–268

[SE] Swan, R. G., Evans, E. G.: K-theory of finite groups and orders. Springer Lecture Notes *149*, 1970

[Sp] Speiser, A.: Gruppendeterminante und Körperdiskriminante. Math. Ann. *77* (1916), 546–562

[Sw1] Swan, R. G.: Induced representations and projective modules. Ann. of Math. *71* (1960), 552–578

[Sw2] Swan, R. G.: Periodic resolutions for finite groups. Ann. of Math. *72* (1960), 267–291

[Sw3] Swan, R. G.: Strong approximation and locally free modules, in Proc. 3rd Oklahoma Conference on Ring Theory and Algebra. Marcel Dekker 1980

[Tt1] Tate, J. T.: Fourier analysis in number fields and Hecke's Zeta functions (Thesis 1950), published in "Algebraic Number Theory", Proceedings of the Brighton Conference 1965. Ac. Press, London 1967, 305–347

[Tt2] Tate, J. T.: Local constants (prepared in collaboration with C. J. Bushnell and M. J. Taylor), in "Algebraic Number Fields", Proceedings of the Durham Symposium 1975. Ac. Press, London 1977, 89–131

[Ty1] Taylor, M. J.: Locally free class groups of groups of prime power order. J. of Algebra *50* (1978), 463–487

[Ty2] Taylor, M. J.: Adams operations, local root numbers and Galois module structure of rings of integers. Proc. London Math. Soc. *3* (1979), 147–175

[Ty3] Taylor, M. J.: On the self duality of a ring of integers as a Galois module. Invent. Math. *46* (1978), 173–177

[Ty4] Taylor, M. J.: Galois module structure of integers of relative abelian extensions. Crelle *303/4* (1978), 97–101

[Ty5] Taylor, M. J.: Galois module structure of rings of integers. Ann. Inst. Fourier *30*, 3 (1980), 11–48

[Ty6] Taylor, M. J.: A logarithmic approach to class groups of integral group rings. J. of Algebra *66* (1980), 321–353

[Ty7] Taylor, M. J.: On Fröhlich's conjecture for rings of integers of tame extensions. Invent. Math. *63* (1981), 41–79

[Ty8] Taylor, M. J.: Galois module structure of rings of integers in Kummer extensions. Bull. London Math. Soc. *12* (1980), 96–98

[Ty9] Taylor M. J.: Monomial representations and rings of integers. Crelle *324* (1981), 127–135

[Ty10] Taylor, M. J.: Fröhlich's conjecture, logarithmic methods and Swan modules, Integral Representations and Applications", Proc. Oberwolfach 1980 (Ed. K. W. Roggenkamp) in Springer Lecture Notes 882 (1981), 207–218

[U1] Ullom, S. V.: A survey of class groups of integral group rings, in "Algebraic Number Fields". Proceedings of the Durham Symposium 1975. Ac. Press, London 1977, 497–524

[U2] Ullom, S. V.: Character action on the class group of Fröhlich, preprint 1977, to appear in "Algebraic *K*-theory" in Springer Lecture Notes

[U3] Ullom, S. V.: Galois module structure for intermediate extensions. J. London Math. Soc. (2), *22* (1980), 204–214

[U4] Ullom, S. V.: Ratios of rings of integers as Galois modules, in "Integral Representations and Applications", Proc. Oberwolfach 1980 (Ed. K. W. Roggenkamp), Springer Lecture Notes *882* (1981), 240–246

[Wa1] Wall, C. T. C.: On the classification of Hermitian forms IV, Adele rings. Inv. Math. *23* (1974), 241–260

[Wa2] Wall, C. T. C.: On the classification of Hermitian forms V, Global rings. Inv. Math. *23* (1974), 261–288

[Wa3] Wall, C. T. C.: Norms of units in group rings. Proc. London Math. Soc. (3) *29* (1974), 593–632

[We1] Weil, A.: Basic Number Theory. Springer 1967

[We2] Weil, A.: Über die Bestimmung Dirichletscher Reihen durch Funktionalgleichungen. Math. Ann. *168* (1967), 149–156

[We3] Weil, A.: Dirichlet series and automorphic forms. Springer Lecture Notes 189, 1971

[Wg] Wang, S.: On the commutator group of a simple algebra. Am. J. of Math. *72* (1950), 323–334

[Wi1] Wilson, S. M. J.: Reduced norms in the *K*-theory of orders. J. Algebra *46* (1977), 1–11

[Wi2] Wilson, S. M. J.: Some counterexamples in the theory of the Galois module structure of wild extensions. Ann. Inst. Fourier *30*, 3 (1980), 1–9

[Wo] Wolf, P.: Algebraische Theorie der Galoisschen Algebren. Deutscher Verlag der Wissenschaften, Berlin 1956

List of Theorems

Some Further Notation

Index

~~1100~~ Ø I

60⁰⁰

#114575 · MATH